AF380212

Design And Modeling for 3D ICs AND INTERPOSERS

WSPC Series in Advanced Integration and Packaging

Series Editors: Avram Bar-Cohen *(University of Maryland, USA)*
Shi-Wei Ricky Lee *(Hong Kong University of Science and Technology, ROC)*

Published

Vol. 1:   Cost Analysis of Electronic Systems
by Peter Sandborn

Vol. 2:   Design and Modeling for 3D ICs and Interposers
by Madhavan Swaminathan and Ki Jin Han

WSPC Series in Advanced Integration and Packaging — Vol. 2

Design And Modeling for 3D ICs AND INTERPOSERS

Madhavan Swaminathan
Georgia Institute of Technology, USA

Ki Jin Han
Ulsan National Institute of Science and Technology (UNIST), Korea

World Scientific

NEW JERSEY · LONDON · SINGAPORE · BEIJING · SHANGHAI · HONG KONG · TAIPEI · CHENNAI

Published by

World Scientific Publishing Co. Pte. Ltd.

5 Toh Tuck Link, Singapore 596224

USA office: 27 Warren Street, Suite 401-402, Hackensack, NJ 07601

UK office: 57 Shelton Street, Covent Garden, London WC2H 9HE

British Library Cataloguing-in-Publication Data
A catalogue record for this book is available from the British Library.

WSPC Series in Advanced Integration and Packaging — Vol. 2
DESIGN AND MODELING FOR 3D ICs AND INTERPOSERS

ISBN 978-981-4508-59-9

Printed in Singapore by Mainland Press Pte Ltd.

This book is dedicated to my dear sister S. Malathi.

— Madhavan Swaminathan

To my dear mother, Hoon Yong Eom.

— Ki Jin Han

Preface

It appears that 3D integration with through silicon vias (TSV) is finally here! What we are referring to here are the memory-on-logic and logic-on-logic applications where the improvement in performance, transistor density and form factor enables the electronics industry to continue the miniaturization of systems. The buzz in this area first started around 2008 which continued until early 2010. During this period several conferences started high lighting the importance of this area through keynote addresses, focused workshops and tutorials. I (MS) remember being at a high profile workshop in Korea where companies were presenting on manufacturing a stack with two dies for mobile applications. From industry presentations, it appeared that most technical challenges had been solved and the major issue was cost. Company executives spoke about 3D integration being the next semiconductor revolution. Universities geared up their research in this area, standards were being defined and then the buzz vanished. Obviously there were technical and business related challenges. Later in 2011, the buzz picked up again with companies demonstrating that a significant improvement in bandwidth, power and form factor can be achieved by stacking dies on each other using TSVs. The targeted application was LPDDR2 for mobile phones. This generated a huge excitement in the electronics industry. Wide I/O became the killer application for the use of this technology. Sematech with help from SRC (Semiconductor Research Corporation) started significant activity in this area. Universities started defining design exchange formats (DEF). The industry forecasted that the first product with 3D ICs would be introduced in 2013. We are now in 2013, with no 3D products in the market place yet! Introduction of new technologies in products has to make business and economical sense and it appears that there are still issues with 3D integration. LPDDR2 using package on package technology still is a cheaper solution with many companies beginning to question wide I/O as the next killer application for 3D. Meanwhile, two major applications are emerging where TSV technology is making all the difference both from a business and technical stand point. First is Field Programmable Gate Arrays (FPGA) where benefits of partitioning a chip using silicon interposer with TSVs are making a lot of economical sense. The industry is calling this 2.5D integration. With a strong supply chain, Xilinx announced its first product in volume in 2012. Introducing the product took Xilinx 6 years from concept to manufacturing, an indication of the complexity of this technology. The second emerging application is in the memory area where memory density can be improved significantly by stacking ICs using TSV technology. I (MS) had the opportunity of moderating a panel at the Design Automation Conference (DAC) in 2013 on 2.5D and 3D integration with executives from Xilinx, Qualcomm and Globalfoundries. When I posed the question "Is 3D finally here", the resounding answer was "Yes". We therefore hope that the timing of this book is about right.

This book has been written to cover material at the basic, moderate and expert levels. With six chapters covering diverse topics such as system integration trends, design exchange formats, parasitic extraction, multiple technologies, signal integrity, power distribution and thermal management in the context of 3D ICs and interposers, we hope that this book is useful to the readers.

Chapter 1 consists of three parts related to 3D integration namely, 1) an introductory part on system integration trends and miniaturization, 2) the importance of modeling at an early phase also called path finding and 3) the need for design exchange formats (DEF). System integration and miniaturization require advances in both IC and package integration which are discussed in the context of Moore and More than Moore scaling. The need for 2.5D and 3D integration using TSVs is explained along with three embodiments of its implementation namely die stacking, package enabled stacking and interposer based solution. Two current interposer technologies being pursued by industry namely, silicon and glass have been compared for their electrical performance. The section on modeling compares the challenges associated with both full wave electromagnetic modeling and physics based (analytical) modeling, from a user perspective. Interpreting the results from port definitions has been highlighted as one of many issues in a full wave solver for a non-expert user, an issue that was raised as a question for discussion at a panel titled "Key Challenges in High Speed Signaling" at the Electrical Performance of Electronic Packaging and Systems conference in 2010, where I (MS) had the pleasure of serving as the moderator. We hope the readers appreciate the role of path finding in 3D integration and the need for electronic design automation (EDA) tools in this area, as discussed in this chapter. Finally, the need for DEF is illustrated through two examples related to DC drop and thermal analysis. This part draws on the methods discussed in later chapters and is targeted towards the expert readers who work on these kinds of analysis on a regular basis. The objective of this section is to illustrate the complexity involved in designing 3D ICs when all of the design information is unavailable.

The focus of chapter 2 is on parasitic extraction and simulation. This chapter delves deep into numerical modeling for ICs and packages that use interconnections with circular cross section. These structures are very important in 3D integration especially when wirebonds or vias are used to connect between IC stacks or between packages (package on package), respectively. Graduate students or EDA companies may find this chapter useful, since the intricacies related to modeling are covered here. Specifically, new basis functions that approximate current and charge density distributions are introduced to simplify the modeling of these structures, which are then combined with the more traditional methods for analyzing planar interconnections.

A natural extension to chapter 2 is chapter 3 where the numerical methods are extended to analyze TSVs. Chapter 3 starts with a discussion on the electrical behavior of TSVs and the reason for such a response. Both physics based and rigorous numerical modeling methods are covered in this chapter to make it useful to both a beginner and an expert reader. The effect of biasing the silicon substrate on the electrical response of TSVs is discussed using full depletion analysis and detailed numerical modeling. The basis functions derived in chapter 2 are extended to cover TSVs to account for polarization current and biasing as well.

Most of chapter 4 relates to a person using EDA tools for design rather than one creating the tool itself. Hence, the material covered in chapter 4 is quite different as compared to chapters 2 and 3. The chapter begins with a focus on parameter sweeps to illustrate the role of physical and material parameters on the electrical response of TSVs, for both biased and unbiased substrates. Unlike chapter 3 where the analysis is restricted to two TSVs, in chapter 4 TSV arrays are analyzed to assess cross talk and their effect on electrical performance. The criticality of excessive cross talk in silicon interposer designs with no ground plane is illustrated through time domain waveforms. An important element of this chapter is the comparison between silicon and glass interposers for achieving signal integrity, an area that is of utmost importance in high speed designs. The concept of return path discontinuity (RPD) is introduced for the first time in this chapter. Since signal and power integrity are related to each other, the response of power and ground planes for both the silicon and glass interposer are compared to determine which one is better, from an electrical performance perspective.

We believe that there are two major technical hurdles that need to be addressed for 3D technology to be viable, especially for logic on logic applications. The first is power delivery and the second is thermal management. Since these topics are inter-related, they are combined in chapter 5. In the interposer and package, one major cause for power supply noise is return path discontinuities (RPD), and therefore this topic is covered in this chapter as well. The difficulty in meeting the target impedance for 3D ICs and the futility in minimizing simultaneous switching noise are illustrated through two examples. The chapter then covers signaling between ICs through the interposer and package containing voltage and ground planes, to illustrate the occurrence of RPDs and their effect in corrupting the eye diagrams. As shown in the chapter, this effect can be mitigated using decoupling capacitors. However, systems today use a large number of capacitors integrated in the chip, embedded in the package and assembled on the board. These components often times increase the cost of a system due to an increase in layer count and board dimensions. This aspect of the problem is highlighted in this chapter. The last part of the chapter addresses the relationship between DC drop and temperature gradient. The numerical aspects of the problem and its application to both silicon and glass interposer technologies are addressed using conduction, convection and fluidic cooling. The numerical methods described in this chapter are also used in chapter 4 for analyzing the effect of temperature on the electrical response of TSVs.

Finally chapter 6 covers alternate methods for power distribution using power transmission lines. This chapter is controversial since it argues that eliminating voltage planes in the package and board can help eliminate RPDs, reduce layer count and reduce the number of decoupling capacitors required. Decoupling capacitors are of two types namely, 1) the ones that supply charge to the switching circuits and 2) the ones used to mitigate RPDs. By using alternate power distribution methods discussed in this chapter, we believe that the second class of capacitors can be eliminated, thereby reducing the total capacitor count. The methods described in chapter 6 are still in their infancy. Nevertheless, it is a different way of looking at the problem and therefore we sincerely hope that these ideas can energize others in the design community to innovate.

Finally we would like to refer the readers to 3D Path Finder (3DPF), a software product from E-System Design (www.e-systemdesign.com), primarily developed for analyzing structures arising in 3D integration, which can be useful to reproduce some of the examples in this book.

Madhavan Swaminathan (MS)
Ki Jin Han (KJH)

Contents

Acknowledgement

I started writing this book in Fall 2011 when I was on my sabbatical in Chennai, India. During the four month period that I spent there, I had the opportunity to spend time with my sister who was suffering from cancer. These four months were the most enjoyable time of my life. It has taken me two years to complete the book. Though my sister is no more, those memorable four months still lingers in my memory. My sister dedicated her life to public service with honesty and integrity. This book is dedicated to her.

First and foremost, I would like to acknowledge my students whose hard work has helped me generate the material for this book. The students who specifically contributed to this book include: KiJin Han, Jianyong Xie, Rishik Bazaz, Biancun Xie, Satyan Telikepalli, David Zhang, Sang Kyu Kim, Suzanne Huh, Tapobrata Bandyopadhyay, Vivek Sridharan and Decao Yang (visiting student). Their contribution over the last five year period has helped make this book possible. I also would like to acknowledge and thank Daehyun Chung (currently with NVidia), whose stay at Georgia Tech helped initiate the 3D activity in my research group. During the course of writing this book I invited my former student KiJin Han to co-author the book with me. I am indeed fortunate that he accepted since his presence has ensured the quality of this book. KiJin's early work on 3D modeling (as part of his thesis) forms the basis for chapters 2 and 3. I also would like to acknowledge the National Science Foundation (NSF), Semiconductor Research Corporation (SRC) and several companies for their support. My special thanks also go to Antonio Ciccomancini of CST for his continuous support over several years.

A very important part of writing a book is to ensure that the material is factually correct, the content is relevant and the information is easy to follow. At the time of writing the book, our goal was to ensure that the book would be useful to both academia and industry. We therefore had each chapter reviewed by two leading experts in the field, one from academia and the other from industry. Their feedback has been valuable in ensuring the quality of the book. We therefore would like to gratefully acknowledge the following people for their reviews: Chapter 1: Sungkyu Lim (Georgia Tech) and Dale Becker (IBM); Chapter 2: Dan Jiao (Purdue) and Zhiguo Qian (Intel); Chapter 3: Tzong-Lin Wu (NTU) and Ivan Ndip (Fraunhofer Institute); Chapter 4: Jose Schutt-Aine (UIUC) and Wendem Beyene (Rambus); Chapter 5/6: Joungho Kim (KAIST) and Woong Hwan Ryu (Samsung).

I had the opportunity to work with Bill Martin, Jim Bruister, Don Klayman, Gene Jakubowski of E-System Design and KiJin Han to develop 3D Path Finder (3DPF). This software product required many late night discussions to develop both the graphical user interface (GUI) and engine. A lot of ideas discussed in the book have been implemented in this product. I am eternally grateful to all of them for working with me.

The editorial staff Chelsea Chin, Yolande Koh and Jason Lim have been very patient working with us to ensure that the book is of high quality. Our thanks go to them. We also would like to thank Zvi Ruder and Avram Bar-Cohen for inviting us to write this book.

I believe this book would not have been possible without the support of my loving family. My special thanks go to my wife Shailaja, daughter Sharanya, mother Rajam, sister Mythili and brother in law Bhaskaran.

Madhavan Swaminathan

Firstly, I would like to thank the main author of this book, Prof. Madhavan Swaminathan, who was my PhD advisor at Georgia Tech. He encouraged me to start working on modeling TSVs for 3D ICs, which was at the beginning stage at that time. Since then, 3D ICs and TSVs have become the most important research topic in semiconductor and packaging areas, which is currently driving 3D integration based on TSV technology into commercialization. Without Dr. Swaminathan's insightful guidance, I could not even have made small progress in the up-to-date research trend. I deeply appreciate his inviting me as the co-author of this book. Although my contribution is tiny, working with Dr. Swaminathan on this book was an honorable opportunity and a valuable experience.

I would like to express my thanks to those who contributed to this book, including the reviewers of chapters, former and present colleagues at Georgia Tech, and people who are working with E-System Design, whose names have been provided by Dr. Swaminathan. I also would like to acknowledge my MS advisor Prof. Hyun-Kyo Jung with Seoul National University, for his guidance that formed the foundation of my research career.

Finally, I would like to give my sincere thanks to my mother, Ms. Hoon Yong Eom for her love and lifelong support. I hope and believe she can recover her health. I wish all my family health and happiness.

Ki Jin Han

Chapter 1

System Integration and Modeling Concepts

The semiconductor industry has come a long way since Dr. Gordon E. Moore, co-founder of Intel, formulated his empirical law called Moore's law in 1965, almost five decades back. Today, Moore's law is being used by the semiconductor industry for research and planning purposes and has been the primary driver for having more than a billion transistors in an integrated circuit chip today. Over the last two decades, the packaging industry has followed Moore's law but more at the System's level where the focus has been on system miniaturization rather than on IC integration, with the coining of the phrase "More than Moore" leading to the development of System in Package (SIP) and System on Package (SOP) technologies. So, what comes next both in the near and distant future? In this chapter we provide an introduction to IC and System Integration and paint a vision for the future with a primary focus on System Integration using packaging as a platform for integration. Since, the next semi-conductor wave is 3D integration, which is also the focus of this book, we discuss the packaging aspects with a focus on interposer based solutions.

A very important part of this chapter is on the need for modeling and simulation to maximize performance. With 3D integration still in its infancy at the time of writing this book, we discuss some introductory aspects to modeling with details covered in the rest of the chapters of this book.

1.1 Moore's Law

Dr. Gordon E. Moore was awarded the IEEE Medal of Honor in 2008 for "pioneering technical roles in integrated circuit processing, and

1

leadership in the development of MOS memory, the microprocessor computer and the semiconductor industry", an apt recognition to a person who has helped shape the semiconductor industry. In 1965, Dr. Moore predicted that the number of components per integrated circuit (IC) in the future will double every year based on a few data points as shown in Figure 1.1(a) [Moore, 1965]. In 1975, Dr. Moore altered his prediction to the doubling of transistors per IC every two years based on additional data points. Coined as Moore's law around 1970 by Dr. Carver Mead, a professor at Caltech, this law has been the primary driver for the semiconductor industry over the last several decades. Moore's law today is attributed to the doubling of transistors every 18 months and has led to more than a billion transistors on a single chip, as shown in Figure 1.1(b).

As we all know today, packing more transistors in an IC requires fine line lithography, which enables the scaling of the transistor. Interestingly enough, Dr. Moore never indicated anywhere that scaling the transistor would result in a better performing transistor, a concept that was derived later. Today, we know that packing more transistors per unit area (increasing transistor density) leads to better performance, with higher functionality made possible due to the availability of more transistors in the IC. Though empirical, Moore's law has helped shape and drive the

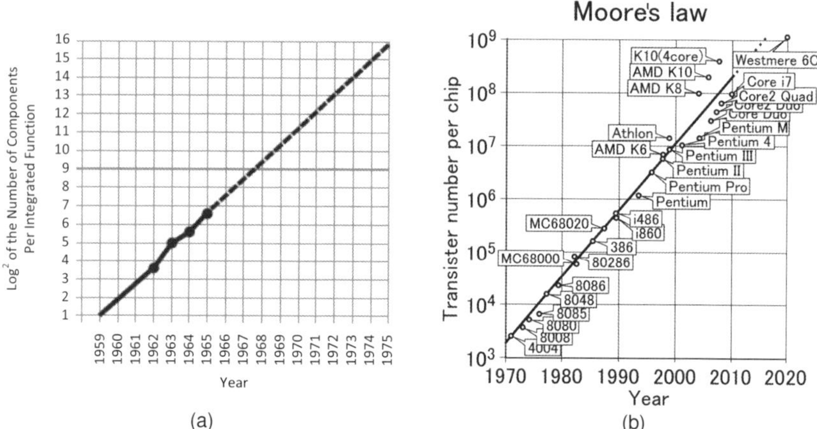

Figure 1.1: (a) Dr. Moore's original prediction in 1965 (components/IC doubling every year) [Moore, 1965] and (b) transistor count in a microprocessor Vs year of introduction.

semiconductor industry. In 2008, Dr. Moore was featured on the cover of IEEE Spectrum Magazine [Perry, 2008] where he was asked "What would you like your legacy to the world to be" to which he replied "Anything, but Moore's law"!

1.2 IC Integration Vs System Integration – What is the Difference?

Over the last four decades the size of systems has reduced exponentially while the functionality that they support has increased dramatically.

This is depicted in Figure 1.2 where the first modern workstations, also called mini computers, introduced by Xerox Palo Alto Research Center (PARC) in 1973 called Xerox Alto were single-user machines with high-resolution graphics and mouse driven graphical user interface. In the late 1970s, minicomputers led to the modern microcomputers with Commodore PET, Apple II and TRS-80 by Radio Shack being introduced in 1977. By 2008, the number of personal computers in use world wide hit 1 billion. Personal computers were desktop computers as opposed to the workstations, where a significant cost reduction led to

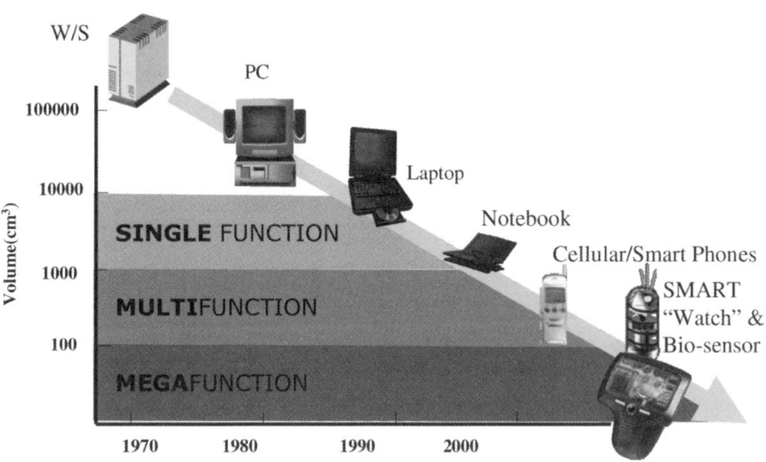

Figure 1.2: System Miniaturization Trend (Courtesy: Prof. Rao Tummala, Packaging Research Center, Georgia Tech).

their usage as a home computer. In the mid 1990s, the modern mobile computers became popular due to the need for portability and lapability, a term meaning the resting of the computer on a person's lap which doesn't exist in the English dictionary. Though subtle in differentiation, the laptops led to notebooks in the mid 2000s through significant weight reduction using a smaller battery, a smaller screen, ultra thin profile, smaller keyboard, removal of the internal floppy drive and with an integrated modem for network connection. Though portable computers have become popular over the last decade, the first portable computer was introduced in 1981 by Adam Osborne, an ex-book publisher. Called as the Osborne 1, it weighed 24 pounds and cost $1795. Along with portability, laptop and notebook computers enabled mobility through the advent of wireless communication for consumer applications in the 1990s, which allowed users to stay connected anywhere in the world. A natural extension of mobility for data through portable computers was voice communication through the introduction of the first commercial cell phone in 1983 by Motorola. Called as DynaTAC 8000x, it had a size of around 19.5cm × 4cm × 8cm, an antenna size of 20cm, an approximate weight of 1.18kg, offered 30 minutes of talk time, had a storage capacity of 30 numbers and cost $3,995. Needless to say, the cell phones have significantly evolved since 1983 where the smart phones of today are mobile, multimedia devices rather than verbal communication tools. The smart phones today are used for surfing the web, checking emails, taking photographs, managing our social status, listening to music, and sometimes for voice communication. As depicted in Figure 1.2, it is interesting to see the evolution of these consumer systems over the last five decades starting with the workstations which were large and bulky to the smart phones of today that are small, light weight and have multi media functionality far greater than what the workstations could support in the 1970s. The multi-function shown in Figure 1.2 refers to voice, video, email, music, internet and other multi media capability available in smart phones today as compared to the workstations that primarily focused on computing in the 1970s. Moving forward, the size of systems will continue to reduce while the functionality supported will only increase, leading to systems with mega-functions. Examples include smart watches and bio-sensor pills that have started appearing in the market today. Given the trend in system miniaturization shown in Figure 1.2, an important question to ask ourselves is: What technologies enable system miniaturization? Are these technologies purely IC driven based on Moore's law or does packaging play a significant role as well? To answer this question, let's compare and contrast the size reduction achieved through IC and package level integration since the 1970s.

Similar to transistor count for IC integration, a measure for package level integration is the component density per square centimeter achievable outside of the IC in the package and printed circuit board. These include the voltage regulator module, interconnections, resistors, capacitors, inductors, substrates, heat sink and other peripheral components outside of the IC required to build a system. Clearly, unless these components are miniaturized, the system as a whole can never be made small and therefore system miniaturization requires both IC and package level integration.

1.3 History of Integration – An Overview

A *qualitative* assessment and comparison of the evolution of IC and Package level integration is depicted in Figure 1.3. To use the same metric, the transistor density and system component density per square centimeter have been used along the two vertical axes. Driven by computing and Moore's law, the transistor density (shown on the left vertical axis) has increased, made possible by a reduction in the feature size from $10\mu m$ used by Intel in 1971 for the 4004 processor to 90nm used in 2002 by Intel for their Pentium M processor.

The world's first single chip microprocessor in 1971 contained 2300 transistors and measured 3mm × 4mm, leading to a density of around 19000 transistors/cm^2. With the usage of dip packages in the early 1970s, the packing density was limited by the pitch of the plated through holes and the line width in the printed circuit board. With a via pitch of 2.5mm, line width of 0.25mm and one line per channel [Tummala et al., 1989], an interconnection density of ~8 lines/cm^2 using two wiring layers (taken as a minimum set for orthogonal wiring) was possible. Given that the printed circuit boards are used primarily for connecting ICs to each other,

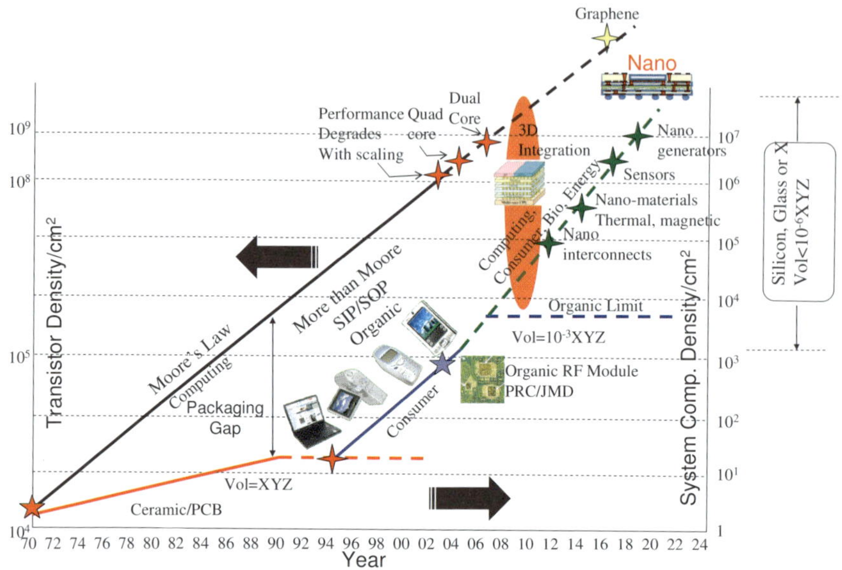

Figure 1.3: Integration trend (Source: Interconnect and Packaging Center, Georgia Tech, Packaging Research Center, Georgia Tech and IEEE Spectrum).

this routing density ultimately determined the size of the system, and is therefore referred to as system component density along the right vertical axis in Figure 1.3 (two different scales have been used for the two vertical axes). With these two data points, the ratio of IC to package level integration was roughly around 2000:1.

Over the next two decades until the early 1990s, the IC integration continued to increase driven by Moore's law while the printed circuit board line widths reduced at the rate of 0.05mm every five years with a dimension of 0.05mm with 6 lines per channel between plated through hole vias on a 2.5mm pitch, translating to an interconnection density of ~48 lines/cm² for two wiring layers. During this period dip packages were replaced by Quad Flat Pack (QFP) packages due to the higher I/O count for the die finally leading to Ball Grid Array (BGA) flip chip packages. In 1990 the Intel 80386 microprocessor was introduced with a transistor count of 275,000 in a chip of size 104mm² leading to a transistor density of ~265,000 transistors/cm². This translates to a ratio of around 5000:1 between transistor and system component density. With Moore's law continuing to drive IC integration, in 1998 Intel introduced the Pentium III with a transistor count of roughly 28.1M and die size of 107mm^2 while the interconnection density in the printed circuit board (2nd level package) continued to stagnate, causing a huge packaging gap (as shown in Figure 1.3) where the ratio of IC to package level integration grew to more than 0.5M:1. Though BGA packages helped reduce the package size, being single chip packages, the system size was limited by the printed circuit board routing density. Hence, several layers in the printed circuit board were necessary to complete all of the wiring to support the system functionality, or in other words, tiny ICs lead to bulky systems. In the late 1980s and early 1990s, IBM's Multichip Module (MCM) technology was prevalent where ceramic substrates were used to connect hundreds of bipolar chips together for mainframe applications. Though this technology helped with a small reduction in the packaging gap, with the transition from bipolar to CMOS technology by IBM for CPUs in the early 1990s, integration levels within the IC increased further.

A comparison of PCB and ceramic substrate technology is shown in Table 1.1 where from the line width and line spacing, the wiring density achievable for two layers is similar for both technologies. The two main advantages of ceramic substrate over PCB wiring were its smaller layer thickness and lower loss tangent, which led to thinner modules and higher performance, respectively.

It is interesting to note that though the semiconductor industry reached a bottleneck around 2004 as shown in Figure 1.3 where scaling beyond 90nm caused significant power leakage, architectural innovations were used to solve these problems through the introduction of the dual core and multi core processors. Hence, the semiconductor industry continued to scale beyond the 90nm to 65nm and below. Since transistor scaling represents the backbone of the semiconductor industry and is required for miniaturization and performance, this can never stop and it is expected that every time the semiconductor industry hits a road block, innovations will emerge, such as for example the introduction of the graphene based transistors integrated into silicon ICs in the distant future.

Table 1.1: Comparison of various packaging technologies [Courtesy: Part Knickerbocker et al., 2006].

Parameters	PCB	Ceramic	Organic/Thin films	Silicon Carrier
Line Width (µm)	127	70	15	1.6
Line Space (µm)	305	326	35	2.9
Line Thickness (µm)	30.5	20	6	1.2
Spacing between Layers (µm)	167	111	9	1.6
Relative Dielectric Constant	4.3	5	3.5	4.0
Loss Tangent	0.025	0.0025	0.013 – 0.004	0.0015
Wiring Density (lines/cm^2)	46	50	400	4444

As shown in Figure 1.3, in the 1970s and 1980s with computing being the primary driver, workstations and personal computers were used where the system though bulky did not require package level integration to the level that was necessary in the mid 1990s when the primary driver shifted to consumer electronics, where mobility became an important requirement.

All of a sudden the need for connecting heterogeneous ICs became necessary and package level integration became a necessity. This was the dawn of the "More than Moore era", a terminology referring to the inability to rely purely on Moore's law for integrating transistors on a single chip. Supporting wireless communication for mobility required the integration of both radio frequency (RF) and digital ICs along with a multitude of passive components that ultimately became the bottleneck for system miniaturization. Two technologies emerged called System in Package (SIP) and System on Package (SOP) [Tummala et al., 2008] where the latter was developed at the Packaging Research Center (PRC), Georgia Tech. The premise behind both these technologies was to mount or embed multiple bare dies using flip chip technology onto substrates containing layers of thin film wiring that connected ICs to each other with embedded passives (resistors R, inductors L and capacitors C) integrated into the thin film layers of the package. A conceptual embodiment of System on Package (SOP) technology is shown in Figure 1.4 where digital, RF and optoelectronic functionally can be embedded into the layers of the package. As shown in the figure, dissimilar bare dies or stacked packages (POP) could be assembled on the system package. With such an integration approach, the level of wiring and surface mount passive components required on the printed wiring board reduces significantly, leading to system miniaturization. Though several processes have been developed to implement both SIP and SOP, one important technology based on organics using Liquid Crystalline Polymers (LCP) was developed by the primary author Madhavan Swaminathan and commercialized through Jacket Micro Devices (JMD), a spin-off company from PRC. As shown in Figure 1.3, a system component density of 400 (at JMD) to 1000 components/cm^2 (at PRC using other polymers) was achieved using this technology for RF Front End Modules which contained switches, Low Noise Amplifiers (LNA) and Power Amplifiers (PA) with a multitude of filters using RLC components embedded into the layers of the package [Swaminathan et al., 2011]. The organic thin film technology have line width and spacing of 15µm and 35µm respectively, leading to a wiring density of 400

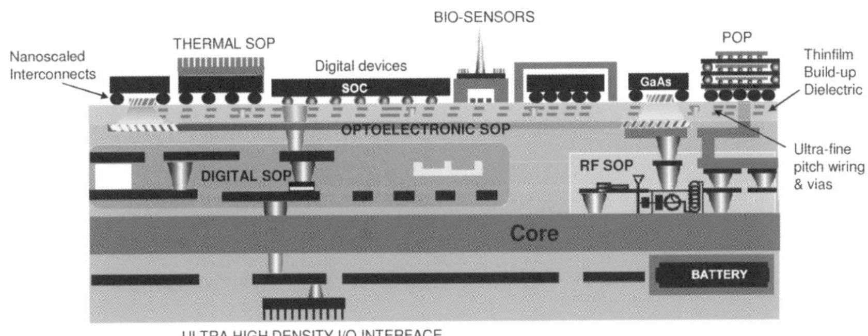

Figure 1.4: System on Package (SOP) (Courtesy: Packaging Research Center, Georgia Tech).

lines/cm^2 based on Table 1.1. In addition, with a small spacing between layers and low loss tangent, thin modules (thickness of 1mm including embedded IC) with high performance was possible. From Figure 1.3, the system component density in mid 2000 increased by a factor of almost 1000 as compared to the 1970s, leading to system volumes that were roughly 1000X smaller as compared to the 1970s, as depicted in Figure 1.2 and Figure 1.3 (Vol $= 10^{-3}$XYZ in Figure 1.3). Though, the system component density still lags behind the transistor density for microprocessors, the packaging industry did make significant progress during the period 1995–2005 for enabling the miniaturization of systems. The density of components is ultimately dictated by the density of interconnections and as shown in Figure 1.3, with organic technology there appears to be a fundamental limit, with the smallest line width and spacing dimensions achievable being of the order of around 10μm. Therefore the question to ask is "What comes next for IC and package level integration" as the semiconductor and packaging industry continues to make progress towards increasing the integration density for both transistors and components.

1.3.1 *3D Integration – Is it the Next Semiconductor Revolution?*

Three interesting trends will shape the semiconductor and packaging industry moving forward namely 1) the need for increased bandwidth between ICs at low power, 2) the ability to work with smaller ICs to minimize cost and improve time to market and 3) heterogeneity for system integration.

The next big wave, called by some as the next semiconductor revolution, is three-dimensional (3D) integration where semiconductor chips can be stacked on each other using short vertical interconnections (z-directed wires), which can be used to communicate between ICs at high speed. The primary driver initially is expected to come from the computing and consumer side as shown in Figure 1.3.

Three embodiments of 3D-integration are shown in Figure 1.5 where the z-directed interconnections are used to communicate between ICs, as opposed to communicating between ICs laterally, as was done

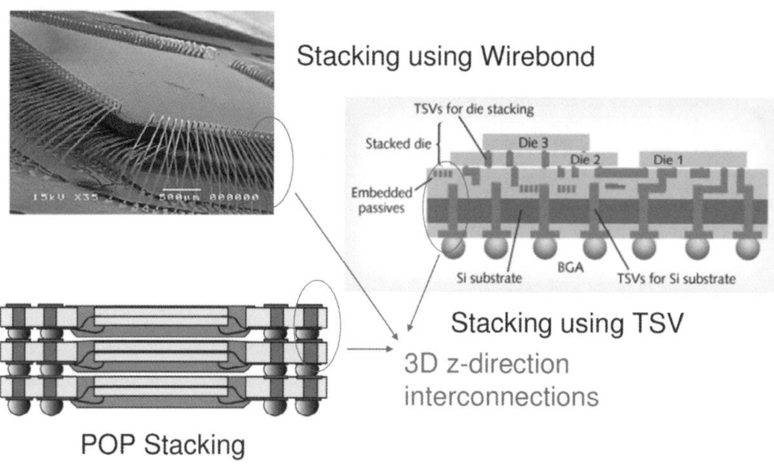

Figure 1.5: 3D Integration using Wirebond, Package on Package (POP) and through silicon vias (TSV) (Courtesy: Packaging Research Center, Georgia Tech).

previously. In the mid 2000s, an important driver was the need to increase memory density which quickly transitioned towards increasing the communication bandwidth between logic and memory for smart phone applications, leading to reduced power. This is an important factor for consumer applications since a reduction in power consumption increases battery life. In Figure 1.5, the early embodiment of 3D integration consisted of stacking dies using wirebonds which provided limited form factor reduction since the wirebonds take up considerable space on the package, as evident from Figure 1.5. Moreover long wirebonds connecting to the dies at the top of the tier limit performance due to increased parasitic resistance and inductance, and hence such stacking was limited to memory applications. Package on Package (POP) that enables the stacking of packages on each other is being used today (2012) for memory on logic applications such as for Low Power Double Data Rate (LPDDR) communication used in mobile phones. Though limited in form factor, LPDDR2 has been shown to provide 3.2Gbps of bandwidth between logic and memory [Kwon, 2011]. As expected, the POP still provides limited integration and bandwidth capability due to the dimensions of the package and pitch of the solder balls in Figure 1.5. Therefore, moving forward, the next big wave is through silicon via

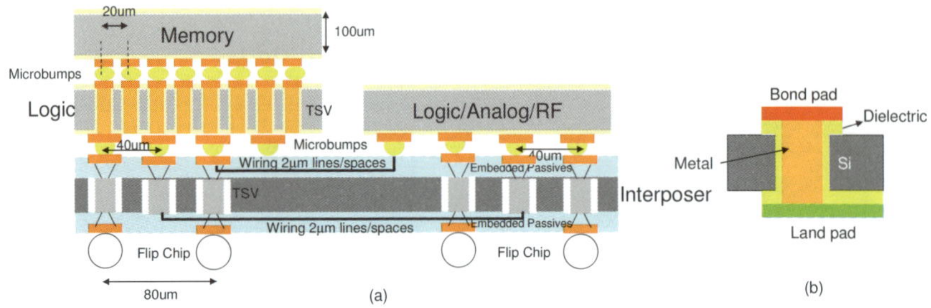

Figure 1.6: (a) Stacked ICs on interposer with embedded passives and (b) through silicon via structure.

(TSV) technology where dies with holes etched in silicon are filled with metal with an oxide liner and stacked on each other using micro-bump technology onto an interposer or substrate, as shown in Figure 1.6. As illustrated in Figure 1.3, the packaging technology is not limited to silicon interposers with TSV but can include through glass vias (TGV) in glass interposer or TXV in a suitable interposer material (where X represents the material used provided it supports high density wiring and good electrical properties).

3D integration provides a marriage between "Moore" and "More than Moore" scaling as never before. Consider Figure 1.6(a), an example of an embodiment showing memory and logic ICs stacked on each other using microbumps on a 20μm pitch bonded onto an interposer with bumps on a 40μm pitch. The interposer contains wiring with 2μm lines and spaces corresponding to a wiring density of 5000 lines/cm^2 for two wiring layers, consistent with the silicon carrier shown in Table 1.1, where thin modules are possible with high performance due to the low loss tangent of the dielectric material used. The interposer is used to communicate between the 3D stack (memory + logic) to another IC that supports logic/analog or RF functionality. The key interconnect structure that enables the stacking of ICs and their communication with each other is through silicon via (TSV) technology, as shown in Figure 1.6(b). With smaller feature sizes in interposers, one would expect that embedded component density (resistors, inductors and capacitors) would be an order of magnitude higher than with organics [Swaminathan et al., 2011]

Table 1.2: Comparison of RLGC parameters for various 3D Interconnect Technologies.

Interconnect Technology	Diameter (um)	Oxide liner thickness (um)	Length (um)	Pitch (um)	Material Properties	R (mΩ)	L (pH)	G (mS)	C (pF)
Wirebond	25	-	1062	60	ε_r=4.3 tand=0.01	130	320	0.01	0.15
POP	200	N/A	500	400	ε_r=4.3 tand=0.01	7	95	0.01	0.19
Die TSV	5	0.1	100	10	ε_r=3.9 σ=10S/m	93	30	1.57	0.35
	6	1.0	100	50		66	47	0.094	0.072
Silicon Interposer TSV	30	1.0	100	100	εr=3.9 σ=10S/m	9.5	21	0.90	0.25
Glass interposer TGV (ENA1)	30	-	100	100	ε_r=5.3 tand=0.004	9.5	27	0.001	0.037

and therefore combined with the wiring density in the interposer, the system component density in the package should increase by a factor of 10–100X in the next decade with 3D integration.

An important comparison of the various 3D interconnect technologies are the parasitics associated with them, as shown in Table 1.2. In the table a pair of interconnects have been analyzed to extract their resistance (R), inductance (L), conductance (G) and capacitance (C) parameters. Since the RLGC parameters change with frequency, Table 1.2 represents approximate values that can be used to gauge the relative performance of these technologies.

In the table, typical physical dimensions and material properties are provided for the interconnect technologies compared. For example, the material with permittivity 4.3 and loss tangent 0.01 for the wirebond represents the molding compound used for protecting it. For the TSVs within the die, two different TSV structures are provided which differ in diameter, pitch (center to center spacing) and oxide thickness to illustrate the effect of these on the RLGC parameters. The TSVs in the die and interposer assume an oxide of permittivity 3.9, silicon conductivity 10S/m and silicon permittivity 11.9. The material properties for the glass interposer assume ENA1 glass from Table 1.3. It is always good to minimize all the four parasitic parameters namely, R, L, G and C while migrating from one technology to another, which is hard to do, as can be seen from the table. Hence, choosing the appropriate interconnect technology is determined by many factors including the application it needs to support, the available infrastructure for manufacturing, the required supply chain and finally the cost. The remaining chapters provide further details on most of the geometrical and material parameters provided in Table 1.2.

The stacking of memory on logic is consistent with Moore scaling (increased transistor density) while the integration of components and high density wiring in the package to increase the system component density is consistent with More than Moore scaling. One may argue that in 3D integration, the metric should be transistors and system components per unit volume rather than unit area due to the stacking of ICs and packages. It is important to note that dies are typically thinned during stacking and packaged using thin substrate material leading to the total height remaining small and therefore the same metric of transistor and component density per square centimeter can be used as in Figure 1.3 for 3D integration as well.

1.3.2 *What Comes Next?*

In addition to computing and consumer, two major drivers are emerging in the area of bio-technology and energy. The systems of the future will continue to follow the path of convergence where digital, RF, optical, analog and sensor functionality on a single integrated package will become necessary, similar to Figure 1.4, but on a nano-platform. This requires the emergence of new nano-devices and materials in the die providing increased transistor density with better performance (Moore Scaling) coupled with new system components at nano-scale in the package (More than Moore scaling) which combined supports the superior electrical, mechanical and thermal functionality required. With nanotechnology enabling the engineering of functional systems at the molecular scale, carbon based nano-electronics is emerging as a leading candidate for the replacement of Silicon based electronics. Several devices based on carbon nano-electronics, classical/non classical CMOS technologies, quantum and molecular devices and nano wires have started emerging.

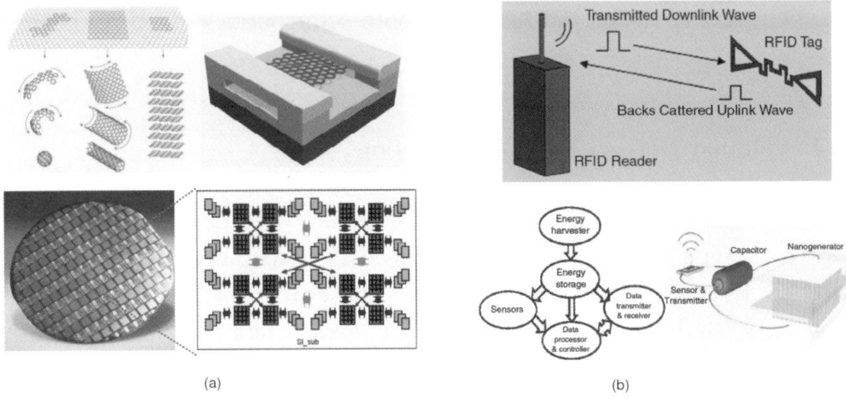

Figure 1.7: (a) Energy efficient multi-core processors and (b) Sensor networks.

As the carbon nano-electronics world unfolds, major advances will be made in packaging as well leading to new nano-interconnects, nano-materials, nano-components and nano-thermal structures in the package as shown in Figure 1.3, enabling the design of convergent heterogeneous nano-systems. We envision nano-systems of the future to contain nano-devices integrated into silicon assembled through nano-scale fine pitch interconnects onto a package containing embedded functional components with suitable heat removal structures with an integrated battery generating energy through harvesting from the environment (nano-generators) and supporting sensing, computing, communication and biological functions, similar to the SOP technology in Figure 1.4. Two possible applications in the future are shown in Figure 1.7 consisting of energy efficient multi-core systems based on carbon nano-electronics and self powered sensor systems (where nano-generators apply) that cover a broad frequency range beyond mm-wave into the terahertz frequencies. With such integration, the move towards nano-modules or systems that are 10^{-6} (Vol $= 10^{-6}$XYZ in Figure 1.3) smaller than the systems of the 1970s should be possible, as shown in Figure 1.3. Hence the future holds a lot of excitement both for IC and package level integration!

In this book our main focus is on the electrical design and modeling for 3D integration primarily for *digital applications*, a technology that we believe will shape the future for both semiconductor and packaging for the next decade.

1.4 Primary Drivers for 3D Integration

3D ICs and interposers improve integration at the system's level, provides superior electrical performance and cost benefits which are tied to the applications it supports. These relate to addressing the communication bandwidth between logic and memory, the ability to use smaller ICs through partitioning to build a system and heterogeneous integration. These issues are elaborated in this section.

1.4.1 *Thirst for More Bandwidth at Low Power*

Logic such as in a processor functions at speed provided it has data available to it. The communication bandwidth between logic and memory is dictated by the number of input/output (I/O) terminals between the two chips used to transfer data. As the number of I/O terminals increases, the communication bandwidth increases as well. Sending data back and forth between the two chips requires power, a large fraction of which is dictated by the parasitics of the wire (or interconnection) used to communicate between the two ICs. A long wire has large capacitance and hence large amount of power is expended in making this communication possible. As the wire lengths become shorter, the capacitance scales linearly with wire length and hence the power consumed becomes smaller as the wire length becomes shorter. This is the premise for 3D integration where short wires (through silicon vias) which are roughly 100μm in length replace the much longer wires of length >1cm in a Multichip Module where the ICs are placed next to each other. This enables the following: 1) higher communication speed per bit due to lower parasitics and 2) reduced power per bit to be able to transmit data.

For smart phones, it is expected that power budgeting will increase 10X over the next decade of which 30–50% will be due to I/O power. This increase is due to the need to support more and more graphics for real time video applications which is dictated by the communication bandwidth between logic and memory. Since power management is critical for mobile phones, the energy required per bit or word of communication between logic and memory has to be decreased, which is possible by reducing the capacitance of the wires connecting the two ICs. With TSV technology, the length can be reduced significantly, thereby reducing the capacitance and power levels per I/O. As an example consider the wide I/O application for mobile phones for communication between logic and memory, which is beginning to emerge as a Jedec standard. Based on [Kwon, 2011], 12.8Gbps of data rate will be used to communicate between the two ICs using 512 I/Os. The energy consumed for an 8 bit word is between 2–7pJ as shown in Figure 1.8. Consider the LPDDR2 Package on Package (POP) technology that is currently being used which supports a data rate of 3.2Gbps, where the energy consumed per word is roughly 512pJ. The faster data rate and lower energy consumed per word are enabled due to larger bandwidth made possible through more I/O connections (512 vs 64) between the dies and the reduced parasitics of the interconnections. Assuming dies are placed side by side as in a Multichip module (MCM), given the physical length of the interconnections, the energy consumed per word is roughly

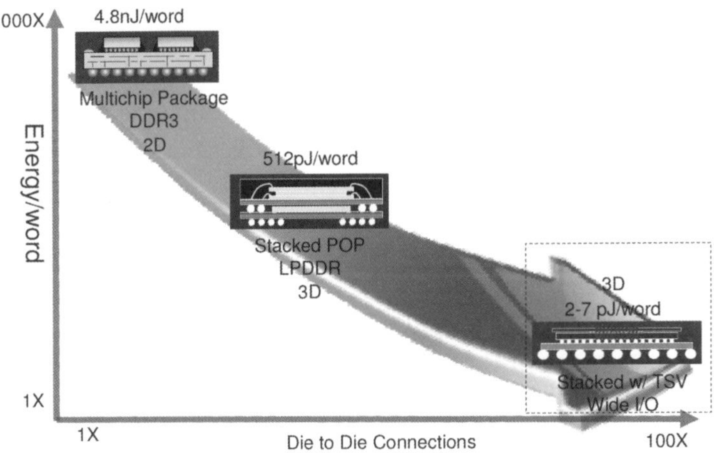

Figure 1.8: Reduction in the energy/word using 3D integration for mobile phones (Source: Part from Dr. Greg Taylor, Intel and Dr. Paul Franzon, NCSU).

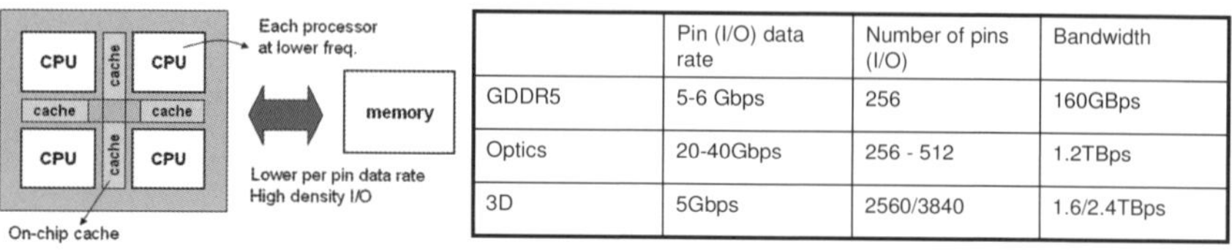

	Pin (I/O) data rate	Number of pins (I/O)	Bandwidth
GDDR5	5-6 Gbps	256	160GBps
Optics	20-40Gbps	256 - 512	1.2TBps
3D	5Gbps	2560/3840	1.6/2.4TBps

Figure 1.9: Tradeoff between I/O speed and channel width.

4.8nJ/word showing the attractiveness of using 3D technology. This is illustrated in Figure 1.8 where as the die to die connections increase from 2D to 3D integration by a factor of 100X, the energy consumed per word decreases by 1000X.

Another important application is the ability to support large data transfers in multi-core computer systems between the microprocessor and memory. The current state of the art in single-ended per pin (I/O) data rate in a system is around 5-6Gbs which is available in graphics memory channels [Bae et al., 2008]. Even with the 5Gbps per pin speed, only 160GB/s bandwidth is achievable with a pin count of around 256. Achieving >1TBps of bandwidth with 256pins can be challenging, since the per pin speed required is around 40Gbps. Very sophisticated optics based technologies may be required to support such speeds. Instead if the per pin speed is maintained around 5Gbps and the pin count is increased to a few thousand, then an architecture can be developed where 1.6TBps or larger bandwidth can be achieved, as shown in Figure 1.9. It has been shown that such high I/O densities can be supported using silicon packaging, where the I/O density can be as large as 10^4–10^6/cm^2, with high wiring density using through silicon via (TSV) and fine pitch interconnection technologies [Knickerbocker et al., 2005]. From Figure 1.6 I/O counts far larger than 2560–3840 is achievable with 3D technology and hence Figure 1.9 is a conservative estimate. The GDDR5 functions at a power level of ~12W which translates to 13.3GBps/watt. From Figure 1.8, if 2pJ/word of energy is consumed using 3D technology, this translates to 3.2W of power assuming 5Gbps and 2560 pins of I/O bandwidth resulting in 0.5TBps/watt for 1.6TBps of data transfer, leading to a bandwidth/watt improvement of 37X. Hence, 3D technology can be very useful in increasing the communication bandwidth between the processor and memory while at the same time reducing power.

1.4.2 *Large Chips Sink Ships*

In wafer processing, the cost of a die after dicing depends on the yield of the wafer, which translates to the number of good dies available per wafer. The yield of a die is proportional to e^{-DA} where D is the defect density and A is the area of the die. For a constant defect density, if the area of the die increases, yield decreases exponentially, leading to fewer good dies per wafer. This translates to a higher die cost.

This is illustrated in Figure 1.10(a) where smaller chips are preferred due to higher yield and lower die cost. Since, semiconductor companies compete on cost in addition to performance, larger chips have always been problematic for companies and hence the terminology "Large chips sink ships". Xilinx has adopted this scheme in their Virtex-7 field programmable gate array (FPGA) where smaller chips that are yield optimized are connected to each other through a silicon interposer using a 65nm back end of the line (BEOL) process, leading to high performance through high bandwidth and low latency interconnections, as shown in Figure 1.10(b).

Though not truly 3D integration, Xilinx uses a passive silicon interposer with high density connections to connect the smaller ICs to each other. Hence, the smaller ICs together form the larger FPGA (like a

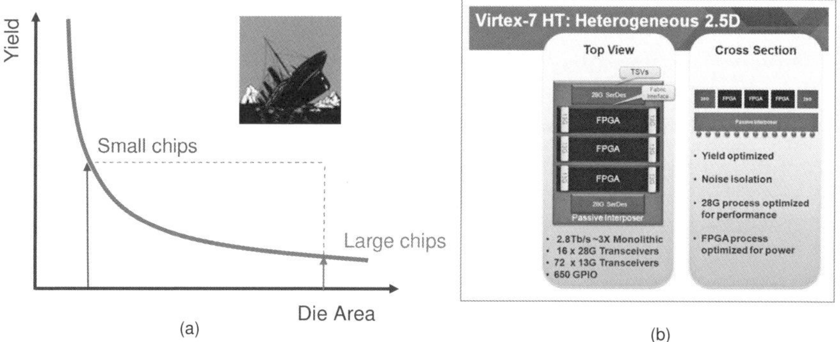

Figure 1.10: (a) Yield Vs Die Area and (b) Virtex-7T (Courtesy: Xilinx).

super chip). This approach has been called as 2.5D integration, a terminology that has been used to differentiate from a Multichip module (MCM), since MCMs in the past have always been an expensive solution and often times is an unpopular acronym to use to describe a new technology.

1.4.3 *Heterogeneous Integration to Continue More Than Moore Scaling*

Heterogeneous integration has several meanings that support a wide range of applications. Heterogeneous integration could mean the connection of dies fabricated through different process nodes (such as 22nm and 65nm), connection of dies fabricated using different materials (such as silicon and silicon germanium), connection of dies from different domains such as RF, digital and optoelectronics (also called mixed signal), to name a few. Clearly, when a wide variety of technologies are required to build a system, integrating these technologies onto a single chip is impossible. Hence, packaging plays a very important role in enabling the deployment of such heterogeneous systems.

As per Figure 1.3, the mid 1990s saw the dawn of the "More than Moore" era where components which could not be integrated into silicon had to be embedded in the package. With wireless communication becoming the primary driver, the need for integrating RF front ends became a necessity. RF front ends are rich in inductors and capacitors, take up considerable space and due to high quality factor requirements are difficult to integrate into silicon [Swaminathan et al., 2011]. Hence, organic based technologies were developed using Liquid Crystalline Polymer (LCP) and other polymer materials (Rogers Experimental Polymer RXP) to embed passives as shown in Figure 1.11. As an example, consider the schematic of a RF Front End Module shown in Figure 1.11(a) with antennas, transmit/receive chain containing a multitude of filters, a switch, matching networks and multiple Power Amplifiers (PA) and Low Noise Amplifiers (LNA). The passive circuitry can be implemented as components integrated into the layers of the package, as shown in Figure 1.11(b) and implemented as in Figure 1.11(c) by mounting the bare ICs (unpackaged) on top of the organic package.

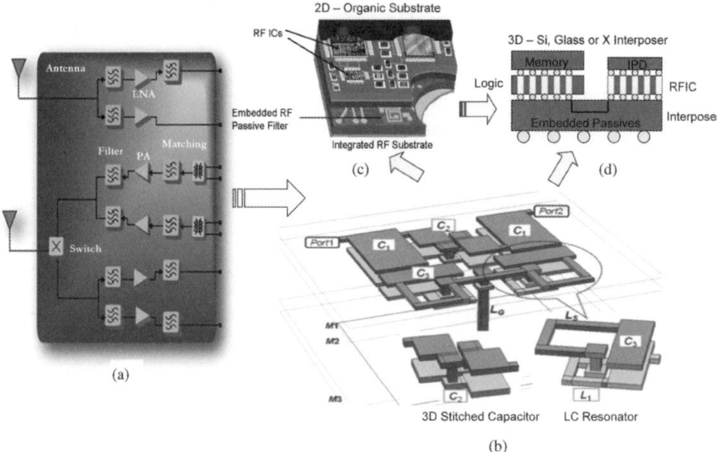

Figure 1.11: (a) RF Front End Module (Courtesy: JMD), (b) Passive Circuit implemented with embedded components, (c) 2D Organic Module (Courtesy: JMD) and (d) 3D Module.

Details of this embodiment are available in [Swaminathan et al., 2011]. As mentioned earlier, organic materials have a line width and spacing limit of around 10µm (which is aggressive in itself). In addition, since the digital (logic and memory not shown) and RF ICs have to be mounted side by side (2D integration), the module or system size can become large. This can be corrected by moving to 3D integration as shown in Figure 1.11(d) where the logic and memory ICs in bare die form can be assembled on top of each other with the RF IC next to the logic/memory stack, which can then be connected to each other using through silicon vias, micro-bumps and fine pitch wiring in the die and interposer. The passive components can be embedded into the layers of the interposer or can be a separate passive IC called IPD (Integrated Passive Device) mounted on the RFIC and connected to the transistors using TSVs and micro-bumps, resulting in a module size that can be far smaller than what has been achieved in the past. Of course, the 3D embodiment shown in Figure 1.11(d) has several issues to be overcome related to thermal management, power delivery and using the right interposer material (such as silicon, glass or X) to name a few. The IPD can be fabricated using glass or other RF friendly materials and provides a possibility for enabling a much lower cost module than embedding the passives into the interposer due to better yieldability and testability.

1.5 Role of the Interposer in 3D Integration

Though it is obvious that the stacking of ICs on top of each other reduces the length of the interconnections and therefore the parasitics, the question to ask is: What is the role of the interposer and how does it affect performance, where performance is loosely defined here to include electrical, thermal and mechanical reliability of the assembled module or system. A module represents a subset of a system which contains ICs and supports specific functionality associated with the system such as communication between logic and memory, wireless communication, analog functionality, power management to name a few. Interposer, as the name implies, is a first level package that is inserted between the die stack and the second level package, as illustrated in Figure 1.12. The interposer supports fine pitch microbumps at the top, coarser C4 flip chip bumps at the bottom, contains multiple ICs in a stack (either on top of each other or side by side) and contains wiring that performs two primary functions namely 1) serves as a space transformer through redistribution by connecting the microbumps to the C4 bumps or 2) provides interconnection between ICs assembled on the interposer as shown in the figure between the stack (IC1 and IC2) and IC3. When the interposer serves as a space transformer, then the number of microbumps on the top surface equals the C4 bumps on the bottom surface and therefore all the interposer does is redistribution by connecting the fine pitch microbumps to the coarser pitch C4 bumps. However, when the interposer is used for wiring by interconnecting ICs side by side, it reduces the wiring in the second level package, where as shown in the figure the 200,000 microbumps reduce to 20,000 C4 bumps, providing an order of magnitude improvement in integration. It is important to note that the via diameter of 10μm in the interposer is an order of magnitude better as compared to the 100μm via diameter in the second level package and supports much higher density wiring as well. In addition the thickness of the interposer is 100μm as compared to 1mm for the second level package, providing an order of magnitude improvement in the thickness as well. These physical dimensions combined provide a large advantage in terms of providing better electrical performance using the interposer as compared to using the second level package for interconnecting the

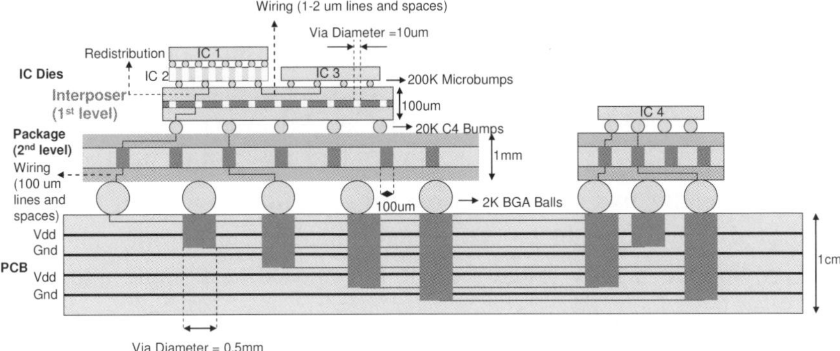

Figure 1.12: Packaging hierarchy showing the role of the interposer.

ICs to each other. In addition, stacking the ICs provides an enormous advantage in increasing the transistor density as well. Notice in Figure 1.12 that the 2000 Ball Grid Array (BGA) bumps at the bottom of the second level package is an order of magnitude lower than the 20,000 C4 bumps at the top of the package and hence the second level package serves as a space transformer. This allows for a large number of C4 bumps to be used for power and ground connections, which is required for 3D integration. Finally, the printed circuit board (PCB) with 1mm via diameter and thickness of 1cm and larger has dimensions which are an order of magnitude larger than the second level package, and hence provides very limited integration capability. It is important to note that the numbers and dimensions provided in Figure 1.12 are approximate values and have been used only for illustration purposes and therefore the actual numbers can vary based on the process and technology used.

The interposer performs the following electrical functions when used for wiring to connect ICs and IC stacks to each other: (1) the short interconnections (of the order of 4–5mm) reduces parasitics and hence improves communication speed and bandwidth, (2) provides better noise management depending on the stack-up used (wiring with voltage and ground planes or just wiring) and (3) improves the ability to decouple the power supply by increasing the effectiveness of the decoupling capacitors mounted on the interposer due to its thinness. The interposer performs several other important functions as well such as (1) improving the heat dissipation depending on the interposer used and hence can serve as an effective thermal management solution and (2) improving the mechanical reliability by matching the coefficient of thermal expansion (CTE) between the IC stack and the first level package, provided silicon is used as the interposer material. From the business side, the interposer provides an attractive solution for reducing the time to market, since ICs fabricated from different processes and domains can be interconnected and provided as a fully tested module as opposed to a single integrated IC.

1.5.1 *Three Embodiments of the Interposer*

The interposer for 3D integration can be used in three ways as shown in Figure 1.13. If the microbump pitch in the 3D stack is compatible with the C4 pitch on the second level package, then the interposer can be removed as shown in Figure 1.13(a) for the wide I/O application from Samsung [Kwon, 2011] where the stacked IC (logic + memory) is assembled directly on the package or PCB. However, as the number of microbumps increases due to its fine pitch the interposer is required to serve as a space transformer and therefore needs to be inserted between the 3D stack and second level package as in Figure 1.12. This is the first embodiment of using the interposer. Typical TSV diameters in the IC are around 5μm and hence are much larger than the transistors which have feature sizes of 22nm or less. Hence, a single TSV can take up considerable space away from the transistors. On the other hand the interposer is a passive structure with no transistors in them and hence using the vias in the interposer to connect ICs to each other in the form of a stack can increase the transistor density. This implementation is shown in Figure 1.13(b) where ICs are mounted on either side of the interposer and connected to each other using the through vias in the interposer. We call this as the 3D enabled with interposer embodiment and this approach can work for some low power applications. It is important to note that this embodiment can reduce the number of the BGA connections and hence can create issues with providing power to the module. In addition, the IC assembled at the bottom of the interposer can create thermal management problems due to the inability of the heat

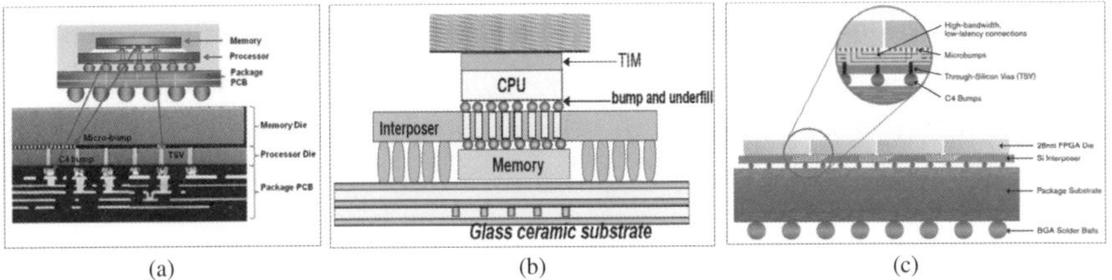

Figure 1.13: Three embodiments of the Interposer (a) 3D with interposer as a space transformer (Courtesy: Samsung [Kwon, 2010], (b) 3D enabled with interposer (Courtesy: Packaging Research Center, Georgia Tech) and (c) 2.5D integration (Courtesy: Xilinx).

to escape easily. The third embodiment is the approach pursued by Xilinx [Madden et al., 2012] where the interposer is used to support wiring between the smaller ICs, as shown in Figure 1.13(c), which is also called as 2.5D integration due to the high density wiring provided by the silicon interposer as compared to the MCM approach used by IBM using glass ceramic technology in the early 1990s [Tummala et al., 1989].

The 2.5D integration used by Xilinx consists of a passive silicon interposer and three active dies: an 8 x 28Gbps transceiver GTZ-IC and two FPGA ICs known as super logic regions (SLR) [Madden et al., 2012]. Derivations of this embodiment include a two IC (GTZ-IC+SLR) and five IC solution (2 GTZ-IC and 3 SLR). The five IC solution consists of 16 28Gbps transceivers and 72 13.1Gbps transceivers with a total bandwidth of 2.78Tbps. The silicon interposer consists of four high density (~1μm pitch) wiring layers, TSVs and microbumps. The silicon interposer lateral routing consists of four layers of metal with one layer for redistribution, two for signal routing and one ground reference layer which separates the two signal routing layers. Two kinds of signals are routed in the interposer namely 1) signals that connect the microbumps to the C4 bumps through the TSVs and 2) signals that connect ICs to each other in the interposer which do not pass through the TSVs. The interposer uses silicon of resistivity 20Ω-cm due to lower losses. The measured eye amplitude for the channel was 923mVp-p with a measured total jitter of 6.25ps at BER 10^{-12}. The measured Random jitter was 230fs. Clearly the 2.5D approach, the first commercial product from Xilinx, is a success and shows the demonstration of using the silicon interposer to reduce cost (due to smaller ICs) and increasing performance. Since the ICs are not stacked on each other, the back side of the ICs is available for thermal management. This is in contrast to 3D IC stacking where heat has difficulty to escape easily.

1.5.2 *Silicon or Glass or X for Interposer*

The interposer can be fabricated using CMOS grade silicon with back end of line (BEOL) process with high density wiring. For example, Xilinx has used the 65nm BEOL process. This approach is very attractive since the silicon wafer processing tools are depreciated and hence passive silicon interposers are easier and cheaper to fabricate. This approach has its advantages and disadvantages. CMOS grade silicon (also called wafer silicon in Table 1.3) has a resistivity of 10Ω-cm and is therefore lossy. Through silicon vias (TSVs) created in such silicon interposers using copper metallization and silicon-di-oxide (SiO_2) lining can cause significant coupling to each other. In addition signals propagating through such TSVs can have significant capacitance due to the high relative permittivity of silicon (11.9) and the existence of slow wave and quasi-TEM modes of propagation (discussed in Chapter 3). In addition, due to the size of the wafers, the cost of processing can be relatively larger. However, with a CTE of 3ppm/K, the coefficient of thermal expansion can be matched to silicon and hence provides a significant advantage. The resistivity can be improved through doping to

Table 1.3: Properties of Silicon and Glass Interposer [Courtesy: Bandyopadhyay, 2011].

Material	Relative Permittivity	Loss Tangent	Resistivity	CTE (ppm/°K)
Interposer				
Wafer-Si	11.9		10 Ω-cm	3
ENA1 Glass	5.3	0.004	10^{13} Ω-cm	3.8
AS01 Glass	6.9	0.011	10^{13} Ω-cm	8.5
O211 Glass	6.7	0.0046	10^{8} Ω-cm	7.4
Dielectric				
SiO_2	3.9	0.001		0.5
ZIF Polymer	3.1	0.011		31
RXP4 Polymer	3.01	0.0043		67

100Ω-cm thereby lowering coupling, but at increased cost. In addition to silicon-di-oxide, a suite of other polymers such as ZIF and RXP4 polymer shown in Table 1.3 can be used as the dielectric for lining the TSV walls and for wiring as well.

An alternative to using the silicon interposer is the glass interposer with similar via diameter, via pitch and high density wiring. Glass as the interposer material has a relative permittivity between 5.3–6.9 and a loss tangent of 0.004–0.011, as shown in Table 1.3. Since, glass is an insulator as opposed to silicon which is a semiconductor, the through glass vias (TGV) can be made with copper without the need for an insulator lining. Since the resistivity of glass is between 10^8–$10^{13}\Omega$-cm, it is much higher than silicon and therefore the coupling between vias is greatly reduced as compared to the silicon interposer. However, glass has a CTE that can be much greater than silicon and hence poses a problem for mechanical reliability. In addition, the thermal conductivity of glass is much lower than silicon, does not help as a heat spreader and hence can create larger hot spots.

With 3D integration in its infancy, the interposer solutions are in their infancy as well. Glass is being seen as a potential solution for replacing silicon while other materials (called X here) are also being pursued. So, the question to ask ourselves is: Which interposer solution is better – Silicon or Glass or X and how do you compare them? In this book, we have tried to compare silicon and glass interposer solutions from the standpoint of electrical and thermal integrity in the context of *digital applications*, meaning the digital communication between ICs through the interposer with emphasis on signal insertion loss, cross talk, power distribution and thermal management. A comparison of silicon and glass interposer for high speed digital applications is shown in Table 1.4. Details that support this comparison are discussed in later chapters as indicated in the comments column of the table.

In Table 1.4, a thumbs-up represents a clear advantage while a thumbs-down is a potential problem. A thumb pointed sideways represents a scenario where there is a problem which can be corrected through suitable engineering of the system. Though it is true that cases with a thumbs-down can also be corrected through adequate engineering, we see these issues as being a major one that could require significant

Table 1.4: Comparison of Silicon and Glass Interposer for Digital Applications.

Parameter	Silicon Interposer	Glass Interposer	Why ?	Comments
Insertion Loss for signal transitioning through via Through Silicon Via (TSV) and Through Glass Via (TGV)	👉 (neutral)	👍 (thumbs up)	Losses in CMOS grade silicon are higher than in glass due to lower resistivity of Si compared to Glass. However, these losses are predictable and hence can be compensated.	See Chapter 3 for details on TSV and Chapter 4 for details on comparison
Cross Talk (w/o Ground Plane) for signal transitioning through via only	👎 (thumbs down)	👍 (thumbs up)	Coupling between vias in silicon is a big problem due to the low resistivity of silicon and is a function of the oxide liner thickness	See Chapter 4 for details
Cross Talk (w/ Ground Plane) for signal lines transitioning through vias	👉 (neutral)	👉 (neutral)	Cross talk is similar since the damped cavity modes in silicon compensate for the increased coupling between Vias. For glass the cavity modes increase the coupling though the direct coupling between vias is small	See Chapter 4 for details
Temperature and Thermal Management	👍 (thumbs up)	👉 (neutral)	Temperature improves cross talk for Si interposer. Si interposer also has better thermal conductivity which helps spread the heat. This may or may not be a problem for glass interposer depending on the power levels.	See Chapter 4 for temperature effect on cross talk in silicon interposer and thermal analysis comparison in Chapter 5
Overall Signal Integrity	👉 (neutral)	👉 (neutral)	Si interposer minimizes the effect of RPDs which can be a very large benefit. However it also introduces additional RC effect. Glass interposer reduces insertion loss but introduces RPDs which need to be managed and can be problematic.	See Chapter 4 for details
Overall Power Integrity	👍 (thumbs up)	👎 (thumbs down)	Si Interposer damps resonances in the Power Distribution Network. This can be a large advantage especially with decoupling solutions, which indirectly improves signal integrity by eliminating RPDs.	See Chapter 4 for details

engineering resources. As seen from the table, our focus is primarily on electrical performance with some aspects of thermal management. A very important item that is neglected here is the mechanical reliability but we assume here that since the coefficient of thermal expansion (CTE) of the silicon interposer is better matched to silicon, this provides a better solution than glass interposer for mechanical reliability.

Due to the low resistivity (10Ω-cm) of CMOS grade silicon, the insertion loss of signal lines in the silicon interposer can be a problem. This issue can be corrected either by shielding the signal lines from the silicon substrate by introducing a ground plane or by increasing the resistivity of silicon (as done by Xilinx using a 20Ω-cm silicon substrate). More expensive ways of improving insertion loss for the silicon interposer is by using active or passive equalization, which is not discussed in this book. Since glass is a very good insulator, the insertion loss of the signal lines will always be superior as compared to silicon, assuming they are matched to 50Ω. The low resistivity of CMOS grade silicon also provides a major problem with cross talk due to the leakage problem, especially between the vias. Surrounding the signal TSVs with ground TSVs doesn't help much since the silicon substrate acts as the coupling mechanism. The RC effect of silicon leading to the long tail in the coupled waveform (explained in Chapter 4) can be a problem with the silicon interposer which is non-existent for glass due to its insulating properties. This comparison assumes that there is no reference planes (voltage and ground) in the silicon interposer. When these planes are added, since the cavity modes are dampened between the voltage and ground planes for the silicon interposer, this effect helps reduce the overall cross talk even though the via-via coupling is larger for silicon as compared to the glass interposer. On the other hand, for glass, the cavity modes can be large which increases the overall cross talk even though the via-via coupling is small and hence the cross talk levels become comparable with the silicon interposer for similar dimensions of the signal lines, with details in Chapter 4. We therefore provide a sideways thumb for both the silicon and glass interposer when voltage and ground planes are introduced in the interposer since one effect may over shadow the other depending on the dimensions used. Since, cross talk is a predictable phenomenon (meaning that it depends on the physical dimensions and spacing of the signal lines), they can be managed through suitable cross talk cancellation techniques, using appropriate shielding or signaling methods. Thermal management is another major issue that needs to be tackled. With silicon having a thermal conductivity of 148W/mK as compared to glass which is at 1.14W/m, silicon is a better spreader of heat which can help in decreasing the chip temperature by 4–5 degrees centigrade for similar conditions. This can be a very large benefit for silicon. In addition, since the conductivity of silicon decreases with increasing temperature, silicon becomes a better insulator at high frequencies and therefore coupled noise can decrease for signals with fast transitions. Though a secondary effect this indicates the need for understanding the temperature distribution for computing signal integrity accurately, since the reduction in cross talk can be significant as illustrated in Chapter 4. The sideways thumb for glass interposer in Table 1.4 is because the thermal issues, though important, may not be a show stopper but requires careful evaluation of the thermal management solution used. Finally, taking all of these effects into account, we provide a sideways -thumb for achieving signal integrity in the silicon interposer for high performance systems where reference places (voltage or ground or both) are used since though the return path discontinuity (RPD) effects are minimized, the channel is still lossy due to the dominant RC effect, if the channel includes TSV transitions (see Chapter 4). In contrast, the glass interposer solution can pose problems for signal integrity due to the presence of RPDs due to the cavity modes in between the voltage and ground planes, which can be corrected using decoupling capacitors or by changing the stack-up, resulting in the sideways thumb for this technology. Since, cavity modes are unpredictable, compensating for RC effect in the silicon interposer is a lesser problem as compared to mitigating the cavity modes in the glass interposer. Finally for managing power integrity, the silicon interposer gets a thumbs up for four reasons namely, (1) reduction in the amplitude of the resonances between voltage and ground planes which suppresses power supply noise, (2) its indirect influence on signal integrity by eliminating return path discontinuities, (3) better management of chip-interposer resonance due to the interaction between the two which though not covered in this book is discussed in [Swaminathan et al., 2007] and (4) better thermal management solution since power is ultimately converted to heat and because silicon is a better heat spreader and reacts favorably to temperature increase at high frequencies. Managing power integrity has always been problematic with any insulating material and therefore the glass interposer gets a thumbs-down which can be corrected using more expensive decoupling solutions.

Finally, in our humble opinion, the fundamental advantage that silicon interposer provides (even though it is a lossy material) is that it is a semiconductor, which in the future provides the possibility of integrating large size transistors using a low cost process. This provides the ability for integrating voltage regulator modules (VRMs) and other transistor level active circuits into the silicon interposer some day (a dream for most package designers) making it an active interposer as opposed to a passive one. In addition, technologies are available to embed trench capacitors into the silicon interposer with large capacitance density, which can significantly enhance power distribution [Dang et al., 2010]. On the contrary, the glass interposer due to its low loss properties can be a potential solution for RF applications where passives (inductors and capacitors) need to be embedded. This aspect of the problem is not covered in this book. In general, we recognize that though glass can be replaced with any other insulating material X with similar or better electrical properties, we still see a clear advantage for silicon as compared to X, especially for digital applications.

Bell Laboratories and IBM have developed silicon module technology in the early 1990s, and hence this technology by itself is not new. During that period, the primary author Madhavan Swaminathan, while at IBM, had the opportunity to compare and contrast silicon based MCM technology with other competing packaging technologies (was called Silicon on X or SOX technology internally within IBM) [Iqbal et al., 1994]. At that time, the only reason why the silicon MCM technology got a non-favorable response was because the input/output (I/O) terminals were along the periphery of the module which restricted the total number of I/Os that could be supported, thereby limiting the power that could be supplied to the ICs. The edge I/Os also increased the parasitics of these connections. With the advent of TSV technology, these issues no longer exist and therefore from an electrical standpoint, the silicon interposer should provide for a superior solution. Therefore both for 2.5D and 3D integration, the authors strongly believe in a silicon interposer based solution with some of the comparisons in Table 1.4 supporting this choice.

1.6 Modeling and Simulation

Designing 3D systems can be complex. One reason is because new technologies such as TSVs, fine pitch microbumps and others are only now beginning to become available. As the processes mature, uncertainty in their manufacturability will become less, making the system architects more comfortable with the usage of such new technologies. This is

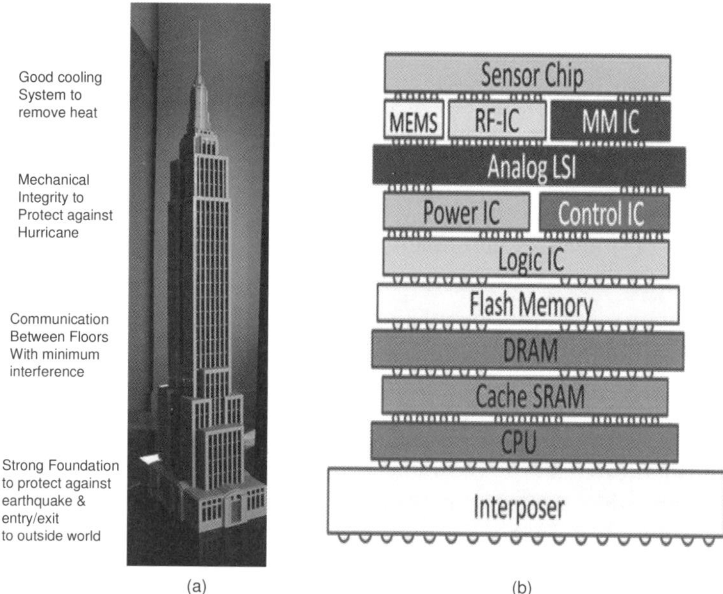

Figure 1.14: (a) Empire State Building and (b) 3D Stack.

beginning to happen today with companies such as TSMC (Taiwan Semiconductor Manufacturing Corporation) defining standard processes such as the CoWoS (Chip on Wafer on Substrate) [Goel, 2012] process and providing foundry service to any company who wants to use the technology. In addition, companies such as Samsung have internal foundries for their own products as well. As with any system, the availability of new technologies alone is insufficient for their implementation and often times electronic design automation (EDA) tools are necessary for supporting the design process. As systems shrink with the advent of 3D technology, it is expected that the interaction between multiple domains (which had minimum interaction with each other before) will grow and therefore the EDA tools need to support these interactions.

This is best explained using Figure 1.14 which shows the analogy between building a sky scraper such as the Empire State Building and stacking dissimilar ICs on each other. The mega functional 3D stack shown in Figure 1.14(b) is a grand vision which though not possible today, can certainly happen in the future as the technologies and know how for 3D integration mature. With any sky scraper, the foundation is very important for two reasons namely 1) it protects the building against an earthquake and 2) it serves as the entrance and exit point to the outside world. In countries such as Japan special care is taken to ensure that tall buildings are constructed to withstand earthquakes. In the case of 3D integration, the interposer plays such a role since it holds the stack together by ensuring that any reliability issues (such as CTE mismatch) with the second level package and PCB is not propagated into the stack. In addition, due to its proximity to the 3D stack, it provides the necessary input/output connections to the rest of the system. The TSVs in the 3D stack provide the communication between dissimilar ICs (such as between the CPU and DRAM) similar to the elevators in a sky scraper that connects different floors together. As the number of elevators increase, the waiting period for accessing these elevators decreases especially during the peak period, which is very similar to the concept of increasing bandwidth in a 3D stack by increasing the number of TSVs that support communication of data in parallel. In places such as Haiti, tall buildings are often constructed such that they withstand hurricanes, which is similar to managing the mechanical integrity of a 3D stack by controlling the stresses in TSVs and between ICs due to CTE mismatch and other effects. Finally, in a sky scraper, the cooling system has to have high efficiency. Since hot air rises, the cooling needs of each floor are very different. In a 3D stack a major problem is the ability to remove heat since it gets trapped between ICs. For mobile applications, where the heat flux levels are low and where no more than two ICs are stacked on each other, thermal management does not pose a major problem. But as the number of ICs in the stack grows and the heat flux level of each IC increases (such as in logic on logic applications), thermal management can be a major show stopper and unless methods are developed to remove heat, 3D integration will have limited applications. Since the sources of heat are the switching transistors, a major interaction occurs between the power delivery network and thermal dissipation network. In fact, the interaction between these two domains can be a major issue.

For 3D integration, the interaction between the three domains namely, electrical, mechanical and thermal requires multi-physics modeling

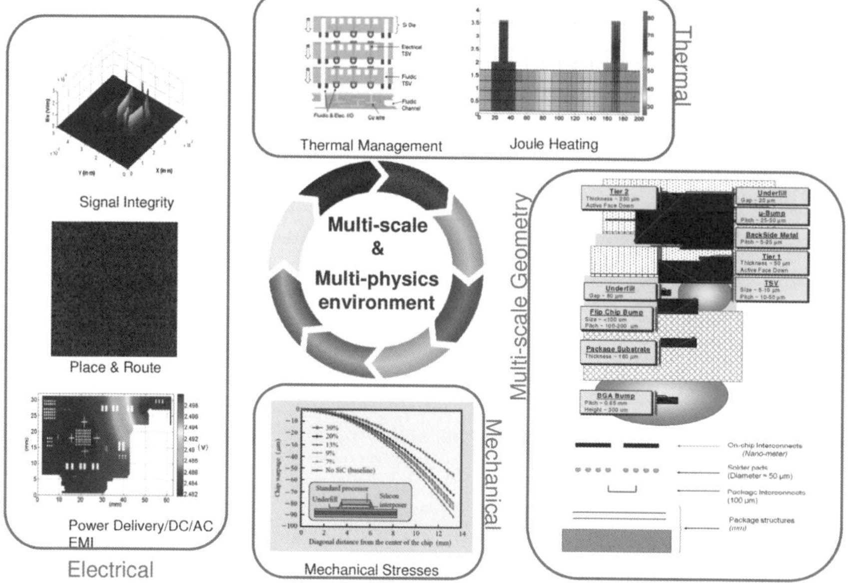

Figure 1.15: Multiphysics and Multiscale Modeling for 3D ICs and Interposers.

where appropriate information needs to be transferred between these domains to ensure that all the required specifications are met. In addition, the geometries of the interconnections have multiple scales ranging from the nanometer to the millimeter range, making the modeling even more complex. This is illustrated in Figure 1.15.

As an example consider the electrical response of TSVs which are defined by their physical dimensions such as diameter, height, oxide thickness, pitch, shape (cylindrical or tapered) and by the materials used (copper, tungsten, oxide and silicon). These structural and material properties not only affect their electrical response but also affect their mechanical and thermal behavior. The position of the TSVs, their density and material properties determine the thermal profile and gradients across the 3D stack based on the power maps of the switching transistor circuits. These power maps can either be static (steady state) or dynamic (transient), which in turn determine the thermal hot spots created in the stack. Hence, thermal modeling is required to ensure that appropriate junction temperature for the transistors is maintained. Unfortunately, this process gets complicated due to Joule heating where the heat generated from the conductors and dielectrics alters the thermal and electrical conductivity of metal and silicon, which in turn alters its thermal profile. The change in electrical conductivity with temperature affects the electrical properties requiring electrical modeling that captures the interaction between the electrical and thermal domains. Since electrical modeling by itself consists of assessing the DC IR drop, computation of current densities to estimate electro-migration limits and high frequency modeling to evaluate signal integrity (insertion loss, cross talk, matching, biasing) and power integrity (power supply noise, simultaneous switching noise) effects, this process in itself can be complex due to the multi-scale dimensions associated with the TSVs and other interconnections where the oxide thickness and aspect ratio along with their density can make modeling difficult. The temperature gradients influence the coefficient of thermal expansion (CTE) of metal and silicon which along with the CTE mismatch between metal and silicon for TSVs causes stresses around them, resulting in the interaction between the thermal and mechanical domains. These stresses can be managed by creating keep out zones (KOZ) which represent regions where the metal densities are reduced, which alters the spacing between signal lines and between TSVs, thereby altering its electrical response (such as insertion loss and cross talk). *In this book, the primary focus is on electrical modeling where the interaction between the electrical and thermal domains have been discussed to some extent in the context of packaging, with importance given to the TSVs and interposer rather than to the transistor level devices. Though very important, only a cursory discussion is provided to address the interaction between the electrical and mechanical domains by defining KOZs, where its effect on the electrical response is discussed. The stresses from TSVs also affect the device behavior by altering the electron and hole mobility for ICs, which is not covered in this book.*

For modeling and simulation, we introduce two concepts here namely the 3D Path Finder (3DPF) and design exchange formats (DEF) which can be useful for 3D integration. These concepts use the simulation methods described in this book, but provide a framework for decision making early in the design process, which may be necessary for 3D integration to be successful.

1.6.1 *Electrical Modeling and 3D Path Finder*

As a system architect, the technologies available for 3D integration are enormous. Though TSVs and microbump technology provide high density integration capability, often times mixing of technologies may be required to reduce the time to market, provided it supports the required performance.

As an example, consider the possible embodiments of memory on logic implementations shown in Figure 1.16(a) [Kumar et al., 2011] where the ICs are stacked using wirebonds, packaged and stacked wirebonded memory ICs are assembled onto a packaged logic IC using package on package format, ICs are either bonded face to face or front to back using TSVs to name a few. Other implementations include the use of glass or silicon interposer to enable stacking, similar to Figure 1.13(b). Clearly, there are a multitude of such embodiments possible. For each of these embodiments, an interconnection path similar to Figure 1.16(b) [Kumar et al., 2011] needs to be evaluated for its electrical performance such as say its insertion loss and timing. Since the whole concept of 3D integration for computing applications is to increase the throughput between ICs (product of I/O speed and number of I/Os), analyzing a single interconnection path may not be sufficient. Hence, multiple paths need to be evaluated to address cross talk and simultaneous switching

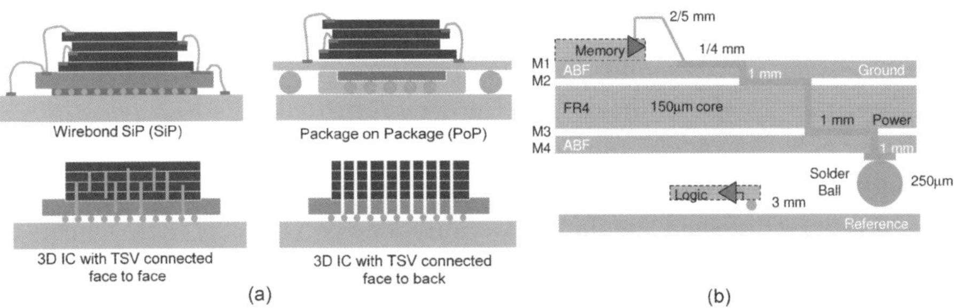

Figure 1.16: (a) Various implementations of memory on logic and (b) interconnection path for Package on Package [Kumar et al., 2011].

noise. In addition, for many of these implementations, the physical dimensions and properties of the materials used can change. As an example, for TSVs, the effect of changes in silicon conductivity and oxide thickness on electrical performance needs to be evaluated, along with the signal to ground ratio to ensure that channel loss and cross talk levels are managed. Since the TSVs interact with the redistribution layers (RDL), this effect needs to be addressed as well. Since the silicon substrate is lossy, a major decision to be made is whether to include a ground reference in the stack-up which can increase the layer count in the interposer. Maybe an alternative is to move to a glass interposer solution that minimizes losses. At an early stage of the design cycle, this exercise of evaluating the various options prior to implementation, to say the least, can be daunting.

The role of the 3D Path Finder (3DPF) as an electronic design automation (EDA) tool is to provide the necessary design and modeling framework for making tradeoffs early in the design cycle where a multitude of structures can be created with ease, the structures are parameterized for easy analysis, has the necessary accuracy for making decisions on the appropriate technology combinations to use and provides the appropriate direction without requiring elaborate interpretation of the results. The word "accuracy" can mean different things to different users. Often times, we relate accuracy with full wave electromagnetic analysis, where unless Maxwell's equations are solved in its entirety, we tend to doubt the results. A person trained in electromagnetic analysis understands the usage of such full wave EDA tools but most users of these tools at an early stage of implementation are not adept with such tools and therefore can make mistakes. One of the questions to therefore ask ourselves is "Are full wave electromagnetic tools always accurate?" At the other extreme, system architects and circuit designers like to work with parasitic elements such as resistance (R), inductance (L), conductance (G) and capacitance (C) to represent structures. For regular and simple geometries, these parameters can be extracted analytically and hence one possibility is to rely on these analytical equations for extraction, using which spice netlists can be created for simulation. This approach is in contrast to full wave analysis where simplifications in the extractions are used for improving the computational speed and also to obtain physical insight into the problem. Though this approach works in certain specific cases, in general for 3D integration, the number of parasitic elements to be considered can be enormous and therefore this method may lead to inaccurate results, especially when complex structures need to be evaluated, since most structures have frequency dependent behavior. Therefore, a compromise is required between full wave analysis and simple analytical based solutions, which could be used as the simulation engine for a tool like the 3D Path Finder. In this section, we elaborate on some issues related to full wave electromagnetic analysis or relying purely on analytical results and paint a high level picture of the 3D path finder.

1.6.1.1 *Full Wave Electromagnetic Analysis*

Full wave electromagnetic tools discretize either the integral or differential form of Maxwell's equations and solve them using several numerical methods with the more popular ones based on the Finite Element Method, Method of Moments and Finite Difference Method in either the time or frequency domain. It is important to note that all these methods had their origin in the microwave area and therefore sophisticated algorithms were developed to analyze the complex wave phenomena occurring in these structures at high frequencies. The users of these tools, in general, have a high level of expertise in electromagnetic analysis or microwave circuit design or both. In the microwave area, understanding the multiple modes of signal propagation is extremely important in determining the electrical performance of the structure. Often times, this analysis is restricted to a few signal and power interconnections. In contrast, for digital applications, the number of interconnections to be analyzed can be enormous (such as in 3D integration) and the mode of signal propagation can be assumed to be quasi-TEM (Transverse Electromagnetic). This assumption enables us to analyze the interconnections as transmission lines and be able to represent the discontinuities along the signal path as R, L, G, C parameters. With system architects, circuit designers, layout engineers, process engineers and more recently signal integrity engineers helping in the design of such complex systems, the electromagnetic know how on

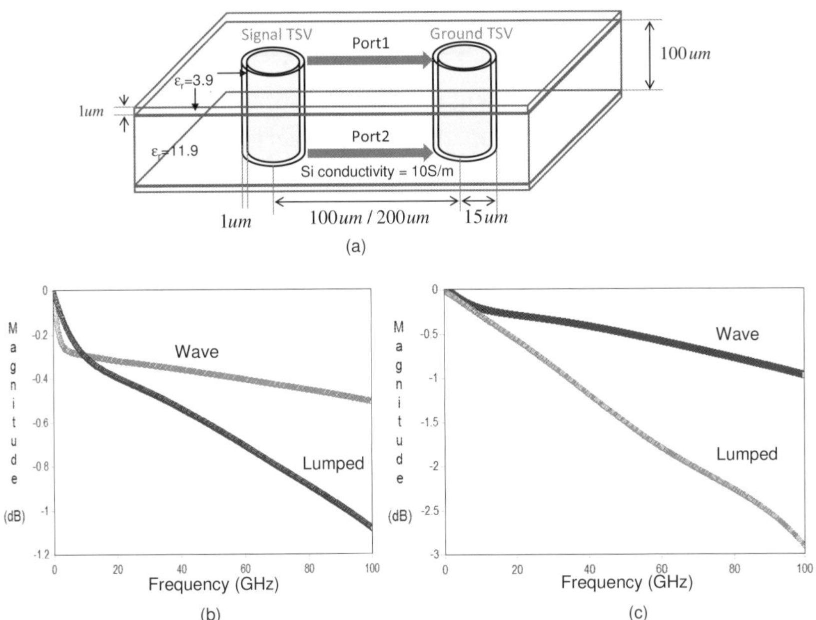

Figure 1.17: (a) TSV signal and ground pair, (b) 100μm pitch and (c) 200μm pitch.

the usage and interpretation of these full wave tools can at times be difficult and lead to misleading results. We illustrate this point using two simple examples related to TSVs by only looking at the results without questioning either the accuracy of the formulation or the numerical method used.

Consider the TSV pair shown in Figure 1.17. In full wave tools port definitions are important and to a user ports represent points where the structure can either be excited or measured. Two types of ports are often used in full wave analysis namely, lumped port and wave port. Lumped ports are used for TEM like modes while wave ports deliver better match to the mode pattern and provide higher accuracy while computing Scattering parameters. The physical dimensions for a signal-ground TSV pair are shown in Figure 1.17(a) where the TSV length, oxide thickness, radius and pitch are 100μm, 1μm, 15μm and 100/200μm respectively. The silicon conductivity and relative permittivity are 10S/m and 11.9 respectively while the relative permittivity for the oxide is 3.9. For such a simple geometry, the response for lumped and wave port can be very different as shown in Figure 1.17(b) for 100µm pitch and Figure 1.17(c) for 200µm pitch, for a frequency bandwidth of 100GHz. These results were generated using a commercial full wave tool. Depending on the dimensions of the vias, the difference between the lumped and wave port results can be significant at lower frequencies as well. Both these results are correct in the context of electromagnetic analysis since a better match between lumped and wave ports can be obtained as the dimension of the lumped port is decreased. Therefore a 60µm pitch will produce a better correlation between lumped and wave port.

Let's next consider reducing the port size (defined as the distance between signal and reference) by adding pads onto the TSVs as shown in Figure 1.18(a) where the structure is identical to Figure 1.17(a). As the dimensions of the pads increase resulting in a reduction in port length, the insertion loss appears to improve for the 100µm pitch, as shown in Figure 1.18(b). To minimize the pad effect, a perfect electric conductor (PEC) can be used for the pads. The results become even more complex for a signal TSV sharing multiple ground TSVs where PEC straps need to be used to short the ground vias together and keep them at the same potential. Considering that these examples are simple, interpreting the results for even such simple geometries can become challenging leading to the following questions: a) how large should the pads be and what are their effects on the over all response, b) is the lumped port small enough to produce appropriate matching, c) do we need to de-embed (or remove) the pad effects and d) is the insertion loss that sensitive to via pitch since one would expect the TSV parasitics to be small due to its very short length. So, the questions to be answered can be endless, requiring significant expertise. So, the assumption that full wave electromagnetic tools always produce the correct results can be at times misleading, since the accuracy depends to a large extent on interpretation. This can be a problem at an early design phase where decisions have to be made fairly quickly on the relative merits between technology options. It is our belief that full wave tools in the context of signal and power integrity analysis still has lots of room to grow, before they can be applied seamlessly by non-electromagnetic engineers for the analysis of designs.

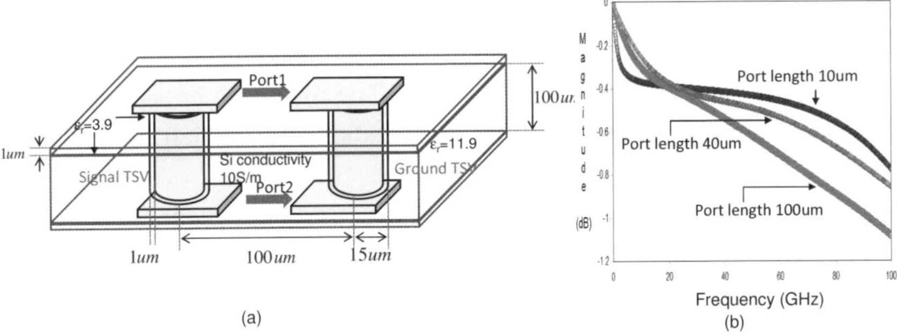

Figure 1.18: (a) TSV signal and ground pair with pads and (b) Insertion loss for different port lengths.

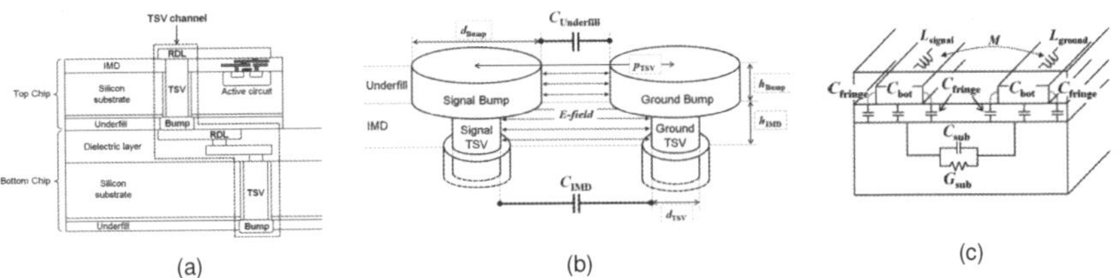

Figure 1.19: (a) TSV channel, (b) parasitics of the bumps and (c) parasitics of the RDL [Kim et al., 2011].

1.6.1.2 *Physics Based Analytical Models*

Designers often times like to extract models based on the physical structure by identifying the parasitics involved. An excellent paper based on this approach is by [Kim et al., 2011] where every parasitic in the interconnection path is first identified, analytical models are used to extract the parasitics and spice netlists are then generated to simulate the response by connecting the parasitic elements together. Through the use of analytical models, the critical dimensions can be parameterized and therefore the trends in the response can be evaluated by changing the physical dimensions. This is ideally suited for fine tuning a process where the critical parameters can be determined for process optimization.

An example of a structure analyzed by [Kim et al., 2011] is shown in Figure 1.19(a) where the TSV channel consists of an RDL (redistribution layer) connected to a TSV with a microbump at the bottom. The microbump connects to another RDL which then connects through a pad to the TSV in the die at the bottom. The material cross section is inhomogeneous with IMD (Inter metal dielectric which is typically silicon oxide), silicon substrate and underfill material. The parasitics between microbumps and pads are represented as equivalent capacitances $C_{underfill}$ and C_{IMD} respectively in Figure 1.19(b) assuming a constant electric field in the region. A similar approach is used to model the RDL layers as well, as shown in Figure 1.19(c) where the parasitic elements to be extracted become more complex. With all these parasitic elements computed an extensive spice netlist is then created where model to hardware correlations show that such an approach indeed produces very good accuracy.

Unfortunately, developing such physics based analytical models for the multitude of technology options available for 3D integration can be difficult for several reasons such as, a) determining all the structures for which parasitic elements need to extracted can be over whelming, b) for certain structures the development of analytical models may not be possible and therefore other tools may be required, c) determining the manner in which the various elements should be connected to each other may not be straightforward and d) the user is continuously guessing on the accuracy of the results since all of the coupling elements may not be included and therefore requires frequent correlation with other full wave electromagnetic tools for a sanity check. However, analytical models do have a place in any analysis since they provide physical insight into the parasitics associated with the structure, which can help in minimizing their effect.

1.6.1.3 *3D Path Finder*

In this book a conceptual tool called the 3D Path Finder (3DPF) is introduced with some elements integrated and available at www.e-systemdesign.com. The graphical user interface (GUI) of 3DPF is shown in Figure 1.20. As illustrated in the figure, the ability to mix and match technologies such as wirebond, TSV, glass interposer, C4, to support applications related to digital, analog, memory and mixed

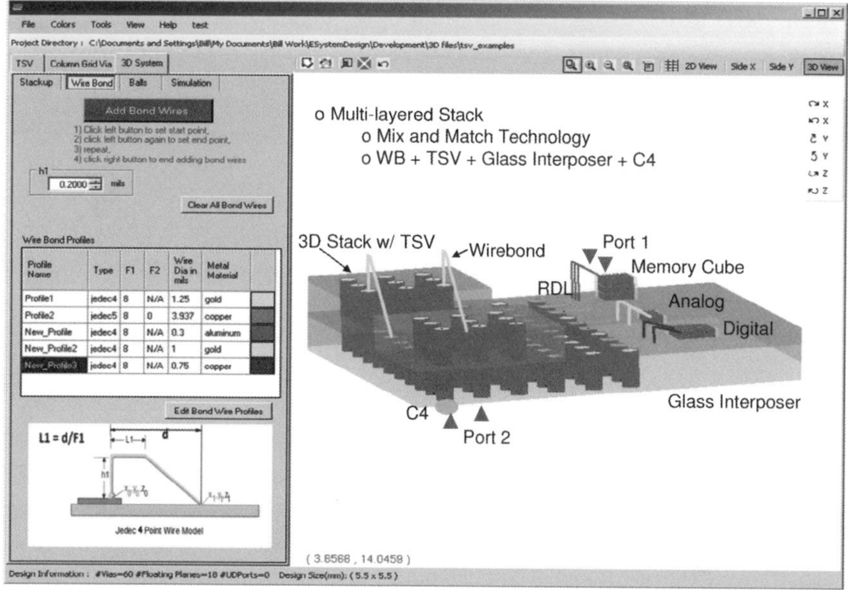

Figure 1.20: 3D Path Finder (3DPF) Graphical User Interface (Courtesy: www.e-systemdesign.com).

signal to name a few, and evaluate the overall electrical response at specific I/O terminals (shown as ports in Figure 1.20) at an early design phase is the role of the tool. The mechanical and thermal domains are equally important and needs to be covered as well (which is not shown in Figure 1.20).

In 3D integration, the authors believe that the vertical interconnections are the most important since they ultimately determine the electrical performance achievable. The lateral interconnections such as the redistribution layers provide the ability to connect to the vertical interconnections, and are often times kept short to minimize the parasitics along the signal path. Unlike the lateral interconnections, the vertical interconnections consisting of wirebonds, through silicon vias, glass vias, microbumps, C4s and ball grid arrays, to name a few, either have cylindrical or spherical cross section. This is in contrast to the lateral interconnections that have rectangular cross section. This information has been taken into account for analyzing both vertical and lateral structures using a combination of electromagnetic and circuit analysis in the frequency domain [Han et al., 2013], with details provided in Chapters 2 and 3. Using circuit concepts with the partial element equivalent circuit (PEEC) as the base, port definitions are suitably defined to extract the frequency domain response. These ports can either be defined with a local reference or infinite reference, providing a path for connecting other components to the structure both for signal and power integrity analysis. This can then be converted into a macro-model and synthesized into a spice netlist for time domain modeling as well, with details on macro-modeling available in [Swaminathan et al., 2007]. This approach has been used to analyze several examples provided in this book.

1.6.2 *Design Exchange Format*

Consider next the design of a system consisting of memory and logic ICs, as an example. In a 2D implementation, the communication between ICs occurs through the package and printed circuit board. Assuming the ICs are stacked on each other, the communication between them occurs through the vertical interconnections such as TSVs and microbumps. In today's industry, defining a supply chain is important for developing a product since few companies are vertically integrated. For example, Xilinx had to partner with TSMC and Amkor for their foundry and assembly services respectively, to develop the 2.5D interposer solution. This was doable since all of the ICs were designed by Xilinx and were assembled side by side, meaning that the interaction between the ICs occurred through the silicon interposer. Consider next the scenario where ICs are stacked in bare die form on each other where each IC is from a different vendor. With companies battling over patents and copyrights, it is unlikely that these vendors will be willing to share circuit level IP, detailed layout information, and other details such as the material stack up of their ICs with each other. This poses a huge problem in the design of the 3D stack since ICs interact with each other (good examples are in power delivery and thermal management), leading to the need for a design exchange format (DEF), which requires that suitable information be passed between ICs at the design phase such that when these ICs are simulated, designed, fabricated, assembled and tested, that they are still functional. The first step towards developing a DEF is to understand the information that needs to be passed between ICs to minimize error in the results. So, though exact modeling and simulation is not possible due to missing information, the goal is to minimize error such that the results provide accurate enough information to be able to design the individual ICs. This is covered in the context of steady state IR drop and temperature gradients [Bazaz et al., 2013] in this section due to their importance and inter-relationship through Joule heating (discussed in Chapter 5), followed by a short write-up on the progress being made for developing DEFs by the engineering community. Though this section uses rigorous analysis and is based on the methods described in Chapter 5, we have included this section in Chapter 1 for completeness and to amplify the need for exchanging detailed structural and material information between dies for ensuring the success of 3D integration.

1.6.2.1 *Example of a Two Die Stack*

Consider Figure 1.21(a) which consists of two dies stacked on each other, assembled on an interposer and mounted on a package or printed circuit board (PCB). The dies communicate with each other and the interposer through TSVs. Each die and interposer has its redistribution layers (RDL) as well. The power to the ICs is supplied from the power supply mounted on the PCB through voltage and ground planes (not shown).

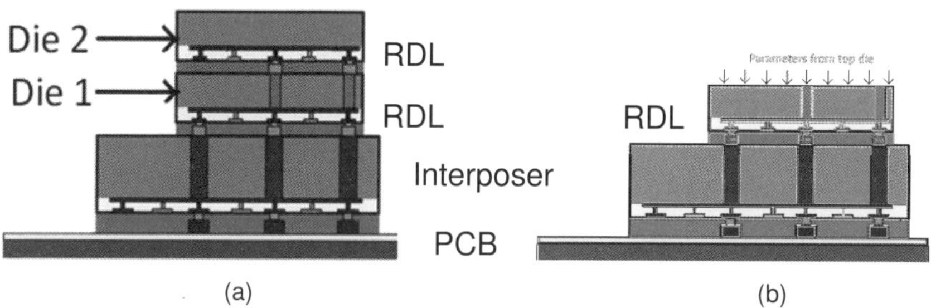

Figure 1.21: (a) Two die exact simulation and (b) Simulation of Die 1 with some parameters passed from Die 2 [Bazaz et al., 2013].

Assuming all of the details are known for Die 1 and Die 2, the entire system can be simulated for calculating the IR drop and temperature gradients. However, with each die being fabricated by a different vendor, not all the information is available. Therefore, for designing Die 1, it is important to determine the parameters that need to be passed from Die 2 (and vice versa), as illustrated in Figure 1.21(b). Examples of these parameters include the total power consumption of Die 2, power map of Die 2, TSV distribution of Die 2 to name a few. The objective here is to minimize the error in the simulated results between Figure 1.21(a) and Figure 1.21(b) such that the data from Figure 1.21(b) can be used to appropriately design Die 1, where error is defined as:

$$Error = \frac{\Pr(Die1 + Die2) - \Pr(Die1)}{\Pr(Die1 + Die2)} \qquad (1.1)$$

with $\Pr(Die1 + Die2)$ being the parameter simulated for Figure 1.21(a) when all the information of both dies are available and $\Pr(Die\ 1)$ is the parameter simulated when some information is passed from Die 2 to Die 1, as in Figure 1.21(b). The goal of defining a DEF therefore is to determine the parameters that need to be passed between dies for their individual design.

1.6.2.2 *IR Drop*

Consider the 3D stack shown in Figure 1.22(a) consisting of a two die stack with all layout details available. In this section, we use the methods described in Chapter 5 to compute the IR drop. In Figure 1.22(b), information from Die 2 is passed to Die 1 in the form of current sources. The dies are connected in face- to- back configuration [Xie et al.]. Hence no TSVs are required in Die 2. The interposer and the two dies are connected to each other using micro bumps. A voltage source is used at the bottom corner as shown in Figure 1.22(a). Current sources are used to simulate the active devices. For a given amount of power and input voltage the required current values can be calculated. A 1V voltage source is used as the power supply. Figure 1.22(c) shows the layout of the RDL layers in each die and the interposer. It comprises of two layers

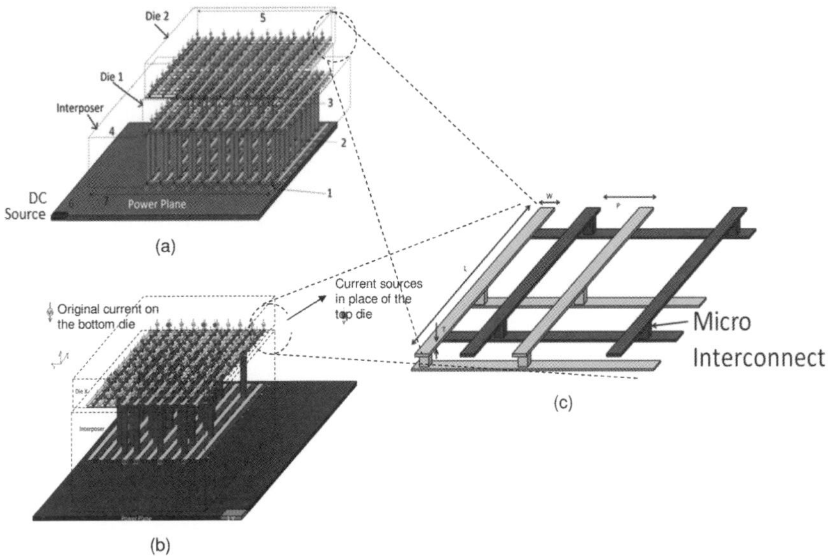

Figure 1.22: (a) Die 1 and Die 2 with RDL layers, (b) Die 1 with RDL and Die 2 represented as current sources and (c) RDL details [Bazaz et al., 2013].

of metal rows placed orthogonally on top of each other. Power and ground rows alternate in each layer. The corresponding power/ground rows in each layer are connected to each other through metal connections which are referred to as micro interconnects in Figure 1.22(c). For simplicity ground vias are not taken into account in this example. The RDL layer at each level gives the flexibility to route the power from one part of the die to the other.

All components including bumps, micro bumps and TSVs are assumed rectangular in cross section for simulation with dimensions as shown in Table 1.5. The RDL layer has length (L) = 3mm for the interposer and 1.1mm for each of the dies. The thickness (T) of power/ground conductors is 1µm. The width (W) is 10 µm and the pitch (P) between power and ground conductor is set to 50 µm. The TSV pitch used is 100µm. The power consumed by the two dies is assumed to be 5W for the top die and 2W for the bottom die, respectively. The error is calculated by taking the voltage drop on Die 1 as the reference when simulating the stack-up in Figure 1.22(a).

Table 1.5: Dimensions of the structures used for simulation.

Index	Component	Connection	Area	Height
1	Bumps	Power plane to interposer	50 X 50 μm^2	50 µm
2	TSV(s)	Interposer to micro bump	30 X 30 μm^2	200 µm
3	TSV(s)	Die 1 to Die 2	10 X 10 μm^2	100 µm
4	Micro Bumps	Interposer to Die 1	20 X 20 μm^2	20 µm
5	Die (1,2)	-	1.1 X 1.1 mm^2	100 µm
6	Power Plane	-	10 X 10 mm^2	30 µm
7	Interposer	-	3 X 3 mm^2	200 µm

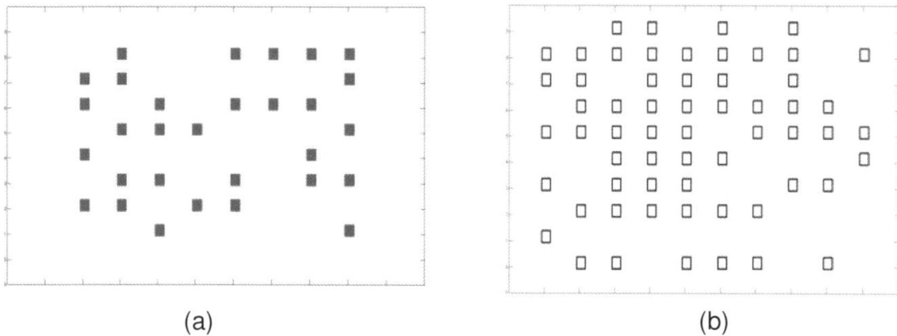

(a)   (b)

Figure 1.23: TSV map from (a) Interposer to Die 1 and (b) Die 1 to Die 2 [Bazaz et al., 2013].

The TSV locations in the interposer and Die 1 are chosen in a random manner (in this example though that is not preferred in a real design) and maintained the same for every simulation, with positions as shown in Figure 1.23(a) for interposer to Die 1 and in Figure 1.23(b) from Die 1 to Die 2. The solid squares denote the TSV positions from interposer to Die 1, while the hollow squares denote TSV positions from Die 1 to Die 2.

The simulations were performed in the following order. First a two die simulation was performed observing the voltage drop on Die 1,

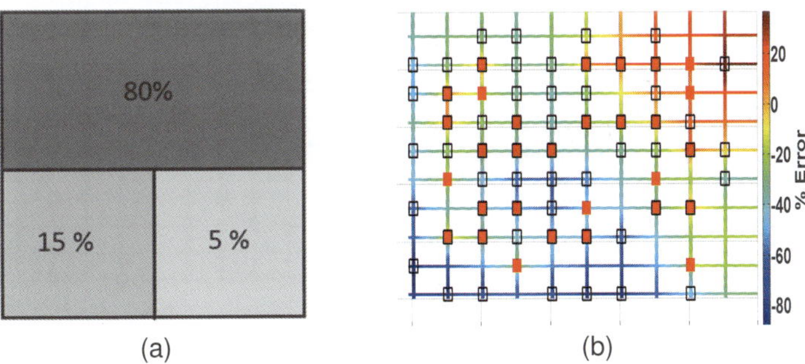

Figure 1.24: (a) Power map for Die 2 and (b) Error in voltage drop on Die 1 [Bazaz et al., 2013].

followed by a single die simulation. Voltage drops were calculated on the RDL layers of Die 1 which is just below the active device. The power on both dies are equal and uniformily distributed in Die 1 but with a power map on Die 2 as shown in Figure 1.24(a). In the second set of simulation only Die 1 is considered. The information that is passed from Die 2 is its total power and power map as in Figure 1.24(a). Using the total power and voltage, individual current sources were placed on top of TSVs (from Die 1 to Die 2) to mimic the current being drawn by Die 2. With power and voltage known, the total current can be calculated. This current was then equally distributed on the TSVs placed within each power map. Rather than plot the absolute voltage drops, the error between the two Die and single Die simulation is shown in Figure 1.24(b). The negative error indicates that the single Die simulation over estimates the voltage drop while a positive error represents the regions where the voltage drop is under estimated. Clearly the error is large between -80 to 20% and with such large errors; it becomes difficult to design Die 1. So, what other information needs to be passed from Die 2 to Die 1 to minimize error?

In Figure 1.24(b), due to the absence of the information pertaining to the RDL layer in Die 2 and the exact position of the current sources for the single Die simulation, the main source of the error ocurs due to inaccuracies in estimating the current being drawns by the TSVs from Die 1 to Die 2. Should Die 2 provide a much finer power map (or current map which can be estimated from the power map and voltage) to Die 1,

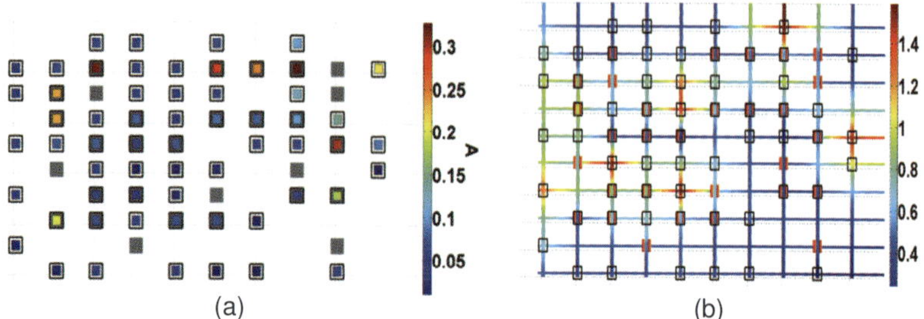

Figure 1.25: (a) Current drawn by each TSV from Die 1 to Die 2 and (b) Error in voltage drop on Die 1 [Bazaz et al., 2013].

then the currents drawn by the TSVs can be better estimated and the error reduced. This is illustrated in Figure 1.25(a), where the current drawn by each TSV is passed from Die 2 to Die 1 (the authors realize that this may be difficult to obtain but is shown here to illustrate the importance of this information in minimizing error) resulting in negligible error in the voltage drop estimation, as shown in Figure 1.25(b). This leads to the use of a Design Exchange Format for IR drop estimations as follows:

(i) Die 2 provides a power map of fine enough granularity that can be used to estimate the current drawn. This needs to be close to the TSV pitch used to connect Die 1 to Die 2, if possble. The coarser the power map, the larger is the resulting error.

(ii) Based on the current profile of Die 2, Die 1 places TSVs connecting Die 1 to Die 2 by ensuring a maximum current limit for each TSV (should there be electromigration concerns which is not discussed in this book).

(iii) Die 1, by relying on the single die simulation, optimizes the voltage drop on its RDL layers by including the TSVs through the interposer.

(iv) Die 1 passes the dimensions and exact location of the TSVs from Die 1 to Die 2 including material properties along with the potential at the bottom of the TSVs based on the single die simulation.

(v) Die 2 uses the information passed from Die 1 to estimate the voltage drop on its RDL layers (not covered in this section).

1.6.2.3 *Thermal Management*

Let's now turn to thermal simulations to determine the temperature gradients in each die for a two die stack, as shown in Figure 1.26. Each die consists of the RDL layers with silicon dioxide as the insulator on a silicon substrate, similar to the IR drop simulations. The interposer and the two dies are bonded together using an adhesive.

A power plane is used in the FR4 board to spread the heat. Power densities are defined for each die to replicate the effect of active devices. The board, substrate, and TSV dimensions are the same as in Table 1.5. The computational methods described in Chapter 5 are used for the thermal simulations in this section, where isothermal boundary conditions are used at the top to simulate an ideal heat sink, with a constant temperature of 30°C. Convection boundary condition with a heat transfer coefficient $10 \mathrm{Wm^{-2}K^{-1}}$ is used, as shown in Figure 1.26.

The conductivity and thickness of the various materials used to simulate the conditions in Figure 1.26 are tabulated in Table 1.6. Power densities of $1\mathrm{W/mm^2}$ are assumed for both dies, which represents a large heat flux, corresponding to logic on logic application. To protect IP, Die 2 only passes its power map to Die 1 as shown in Figure 1.24(a). A constant power map is assumed for Die 1. The objective is to be able to simulate the temperature profiles in Die 1 using this information with minimum error as compared to a two die simulation where all the stack-up details as in Figure 1.26 are available.

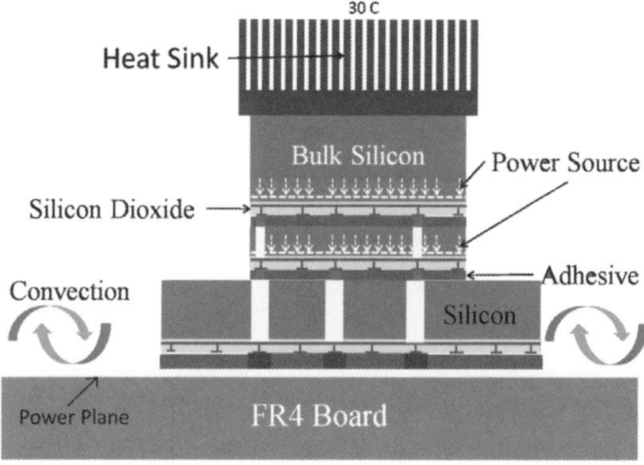

Figure 1.26: Thermal simulations for a two die stack.

Table 1.6: Material thickness and thermal conductivity.

	Material Thickness	Thermal Conductivity (Wm^{-1}K^{-1})
Silicon	100 μm	110
Bulk Silicon	500 μm	110
Adhesive	50/20 μm	0.3
FR4 Board	800 μm	4.3
Silicon Dioxide	10 μm	1.3
Copper (Power Plane)	30 μm	400

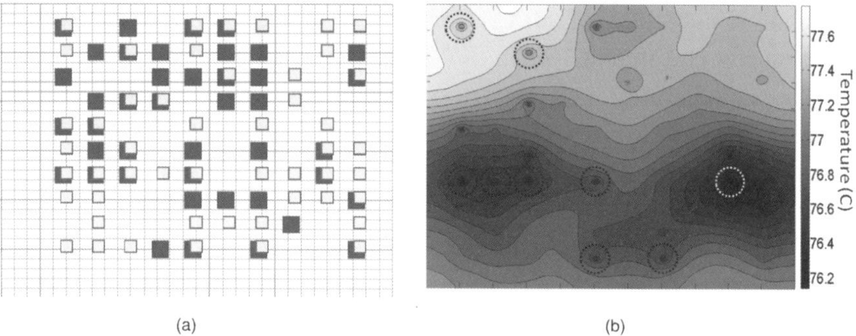

(a)   (b)

Figure 1.27: (a) TSV map and (b) Temperature profile for Die 1 [Bazaz et al., 2013].

Since TSVs provide a low resistance thermal path for heat to flow towards the heat sink, their position and cross sectional dimension are critical to be able to build a good model. The mapping information for both the TSVs in the interposer and Die 1 are available to Die 1, which are shown in Figure 1.27(a) where the dark squares represent TSVs from interposer to Die 1, while the light smaller squares are TSVs from Die 1 to Die 2.

The simulated temperature profile on Die 1 based on a two die simulation is shown in Figure 1.27(b). In Figure 1.27(b) the regions where the temperature is low (indicated by circles) correspond to positions where the TSV in Die 1 is directly above a TSV in the interposer. Since TSVs are the primary sources of heat removal from the die, they conduct the heat away from Die 1 towards the heat sink on the top and the board at the bottom, resulting in reduced temperature at these locations.

Consider next the single die simulation. To be able to simulate the stack in Figure 1.26, the missing information is the RDL layers in Die 2. This information is critical since it is in the direct path of heat flow from Die 1 towards the heat sink and can have a significant effect on the temperature profile. The RDL layers can be modeled using an equivalent thermal conductivity assuming the volume of metal and oxide in the RDL layers along with the thickness can be provided by Die 2. With this information, the equivalent thermal conductivity can be computed as:

$$k_{eq} = x \times 400 \mathrm{Wm^{-1}K^{-1}} + (100 - x) \times 1.3 \mathrm{Wm^{-1}K^{-1}} \qquad (1.2)$$

where x is the percentage volume of metal (copper) with thermal conductivity $400 \mathrm{Wm^{-1}K^{-1}}$ and $(100 - x)$ is the percentage volume of oxide with thermal conductivity $1.3 \mathrm{Wm^{-1}K^{-1}}$. The equivalent thermal conductivity along with the total thickness of the RDL layer and power map can be used to mimic Die 2 in the stackup in Figure 1.26. Unlike IR drop simulation, the fine granularity of the power map in Die 2 passed to Die 1 is not necessary, and a coarse power map as in Figure 1.24(a) is sufficient. The percentage error in the temperature profile of Die 1 as compared to Figure 1.27(b) is shown in Figure 1.28(a) for 10% metal and

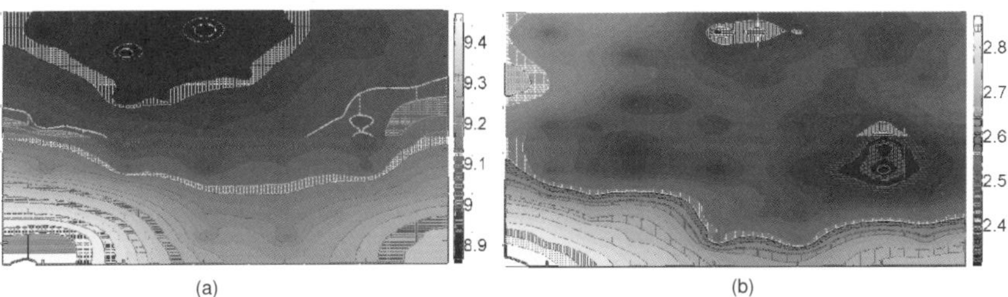

(a)   (b)

Figure 1.28: Temperature profile in Die 1 using single die simulation with (a) 10% metal and 90% oxide for RDL of Die 2 and (b) 5% metal and 95% oxide for RDL of Die 2 [Bazaz et al., 2013].

90% oxide. The error is relatively large (~8–9%) and can be reduced by using 5% metal and 95% oxide as in Figure 1.28(b) where the error is reduced (~3%). Hence the accuracy of the temperature estimated in Die 1 for a single die simulation depends to a large extent on the supply of accurate information on the volume percentage of metal in the RDL layers of Die 2 and its thickness. The positive error indicates that the temperature is being underestimated in the single die simulation.

A possible design exchange format for tempertaure gradient estimations for enabling thermal design can therefore be defined as follows:

(i) Die 2 passes information on the equivalent thermal conductivity of its RDL layers along with the RDL and bulk silicon thickness. A power map of the active circuits in Die 2 is also passed to Die 1.

(ii) Die 1 constructs a thermal model using the thermal conductivity of the material interfaces (such as underfill), microbumps, heat sink, convection heat transfer coefficient, equivalent thermal conductivity of Die 2, bulk silicon and others. The heat sources in the form of power maps of Die 2 and Die 1 are included in the model. Die 1 positions the TSVs in the interposer and Die 1 based on the power map of both dies to enable the least thermal resistance path. This exercise can also involve Die 1 ensuring that its high power area doesn't overlap with a high power area in Die 2, to maximize heat flow.

(iii) Die 1 simulates the stack to estimate its temperature profile and optimizes the placement of TSVs and its power map to minimize its temperature gradients and hot spots.

(iv) Die 2 uses the temperature profile of Die 1 as a boundary condition along with the details of its RDL layers, heat sources and TSV locations to obtain an estimation of its temperature profile (not covered in this section).

For both IR drop and thermal simulations, a tool such as the 3D Path Finder can be very useful since it can be used to estimate the compatibility between dies during the early design phase, before the tape out of such designs occur.

1.6.2.4 *Move Towards a DEF by the Engineering Community*

Thermal and power delivery networks are just two examples of design exchange formats that are required for the design of 3D ICs. These can be expanded to include design exchange formats for partitioning and floor planning, stress analysis, signal integrity management and Design for Test. The DEF described in this chapter for IR drop and thermal simulations is based on early work and by no means represents a comprehensive solution. The authors hope that this preliminary work will amplify the importance of DEF for 3D integration and will lead to refinement in these formats over time.

A working group consisting of representatives from Sematech, Semiconductor Research Corporation (SRC), Si2, several universities including North Carolina State University (NCSU), University of San Diego (UCSD), University of Minnesota (UMN), University of California Los Angeles (UCLA), Georgia Tech, several semiconductor and EDA companies have identified the development of DEF for 3D as an important exercise and are developing standards for it. Excerpts of the general principles for defining a DEF put together by a group consisting of Altera, Samsung, UCLA, IBM, NCSU, UCSD, Global Foundries, Intel, Georgia Tech, UMN and Qualcomm in 2011 were as follows:

(i) The requirements of the standard are the minimum necessary to produce satisfactory answers and no more. Producing such models is a customer support function, not a core engineering function, i.e. it is not generated by the chip design scheme. The role of the DEF is to facilitate the transfer of information through a compact model, not to build one.

(ii) The resulting compact model be verifiable. A verifiable model ideally has the following properties:

- It be human readable. Often a quick scan of the model shows what is wrong.
- Its syntax is verifiable.
- It has some physical basis, though non-physical models can also be acceptable if provided in the appropriate spice format.
- It be verifiable through measurement.

(iii) An active technical community is available, which is willing to drive evolution of the standard as essential needs of the community changes.

Clearly, as the 3D IC world unfolds, the need for Design Exchange Formats and their adoption as a standard by the engineering community will get validated. Only time will tell if such DEFs are even necessary!

1.7 Summary

The electronics industry is beginning to recognize 3D integration as the next big wave for enabling the miniaturization of systems. 3D integration provides benefits both for Moore and More than Moore scaling by providing larger bandwidth, better yield, heterogeneity and most importantly a shorter time to market. Of course, like any new technology, standardization of design formats, processes, assembly and design for test have to evolve over time, supported by a supply chain that can deliver in volume and at low cost. In this chapter, a general overview of 3D integration is provided with emphasis on the need for modeling and simulation at an early design stage, often times categorized as a path finding activity. A high level comparison of two competing interposer technologies is also discussed, with inferences drawn from the remaining chapters in this book. A preliminary discussion on the need for design exchange formats (DEF) is also addressed in this chapter.

With a focus on Design and Modeling for 3D integration, the rest of the chapters are organized as follows:

- In Chapter 2, the numerical methods used to extract and simulate the parasitics of interconnections are discussed in detail. The focus is on using specialized basis functions for analyzing structures with cylindrical cross section and combining it with the more classical methods such as partial element equivalent circuit (PEEC) for computing the electrical response. This chapter is ideally suited for tool developers and others with a strong background in computational electromagnetics and circuit theory.

- Though Chapter 3 is a continuation of the numerical method in Chapter 2 applied to through silicon via (TSV) structures, it starts with a fundamental treatment of TSVs for understanding its electrical behavior using physics based models. This chapter also describes the effect of biasing on the TSV response and the impact of via taper. The methods discussed are supported through numerous examples.

- The focus of Chapter 4 is on design related to Electrical Performance and Signal Integrity. The first part of the chapter can be useful for process engineers and designers for defining optimum parameters for the TSV structures such as oxide thickness, length, metallization, silicon resistivity, radius, to name a few. A significant part of the chapter focuses on cross talk and temperature effect. A comparison of Silicon and Glass interposer in the context of signal integrity is discussed in this chapter leading to some of the conclusions in Chapter 1.

- Chapter 5 addresses the two most important areas for 3D integration namely, power distribution and thermal management. The criticality of managing power supply noise and the role of return path discontinuities (RPD) are addressed. With voltage scaling and power levels increasing, this is a major problem for the electronics industry today. The importance of capturing the interaction between electrical and thermal analysis due to Joule heating and their response in the presence of conduction, convection and fluidic cooling is illustrated in this chapter.

- In Chapter 6, alternate methods for power distribution are discussed for managing power supply noise and RPDs. A new concept based on power transmission lines with alternate signaling methods are introduced with a focus on minimizing RPDs and reducing power supply noise. These methods are applied for chip to chip communication in a printed circuit board with some early results for 3D ICs.

References

1. [Moore, 1965] Gordon E. Moore, "Cramming more components onto integrated circuits", Electronics, Volume 38, Number 8, April 19, 1965.

2. [Perry, 2008] Tekla S. Perry, "Gordon Moore's Next Act", IEEE Spectrum, May 2008.

3. [Tummala et al., 1989] Rao R. Tummala and Eugene J. Rymaszewski, "Microelectronics Packaging Handbook", Van Nostrand Reinhyold, 1989.

4. [Tummala et al., 2008], Rao R. Tummala and Madhavan Swaminathan, "Introduction to System on Package (SOP) – Miniaturization of the Entire System", McGraw Hill, 2008.

5. [Swaminathan et al., 2011] M. Swaminathan, V. Sundaram, J. Papapolymerou and R. Pulugurtha, "Polymers for RF Apps", IEEE Microwave Magazine (Invited Feature Article: Cover: Packed and Connected), pp. 62–77, 2011.

6. [Kwon, 2011] Oh-Hyun Kwon, "Eco-Friendly Semiconductor Technologies for Healthy Living – Plenary Talk", International Solid State Circuits Conference (ISSCC), 2011.

7. [Bae et al., 2008] S. Bae et al., "A 60nm 6Gb/s/pin GDDR5 Graphics DRAM with Multifaceted Clocking and ISI/SSN ‑ Reduction Techniques", *IEEE International Solid State Circuits Conference*, pp. 278–279, 2008.

8. [Knickerbocker et al., 2005] J. U. Knickerbocker, P. S. Andry, L. P. Buchwalter, A. Deutsch, R. R. Horton, K. A. Jenkins, Y. H. Kwark, G. McVicker, C. S. Patel, R. J. Polastre, C. D. Schuster, A. Sharma, S. M. Sri-Jayantha, C. W. Surovic, C. K. Tsang, B. C. Webb, S. L. Wright, S. R. McKnight, E. J. Sprogis, B. Dang, "Development of Next ‑ Generation System ‑ on ‑ Package (SOP) Technology Based on Silicon Carriers with Fine ‑ Pitch Chip Interconnection", *IBM Journal of Res. & Dev.*, Vol. 49, No. 4.5 pp. 725–753, July/September 2005.

9. [Knickerbocker et al., 2006] J. U. Knickerbocker, C. S. Patel, P. S. Andry, C. K. Tsang, L. P. Buchwalter, E. J. Sprogis, Gan Hua, R. R. Horton, R. J. Polastre, S. L. Wright, and J. M. Cotte, "3D Silicon Integration and Silicon Packaging Technology Using Silicon Through Vias", IEEE Journal of Solid State Circuits, pp. 1718–1725, Vol. 41, No. 8, August 2006.

10. [Madden et al., 2012] Liam Madden, Ephrem Wu, Namhoon Kim, Bahareh Banijamali, Khaldoon Abugharbieh, Suresh Ramalingam and Xin Wu, "Advancing High Performance Heterogeneous Integration Through Die Stacking", Proceedings of the European Solid State Device Research Conference (ESSDERC), pp. 18–24, 2012.

11. [Swaminathan et al., 2007] Madhavan Swaminathan and Ege Engin, "Power Integrity Modeling and Design for Semiconductors and Systems", Prentice Hall, 2007.

12. [Goel, 2012] S. K. Goel, "Test Challenges in designing complex 3D chips: What is on the horizon for EDA industry?", International Conference on Computer Aided Design (ICCAD), pp. 273, 2012.

13. [Iqbal et al., 1994] A. Iqbal, M. Swaminathan, M. Nealon and A. Omer, "Design Trade-off among MCM-C, MCM-D and MCM-D/C Technologies" , *IEEE Trans. on*

Components, Hybrids and Manufacturing Technology, Vol. 17, No. 1, pp. 22–29, Feb. 1994.

14. [Dang et al., 2010] Bing Dang, Michael Shapiro, Paul Andry, Cornelia Tsang, Edmund Sprogis, Steven Wright, Mario Interrante, Jonathan Griffith, Van Truong, Luc Guerin, Roger Liptak, Daniel Berger, and John Knickerbocker, "Three-Dimensional Chip Stack With Integrated Decoupling Capacitors and Thru-Si Via Interconnects", Vol. 31, No. 12, IEEE Electron Device Letters, Dec. 2010.

15. [Kumar et al., 2011] Gokul Kumar, Tapobrata Bandyopadhyay, Vijay Sukumaran, Venky Sundaram, Sung Kyu Lim and Rao Tummala, "Ultra-high I/O Density Glass/Silicon Interposers for High Bandwidth Smart Mobile Applications", ECTC, pp. 217–223, 2011.

16. [Kim et al., 2011] Joohee Kim, Jun So Pak, Jonghyun Cho, Eakhwan Song, Jeonghyeon Cho, Heegon Kim, Taigon Song, Junho Lee, Hyungdong Lee, Kunwoo Park, Seungtaek Yang, Min-Suk Suh, Kwang-Yoo Byun, and Joungho Kim, "High-Frequency Scalable Electrical Model and Analysis of a Through Silicon Via (TSV)", IEEE Transactions on Components, Packaging and Manufacturing Technology, Vol. 1, No. 2, pp. 181–195, February 2011.

17. [Han et al., 2013] Ki Jin Han and Madhavan Swaminathan, "Modeling electrical interconnections in three-dimensional structures", US Patent Number: 8,352,232, Issue date: January 8, 2013.

18. [Bazaz et al., 2013] Rishik Bazaz, Jianyong Xie and Madhavan Swaminathan, "Electrical and Thermal Analysis for Design Exchange Formats in Three Dimensional Integrated Circuits", International Symposium on Quality Electronic Design (ISQED), 2013.

19. [Bandyopadhyay, 2011] T. Bandhyopadhyay, "Modeling, Design and Characterization of Through Vias in Silicon and Glass Interposers", Ph.D. Dissertation, Georgia Institute of Technology, 2011.

20. [Xie et al.] Yuan Xie, Jason Cong and Sachin Sapatnekar, "Three Dimensional Integrated Circuit Design EDA, Design and Microarchitectures (Integrated Circuits and Systems)", Springer Publications, 2009.

Chapter 2

Modeling of Cylindrical Interconnections

Interconnections play a very important role in the miniaturization of systems. With a reduction in volume being an important driver for future electronic systems, vertical interconnections are beginning to play an important role in providing the necessary electrical performance. Vertical interconnections enable the 3D stacking of ICs and packages, as illustrated in Chapter 1. These vertical interconnections have an important characteristic that their cross section is cylindrical (or spherical in the case of solder balls) unlike the lateral interconnections that have a rectangular cross section. This is true for wirebonds for the stacking of ICs or column grid arrays for Package on Package (POP) stacking. Unlike the lateral interconnections used for redistribution in a package, the wirebonds can criss cross each other causing electromagnetic coupling. In addition, the frequency dependent variation of the vertical interconnections can affect performance. In this chapter, the electromagnetic modeling of wirebonds, vias and other interconnections with cylindrical cross sections are discussed with an extension to through silicon vias in Chapter 3. A robust method for modeling these interconnections is discussed over a broad frequency range followed by several results for both canonical and realistic structures.

2.1 Introduction

The challenges of multi-functional integration in today's electronic applications can be overcome by using package-based system integration. To reduce volume, instead of placing the ICs next to each other and interconnecting them using a package, the ICs can be placed on top of each other and interconnected using vertical interconnections. The three embodiments of vertical stacking are shown in Figure 2.1 where the ICs can be interconnected using wirebonds, through silicon vias (TSV) or Package on Package (POP) techniques. Unlike wirebond and TSV based stacking, in POP scheme the ICs are first pre-packaged before stacking the packages on each other. Since the length of the wirebonds are always longer than the TSV based approach, it is expected that wirebond based stacking will have limited performance, but provide a lower cost method for integration. Therefore, in terms of electrical performance, the TSV based approach will perform better as compared to POP which in turn will be better than Wirebonds. It is important to note that all three approaches provide viable methods for integration and it is expected that these will be used in a wide range of products being initially driven by consumer and mobile applications.

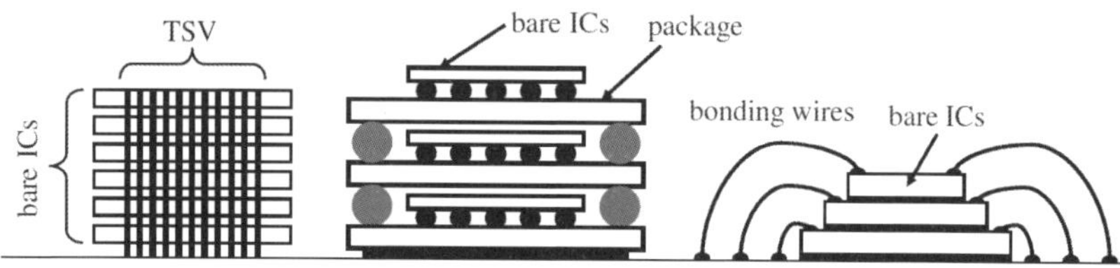

Figure 2.1: Examples of System in Package (SIP) structures with interconnections for 3D stacking [Tummala, 2005].

Since processing the vertical interconnections is a lot more challenging as compared to the planar ones, the electrical and mechanical characteristics of the vertical interconnections can be a bottleneck for achieving the required system performance. Currently, a popular stacking solution is using bonding wires because their processing is very mature and cost effective. Since the original beam lead technology of AT&T, bonding wires have been the preferable technology for chip-to-package interconnections because of their high flexibility, high reliability, and low defect rates [Tummala, 2001]. Bonding wires have been useful for low-frequency and consumer electronic applications, and the technology is still evolving with increasing levels of integration. A method for increasing density is to reduce the wire pitch, which is currently at about 60μm [Milke et al., 2007]. For high-speed applications, optimization of wiring enables the use of bonding wire interconnections, as reported in [Kam et al., 2008] [Kim et al., 2009] [Hu et al., 2008]. The application of bonding wires to 3-D integration was proposed in the late 1990's [Fukui et al., 2000].

Although bonding wire technology is a popular interconnection choice for 3-D integration, its undesirable electrical properties create many challenges. A major problem is the estimation of the electrical behavior of 3-D interconnections since their electrical model can be very difficult to extract. A major difficulty in modeling 3-D interconnections comes from the need to obtain the entire coupling model that includes a large number of 3-D interconnections. In a typical stacked structure composed of several ICs, the number of bonding wires can be close to a thousand [Dreiza et al., 2005], causing coupling between interconnections due to criss-crossing of the wires. Since achieving higher integration density reduces the pitch between interconnections, the number of interconnections will continue to increase, causing stronger electrical coupling which can lead to unwanted electromagnetic interference. Furthermore, for accurate electrical design that requires the stacking of radio frequency (RF), analog, and digital ICs, the interconnection models should cover a large frequency range. This broadband model needs to capture frequency-dependent losses, coupling, and any mismatch, which are contributed by the parasitic elements such as inductance, resistance, capacitance, and conductance. Extracting the frequency-dependent conductor loss and inductive coupling can be especially difficult since they need to be calculated from the current density distribution that is affected by skin- and proximity-effects.

To address the above issues, an interconnection modeling method targeted towards vertical interconnections of cylindrical cross section, is described in this chapter. Until now, existing numerical or analytical methods are not optimized in their accuracy and applicability for modeling such interconnection structures since they have been primarily developed for analyzing planar structures of rectangular cross section. This chapter describes the basis for modeling these structures with cylindrical cross section which include the formulation, comparison of the results for canonical structures, comparison with existing methods and finally their application to realistic examples.

2.2 Specialized Basis Functions

The proposed methodology in this chapter is an integral-equation-based model associated with new types of global basis functions [Han et al., 2013]. The basis functions are developed to capture the physical behavior for interconnections with cylindrical cross section such as bonding wires and Column Grid Arrays (CGA), which are then extended to TSVs in Chapter 3. Two kinds of basis functions are described in this section, namely: a) conduction mode basis function (CMBF) for conductor resistance and inductance extraction and, b) accumulation mode basis function (AMBF) for conductance and capacitance extraction. Polarization mode basis function (PMBF), which is used for TSV modeling, will be discussed in detail in Chapter 3.

When used in integral equations, the proposed basis functions are efficient both in memory and computational performance, as will be discussed in this chapter. Compared to the generic computational electromagnetic methods that are based on spatial discretization, the specialized basis functions described in this chapter are geometrically suited for cylindrical interconnections, and therefore the required number of basis functions can be reduced considerably. In addition, the specialized basis functions can capture high-frequency effects by their very nature, and hence no refined discretization for modeling high-frequency effects is required. In short, the methods described eliminate the need to mesh the geometry and instead the basis functions used capture the necessary physical effects. Therefore, regardless of the method used for solving the matrix equation, the size of system matrix is often small and the simulation time is reduced as well.

2.3 Electric Field Integral Equation (EFIE) with Cylindrical CMBF for Resistance and Inductance Extraction

This section introduces the cylindrical CMBF with its classification and applies the basis functions to the construction of equivalent circuit equations. Several techniques for computing partial resistances and inductances are discussed as well.

2.3.1 *Cylindrical Conduction Mode Basis Functions (CMBF)*

The main feature of the CMBF is that it globally describes the current density distribution in the cross section of a conductor. Using this global nature of the CMBF reduces the required number of basis functions, which can be large when using localized constant basis functions [Ruehli, 1974]. Clearly, the smaller number of bases has merit for reducing the size of the partial impedance matrix, as discussed in the use of the CMBF for rectangular geometries [Daniel et al., 2001].

Examples of cylindrical conductors are shown in Figure 2.1. The cylindrical CMBFs are constructed from the following current density diffusion equation for the structure in Figure 2.1 [Silvester, 1966].

$$\nabla \times \nabla \times \vec{J} + \alpha^2 \vec{J} = 0, \tag{2.1}$$

where $\vec{J}$ is the current density (A/m^2), $\alpha^2 = -j\omega\mu\sigma = -\left(\dfrac{1+j}{\delta}\right)^2$,

$\omega = 2\pi f$ is angular frequency (rad/sec), $\mu = 4\pi \times 10^{-7}$ is the free-space permeability (H/m), σ is the conductivity (S/m), and $\delta = \dfrac{1}{\sqrt{\pi f \mu \sigma}}$ is the skin depth (m). One assumption for deriving (2.1) from Maxwell's equations is that the medium is a good conductor ($\sigma \gg \omega\varepsilon$). The other assumption about the current density is that it flows in the axial direction without any transverse variation. These assumptions are valid for thin conductors used in wirebonds and vias. By solving the diffusion equation, we can find that the solutions, which will be used for basis functions, have the following form as described in the Appendix A.1 [Han et al., 2008a]:

$$\cos\left(n(\varphi - \varphi_0)\right) J_n(\alpha\rho) \qquad n = 0, 1, 2, \cdots, \tag{2.2}$$

where $J_n(\alpha\rho)$ is the nth order Bessel function or Kelvin function [Abramowitz et al., 1965], the asymptotic behavior of which is the exponential function of ρ. For the use of (2.2) as basis functions, a proper classification of the order "n" and the orientation "φ_0" is necessary.

 Design and Modeling for 3D ICs and Interposers

The physical behavior of the bases with different orders classifies the cylindrical CMBFs into skin-effect (SE) and proximity-effect (PE) modes. The SE-mode basis is the fundamental-order ($n=0$) function, which shows the same behavior as that of the skin-effect current distribution in a circular cross section. The PE modes are the remaining higher-order ($n>0$) basis functions, which have sinusoidal behaviors in their angular variations. The collection of the harmonic angular functions in PE modes enables the description of current crowding caused by proximity-effects. Considering that Fourier series expansion can express any periodic function of φ, we classify two orthogonal basis functions for each order of the PE modes. In summary, the cylindrical CMBFs are classified into the following groups for the ith conductor in global coordinates:

Skin-effect (SE) mode ($n=0$):

$$\vec{w}_{i0} = \begin{cases} \dfrac{\hat{z}_i}{A_{i0}} J_0\left(\alpha(\vec{r}-\vec{r}_i)\cdot\hat{\rho}_i\right) & \vec{r}\in V_i \\ 0 & \text{elsewhere} \end{cases}, \tag{2.3}$$

Proximity-effect, direct (PE-d) mode ($n>0$):

$$\vec{w}_{ind} = \begin{cases} \dfrac{\hat{z}_i}{A_{in}} J_n\left(\alpha(\vec{r}-\vec{r}_i)\cdot\hat{\rho}_i\right)\cos(n\varphi_i) & \vec{r}\in V_i \\ 0 & \text{elsewhere} \end{cases}, \tag{2.4}$$

Proximity-effect, quadrature (PE-q) mode ($n>0$):

$$\vec{w}_{inq} = \begin{cases} \dfrac{\hat{z}_i}{A_{in}} J_n\left(\alpha(\vec{r}-\vec{r}_i)\cdot\hat{\rho}_i\right)\sin(n\varphi_i) & \vec{r}\in V_i \\ 0 & \text{elsewhere} \end{cases}, \tag{2.5}$$

where $\vec{r} = x\hat{x} + y\hat{y} + z\hat{z}$ is a point in the ith conductor, $\vec{r}_i = x_{i0}\hat{x} + y_{i0}\hat{y} + z_{i0}\hat{z}$ is the center point of the ith conductor, and A_{in} is the effective area. A_{in} is the constant that normalizes the basis function so that the integration of the function over the cross section equals unity [Han et al., 2009], as shown in the Appendix A.2.

A main advantage of using the cylindrical CMBF is that the orthogonal PE-mode bases automatically capture current crowding in any orientation. This feature makes the proposed method free from pre-constructing the shapes of the basis functions based on conductor geometry or the generation of proximity templates [Daniel et al., 2003]. Thus, we can apply the cylindrical CMBF for more general 3-D interconnection problems, where many conductor segments are located in a complex fashion. For example, Figure 2.2 demonstrates how the linear combination of SE- and PE-mode basis functions describes a specified current density distribution induced by the proximity of nearby conductors.

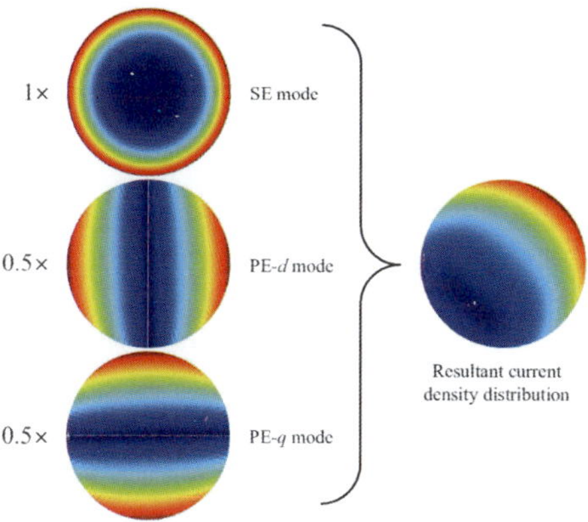

Figure 2.2: Current distribution created using a combination of SE and PE mode basis functions at 100MHz.

2.3.2 *EFIE Formulation*

The cylindrical CMBFs can now be inserted into the volume electric field integral equation (EFIE) to form equivalent voltage equations, which are composed of the modal partial impedances of each conductor. This subsection outlines the formulation procedure, including the calculation of the impedances and the construction of the equivalent network.

2.3.2.1 *Voltage Equation*

As in the classical PEEC method [Ruehli, 1974] and the CMBF-based method for conductors of rectangular cross section [Daniel et al., 2001], the proposed method approximates the volume EFIE in the form:

$$\frac{\vec{J}(\vec{r},\omega)}{\sigma} + j\frac{\omega\mu}{4\pi}\int_{V'} G(\vec{r},\vec{r}')\vec{J}(\vec{r}',\omega)dV' = -\nabla\Phi(\vec{r},\omega) , \qquad (2.6)$$

where $\Phi(\vec{r},\omega)$ is the electric potential (V) and $G(\vec{r}-\vec{r}')= e^{-jk_0|\vec{r}-\vec{r}'|}/|\vec{r}-\vec{r}'|$ is the Green's function. An assumption used in this chapter is that the maximum size of the problem space is much smaller than the wavelength of the maximum modeling frequency, so the retardation term ($e^{-jk_0|\vec{r}-\vec{r}'|}$) is negligible as in the (L_p, R) PEEC method [Ruehli et al., 2001]. Hence, this can be categorized as a quasi-static solution.

Figure 2.3 defines a general *N*-conductor system to be discussed in this section. As discussed in the previous subsection, a major difference of the CMBF-based method from the classical PEEC method is that the integral equation is not discretized to volume filaments but to globally-defined conduction modes. However, each cylindrical CMBF is localized to each conductor, as shown in (2.3) to (2.5). Therefore, for the approximation of current density on a conductor segment *j*, basis functions that belong to the conductor are combined as follows:

$$\vec{J}_j(\vec{r},\omega) \cong \sum_{n,q} I_{jnq}\vec{w}_{jnq}(\vec{r},\omega) . \qquad (2.7)$$

By inserting the approximation (2.7) into the current density term in (2.6) and applying the following inner product based on Galerkin's method:

$$\left\langle \vec{w}_{imd}(\vec{r},\omega),\vec{x}\right\rangle = \int_V \vec{w}_{imd}^*(\vec{r},\omega)\cdot\vec{x}dV , \qquad (2.8)$$

we obtain the voltage equation containing the following partial resistance, partial inductance, and modal voltage difference:

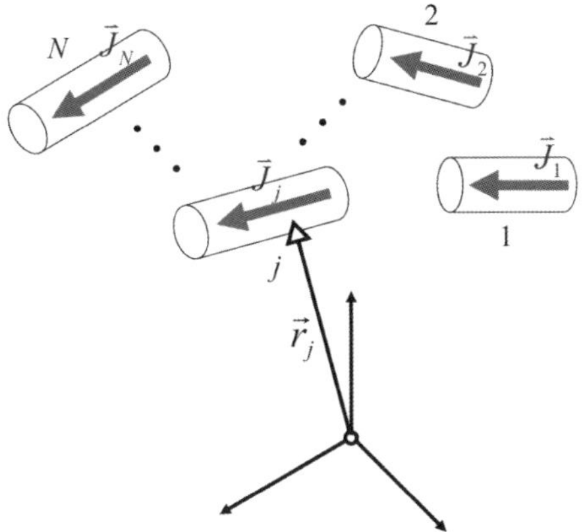

Figure 2.3: Configuration of N-cylindrical conductor system.

$$\sum_{n,q} I_{jnq} R_{imd,jnq} + j\omega \sum_{n,q} I_{jnq} L_{imd,jnq} = \Delta V_{imd}^{j}, \tag{2.9}$$

where

$$R_{imd,jnq} = \frac{1}{\sigma} \int_{V_i} \vec{w}_{imd}^{*}(\vec{r}_i,\omega) \cdot \vec{w}_{jnq}(\vec{r}_j,\omega) dV_i ,$$

$$L_{imd,jnq} = \frac{\mu}{4\pi} \int_{V_i} \int_{V_j} \vec{w}_{imd}^{*}(\vec{r}_i,\omega) \cdot \vec{w}_{jnq}(\vec{r}_j,\omega) \frac{1}{\left|\vec{r}_i - \vec{r}_j\right|} dV_j dV_i ,$$

$$\Delta V_{imd}^{j} = -\int_{S_i} \Phi_j(\vec{r}_i) \vec{w}_{imd}^{*}(\vec{r}_i,\omega) \cdot d\vec{S}_i .$$

After applying the same inner products with other basis functions in the ith conductor, we can combine all voltage equations into the following sub-matrix equation, which represents the interactions between all modes in two conductors i and j.

$$(\mathbf{R_{ij}} + j\omega\mathbf{L_{ij}})\mathbf{I_j} = \mathbf{V_i^j}, \tag{2.10}$$

where

$$\mathbf{R_{ij}} = \begin{pmatrix} R_{i0,j0} & R_{i0,j1d} & \cdots & R_{i0,jNq} \\ R_{i1d,j0} & R_{i1d,j1d} & \cdots & R_{i1d,jNq} \\ & & \ddots & \\ R_{iMq,j0} & R_{iMq,j1d} & \cdots & R_{iMq,jNq} \end{pmatrix},$$

$$\mathbf{L_{ij}} = \begin{pmatrix} L_{i0,j0} & L_{i0,j1d} & \cdots & L_{i0,jNq} \\ L_{i1d,j0} & L_{i1d,j1d} & \cdots & L_{i1d,jNq} \\ & & \ddots & \\ L_{iMq,j0} & L_{iMq,j1d} & \cdots & L_{iMq,jNq} \end{pmatrix},$$

$$\mathbf{I_j} = \begin{pmatrix} I_{j0} & I_{j1d} & \cdots & I_{jNq} \end{pmatrix}^T, \text{ and } \mathbf{V_i^j} = \begin{pmatrix} \Delta V_{i0}^j & \Delta V_{i1d}^j & \cdots & \Delta V_{iMq}^j \end{pmatrix}^T.$$

Finally, all the sub-matrix equations between conductor segments congregate to form the global impedance matrix equation, which contains resistive loss and inductive coupling in the entire conductor system. The size of the global impedance matrix is approximately $(N_c N_m) \times (N_c N_m)$, where N_c and N_m are the number of conductor segments and the required number of modes for a conductor, respectively. The parameter N_m is one when only the SE mode is used and is more than one when additional PE modes are used. For practical structures, the number of PE-mode basis function pairs is typically two or three for accurately describing current crowding in the conductor, and therefore the required memory is considerably reduced compared to that of the classical PEEC method.

2.3.2.2 *Partial Impedances*

The calculation of partial resistances and inductances in (2.9), which involve the computation of a six fold integral with frequency-dependent integrands, can be computationally expensive. This section discusses analytical and numerical integration techniques that can reduce computation time.

For partial resistances, indefinite integrals can be easily found, and the mutual resistances vanish because of the local and orthogonal properties of the cylindrical CMBFs. Therefore, the global matrix of partial resistances becomes diagonal in the form:

$$
R_{imd,jnq} = \begin{cases} \dfrac{\pi\delta^2 \rho_i l_i}{\sigma |A_{i0}|^2} \operatorname{Im}\!\left(\alpha J_0^*(\alpha\rho_i) J_1(\alpha\rho_i)\right) & i=j, m=n=0 \\[3ex] \dfrac{\pi\delta^2 \rho_i l_i}{2\sigma |A_{im}|^2} \operatorname{Im}\!\left(\alpha^* J_{m-1}^*(\alpha\rho_i) J_m(\alpha\rho_i)\right) & i=j,\ m=n\neq 0,\ d=q \\[3ex] 0 & \text{otherwise} \end{cases}
$$

$$(2.11)$$

The partial resistance from the SE mode is actually identical to the analytic internal resistance formula of a cylinder [Paul, 1994].

In contrast to the partial resistances, closed-form expressions of the partial inductances cannot be found; therefore numerical methods need to be used. However, analytical integrations over three variables can reduce the original six fold integral to the following triple integral:

$$
L_{imd,inq} = \frac{\mu}{8\pi} \int_0^{\rho_i}\int_0^{\rho_i}\int_0^{2\pi} \rho\rho' \frac{J_m^*(\alpha\rho)J_n(\alpha\rho')}{A_{im}^* A_{in}} I_{\varphi\Sigma} I_z \, d\varphi\, d\rho'\, d\rho, \quad (2.12)
$$

where

$$
I_{\varphi\Sigma}(\varphi_\Delta) = \begin{cases} 8\pi - 4\varphi_\Delta & m=n=0 \\[2ex] 2(2\pi-\varphi_\Delta)\cos(n\varphi_\Delta) - \dfrac{2}{n}\sin(n\varphi_\Delta)\cos(\varphi_d) & m=n\neq 0,\ d=q \\[3ex] \dfrac{4(-1)^{m+n+1}}{m^2-n^2}\left[m\sin(m\varphi_\Delta) - n\sin(n\varphi_\Delta)\right] & m\neq n,\ d=q\ (\text{PE}-d) \\[3ex] \dfrac{4(-1)^{m+n+1}}{m^2-n^2}\left[n\sin(m\varphi_\Delta) - m\sin(n\varphi_\Delta)\right] & m\neq n,\ d=q\ (\text{PE}-q) \\[3ex] 0 & \text{otherwise} \end{cases}
$$

$$I_z(D,l_i) = 2\left(\sqrt{D^2} - \sqrt{l_i^2 + D^2}\right) + l_i \log\left[\frac{l_i^2 + \sqrt{l_i^2 + D^2}}{-l_i^2 + \sqrt{l_i^2 + D^2}}\right],$$

and $D^2(\rho,\rho',\varphi_\Delta) = \rho^2 + \rho'^2 - 2\rho\rho'\cos\varphi_\Delta$. where $D(\rho,\rho',\varphi_\Delta)$ is the distance between two points on the cross sectional plane. φ_Δ is a new angular variable that is obtained from the following coordinate transformation of φ and φ', which is useful for both reducing computational cost and avoiding the Green's function singularity during numerical integrations [Ortiz et al., 2007].

$$\begin{pmatrix} \varphi_\Delta \\ \varphi_\Sigma \end{pmatrix} = \begin{pmatrix} 1 & -1 \\ 1 & 1 \end{pmatrix}\begin{pmatrix} \varphi \\ \varphi' \end{pmatrix} \tag{2.13}$$

Since the Green's function is not a function of φ_Σ, the indefinite integral over φ_Σ reduces one of the six numerical integrals. In the remaining numerical integrals, singular points of the integrands are concentrated on a line where both $\varphi_\Delta = 0$ and $\rho = \rho'$ hold true. In numerical integration based on adaptive Lobatto quadrature [Press et al., 1995], the singular points are simply avoided by adjusting the starting points as a small value such that $\varphi_\Delta = 0.01\pi$.

For the calculation of the partial mutual inductances, we rewrite the inductance formula in (2.12) with the following frequency-dependent and frequency-independent parts:

$$L_{imd,jnq} = \frac{\mu}{4\pi} \int\limits_{\rho_j,\rho_i} \rho_i\rho_j \frac{J_m^*(\alpha\rho_i)J_n(\alpha\rho_j)}{A_{imd}^* A_{jnq}} I_{z,\varphi}(\rho_i,\rho_j)d\rho_i d\rho_j, \tag{2.14}$$

where

$$I_{z,\varphi}(\rho_i,\rho_j) = \int\limits_{\varphi_j,\varphi_i} \cos(\varphi_i - \varphi_{i,d})\cos(\varphi_j - \varphi_{j,q}) \times \int\limits_{z_j,z_i} \frac{(\hat{z}_i \cdot \hat{z}_j)}{|\vec{r}_i - \vec{r}_j|} dz_i dz_j d\varphi_i d\varphi_j$$

The frequency-independent part $I_{z,\varphi}(\rho_i,\rho_j)$ is composed of analytical integration over (z_i,z_j) and numerical integration over (φ_i,φ_j).

For the calculation of the analytical integral over (z_i, z_j), the distance between two points in two different conductor segments should be formulated with predefined orientations. In Figure 2.4, translation and rotation (based on Euler angles) can be determined from the defined global coordinates. Then, each point in a conductor is specified with the local cylindrical coordinates, which are related to the cylindrical CMBFs. With the calculated distance, we can obtain indefinite integrals for axial variables (z_i, z_j).

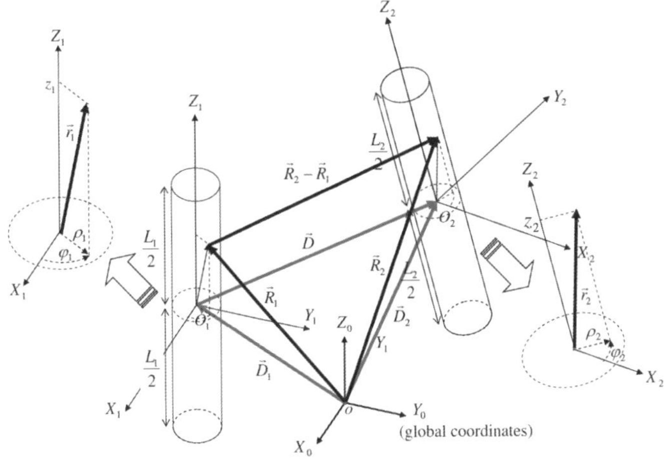

Figure 2.4: Definition of the orientation parameters between two separate cylinder segments.

To reduce total frequency sweep time, the frequency-independent integrals ($I_{z,\varphi}$s) over (φ_i, φ_j) can be computed before the sweep simulation. The pre-computed frequency-independent $I_{z,\varphi}$ is multiplied by the frequency-dependent integrands during the numerical integration over (ρ_i, ρ_j). One issue with this approach is that storing the values of $I_{z,\varphi}$s for every point of (ρ_i, ρ_j) requires a large amount of memory. Fortunately, we can reduce the memory requirement by using the property that $I_{z,\varphi}(\rho_i, \rho_j)$ is a "smooth" bi-variate function. That is, the variations in the frequency-independent part of the integrand are smaller than those in the frequency-dependent Kelvin functions. Therefore, we can use an interpolated value obtained from pre-computed integration values for only a small number of data points. The total number of the sampled data points is determined adaptively according to the relative variation of the Green's function.

For integrals over the remaining two variables (ρ_i, ρ_j), the double integral using adaptive Lobatto quadrature [Press et al., 1995] was used since the algorithm provides improved accuracy and reliability compared to adaptive Simpson quadrature and other higher-order quadratures [Gander et al., 2000].

2.3.2.3 *Equivalent Circuit*

In addition to the calculated partial resistances and inductances, the modal voltage difference should be considered to generate the global impedance matrix equation and the corresponding equivalent circuit. Since $\Delta V_{imd}^j = 0$ when $i \neq j$, the combined modal voltage difference can be reduced as follows:

$$\Delta V_{imd} = \sum_j \Delta V_{imd}^j = \Delta V_{imd}^i \tag{2.15}$$

where ΔV_{imd}^j are modal voltages induced by the jth current density in (2.15). Since the integral over the lateral surface of a cylinder is zero, we can simplify ΔV_{imd} to the integral over the inlet and the outlet planes S_i^- and S_i^+, respectively) as follows:

$$\Delta V_{imd} = - \int_{S_i^+} \Phi(\vec{r}_+)\vec{w}_{imd}^*(\vec{r}_i, \omega) \cdot d\vec{S}_i^+ - \int_{S_i^-} \Phi(\vec{r}_-)\vec{w}_{imd}^*(\vec{r}_i, \omega) \cdot d\vec{S}_i^- \tag{2.16}$$

When the SE-mode basis is involved, the integrals of $\vec{w}_{imd}^*$ in (2.16) are unity since the basis functions are normalized, as discussed in Section 2.2.3.1. Thus, the modal potential difference becomes the actual voltage difference between the two nodes. In the case where the PE-mode bases are involved, the modal potential difference becomes zero since the integral of the harmonic functions in the higher-order bases vanish. In summary, the global impedance matrix equation can be expressed as follows:

$$\begin{pmatrix} \mathbf{Z}_{ss} & \mathbf{Z}_{sp} \\ \mathbf{Z}_{ps} & \mathbf{Z}_{pp} \end{pmatrix}\begin{pmatrix} \mathbf{I}_{s} \\ \mathbf{I}_{p} \end{pmatrix} = \begin{pmatrix} \mathbf{\Delta V}_{i} \\ \mathbf{0} \end{pmatrix}, \tag{2.17}$$

where $\mathbf{Z}_{ss}$, $\mathbf{Z}_{sp}$, $\mathbf{Z}_{ps}$, and $\mathbf{Z}_{pp}$ are partial impedances grouped by SE and PE modes, $\mathbf{I}_{s}$ and $\mathbf{I}_{p}$ are skin- and proximity-effect currents, and $\mathbf{\Delta V}_{i}$ is the voltage difference across a conductor segment.

From the viewpoint of circuit topology, the equivalent circuit generated from the PE-mode basis function forms a closed loop like a shielded conductor, which is inductively coupled with the other circuits. In an example of the equivalent circuit of two conductor segments shown in Figure 2.5, two branches are generated from the SE-mode partial components, and eight loops come from four orthogonal pairs of two PE modes. The number of PE-mode loops will vary according to the strength of the proximity-effect.

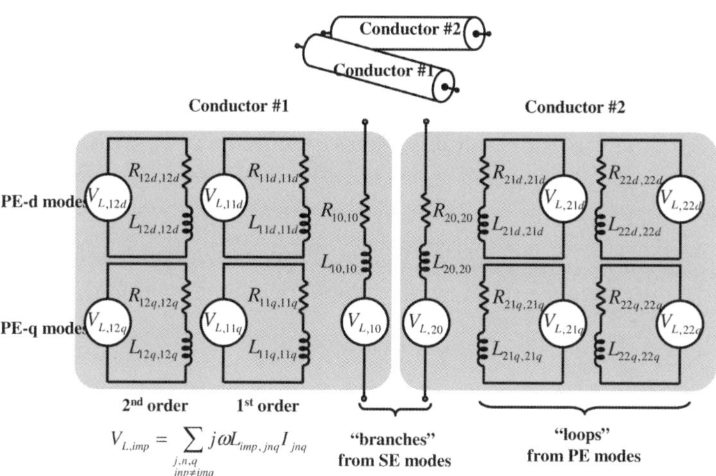

Figure 2.5: Equivalent circuit for two coupled conductors showing SE and two PE modes.

Extending the two-conductor model, Figure 2.6 illustrates a general equivalent network of two coupled 3-D bonding wires. The wires are approximated using connections of several straight conductors, and the physically connected nodes are identical to the circuit nodes of the SE

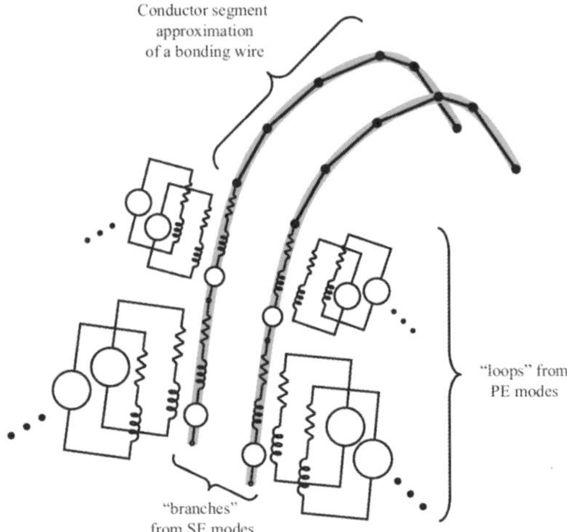

Figure 2.6: Equivalent circuit for two bond wires.

branches. During the approximation of the bonding wires with the conductor segment model, the number of segments is controlled so that the approximate model captures the original curvature of bonding wires accurately. It is important to note that since the proposed method assumes that current flows in the axial direction only, the current distribution may be inaccurate if any sharp edges connecting adjoining conductor segments are present, which is typically not true for bond wires or vias.

2.3.3 *Efficiency Enhancements and Implementation*

The proposed method discussed in the previous sections has the benefit of using the equivalent network's system matrix (2.17), which is much smaller than the matrix using the classical PEEC method. However, the calculation of the partial impedances (2.9) for each frequency step is more complicated than the classical PEEC method. Therefore, for the modeling of 3-D interconnections in stacked ICs, further reduction of computational time is necessary.

2.3.3.1 *Controlling the Number of PE-Mode Basis Functions*

One of the ideas for reducing computational time is to use the number of higher-order (PE-mode) bases differently for each neighboring conductor [Han et al., 2008b]. This is possible by assigning a reduced number of higher-order bases to each conductor because calculations related to the higher-order basis functions are not necessary when the distance between two conductors is sufficiently large or when the coupling coefficient is small. This can be explained more clearly with an example shown in Figure 2.7. Here we select an arbitrary conductor (e.g., conductor i) and group its neighboring conductors according to their coupling levels to the conductor i. For the group of conductors that is within close proximity to the conductor i, such as conductor j in Figure 2.7, the computation of PE-mode interactions is required up to the second order.

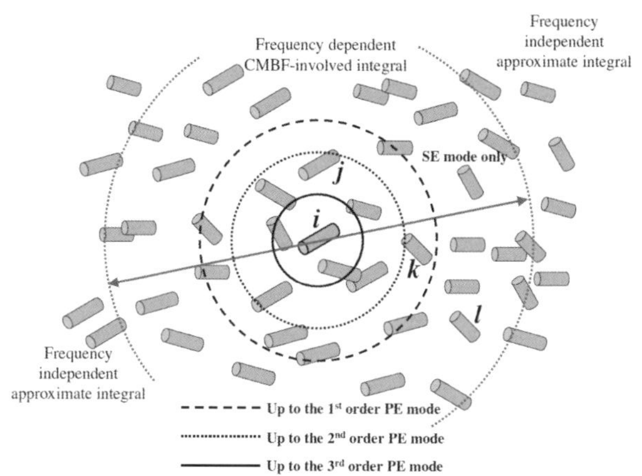

Figure 2.7: Diagram showing two efficiency enhancement schemes.

However, for those that are more distant from conductor i, such as conductor k, only the computation of the first order PE-mode interaction is required. Then, when generating the matrices involving PE-mode bases, only partial mutual inductances between the required PE modes are computed and filled, and other elements are set to zero, as shown in Figure 2.8. Therefore, besides reducing the time to compute modal

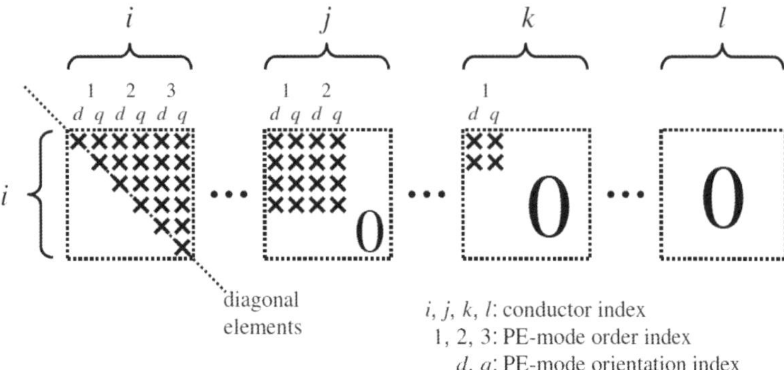

Figure 2.8: Matrix (Z_{pp}) involving conductor i in Figure 2.8. 'X's and '0's represent computed non-zero elements and zeros, respectively.

mutual inductances, we can save memory for storing non-zero elements because such grouping enables the higher-order submatrices ($\mathbf{Z}_{sp}$, $\mathbf{Z}_{ps}$, and $\mathbf{Z}_{pp}$) of the partial impedance matrix to become sparse.

In the actual calculation of the required number of PE-mode basis functions in each group, we need to consider two key parameters. One is the initial coupling coefficient obtained with SE-mode bases only, and the other is the aspect ratio of the diameter to length of a cylinder. These parameters have the following important characteristics. The higher the initial coupling coefficient and the larger the aspect ratio, the more higher-order PE-mode basis functions are required. We can show these characteristics by drawing the boundaries of the required number of basis functions under a defined error bound (10^{-3} for example) in Figure 2.9, which are obtained by computing relative errors in the resultant coupling coefficients for various aspect ratios and initial coupling coefficients. The numerical experiments of Figure 2.9 are based on a simplified case where the two conductors are parallel and have identical shape (and aspect ratio). Therefore, one might expect that more rigorous evaluation covering other possible cases where the two conductors have different orientation and shapes from each other is necessary. However, the proximity-effect arising from parallel conductors is the maximum compared to other cases of arbitrarily oriented conductors; hence the

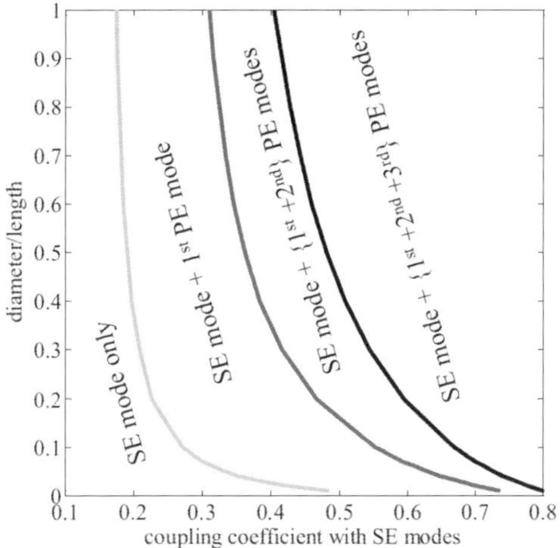

Figure 2.9: Relative error boundaries that define the required cylindrical CMBFs at 10GHz (from copper based two-parallel-conductor structure).

required PE modes are maximized as well. As for the issue of different shapes of cylinders, we can select the largest aspect ratio among cylinders when determining the required number of PE modes.

2.3.3.2 *Multi-Function Method (MFM)*

In addition to the PE-mode order reduction, we can use simplified approximate integrals to reduce the computational cost of generating $\mathbf{Z}_{ss}$. Since the approximations are frequency-independent, the number of $\mathbf{Z}_{ss}$ elements to be calculated is reduced during a frequency sweep. Therefore, the computational effort of generating the dense matrix becomes that of generating a banded matrix.

When the two conductors are sufficiently separated as in Figure 2.7, the variation of current density in conductors is negligible. Thus, the following thin-filament approximation can be used instead:

$$L_{i,j} = \frac{\mu}{4\pi} \int\limits_{z_i} \int\limits_{z_j} G(\vec{r}_i, \vec{r}_j) dz_j dz_i \tag{2.18}$$

The integrand in the above double integral does not contain frequency-dependent CMBFs and can be calculated analytically for any orientation of two straight conductor segments [Ray et al., 1999]. The accuracy of the thin-filament approximation is ensured when the distance between conductors is sufficiently large. The numerical experiments of two parallel cylinders with various dimensions show that the relative error of the thin-filament approximation from the exact integral depends on the aspect ratio of diameter to length of a cylinder, as in the case of the PE-mode order reduction. Figure 2.10 shows the boundary where the thin-filament approximation maintains the relative error less than 10^{-3}. The threshold pitch of using the frequency-independent approximations is usually higher than that of the controlling higher-order bases.

For a conductor system occupying a very large dimension, the following center-to-center approximation [Antonini et al., 2003] is also available:

$$L_{i,j} = \frac{\mu}{4\pi} \frac{l_i l_j}{R_{ij}},$$ (2.19)

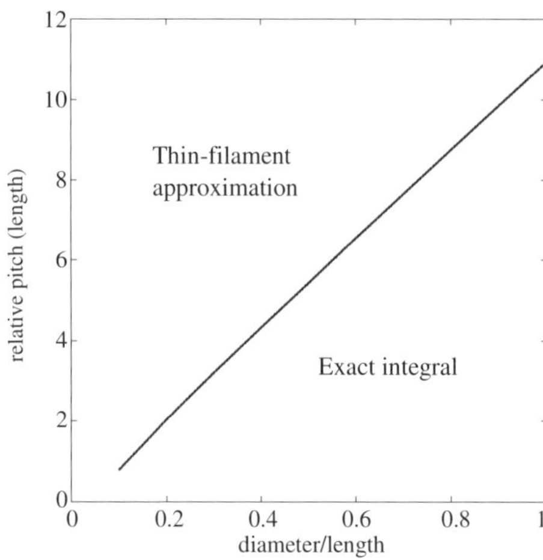

Figure 2.10: Error thresholds of using the thin-filament approximation as a function of diameter per unit length at 10GHz.

where l_i, l_j, and R_{ij} are the length of the ith and the jth conductor, and the distance between the centers of the two conductors, respectively. From a similar numerical experiment, the relative pitch (w.r.t. length) where the center-to-center approximation is available is about 9.12, regardless of the cylinder lengths.

2.3.4 *R-L Extraction Example: Comparison with PEEC Method*

This subsection demonstrates a simple three-conductor problem for evaluating the accuracy of the approach discussed in the previous subsections. The test structure is shown in Figure 2.11(a), where two conductors (1 and 2) are connected at the far end, with the other conductor being grounded. To examine the accuracy of the proposed method for arbitrary orientations of conductor segments, we applied variations in yawing angle (θ_Y) for conductor 1. Simulation tool used for comparison study was FastHenry [Kamon et al., 1994], which is an inductance extraction tool based on the PEEC method combined with the fast multipole method. Since FastHenry uses brick-type filaments for modeling interconnection geometry, we constructed an approximate cylinder model with the brick elements, as shown in Figure 2.11(b).

Figure 2.12 shows that the loop resistances and inductances obtained from the proposed method and FastHenry are well matched for all geometric variations. For several test cases, the proposed method required at least 8 times less simulation time than FastHenry mainly because the number of basis functions used was considerably smaller. The large simulation times of FastHenry are due to the approximate discretization of the circular cross section, which requires the solution of a 378 by 378 impedance matrix. By contrast, the number of required basis functions in the proposed extraction method is up to seven (one SE mode and six PE modes), resulting in a 7 by 7 impedance matrix. Additional examples with the three-conductor configurations can be found in [Han et al., 2009].

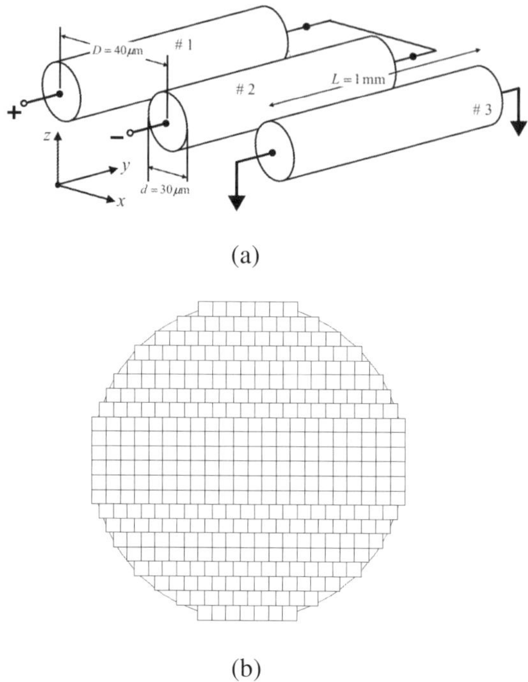

Figure 2.11: Three conductor example to validate the proposed method. (a) Geometry (b) Discretized approximate model of a cylindrical conductor in FastHenry.

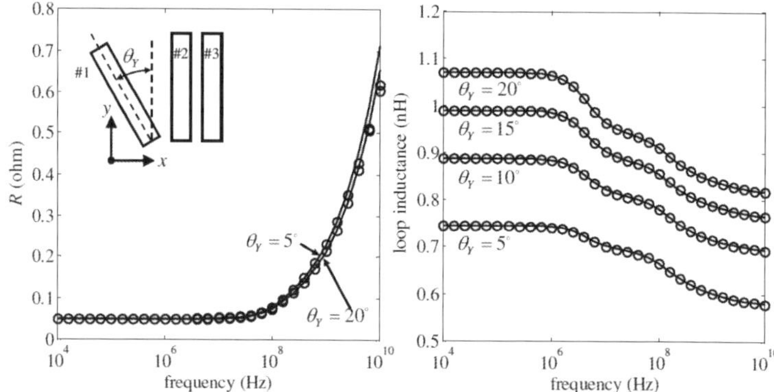

Figure 2.12: Loop resistances and inductances of cylindrical conductors with increased yawing angle of conductor 1. (circles: FastHenry, lines: proposed method).

2.4 Scalar Potential Integral Equation (SPIE) with Cylindrical AMBF for Conductance and Capacitance Extraction

Inductance and resistance of 3-D interconnections, which were discussed in the previous section, are major parasitic elements affecting the electrical behavior of interconnections at low and medium frequencies. As frequency increases, however, the effect of capacitive coupling between interconnections becomes significant, so the interconnections behave like electrically long transmission lines. Moreover, when interconnections are surrounded by lossy dielectric materials, the loss tangent of the material causes signal attenuation as well. Thus, interconnection models should be constructed with accurate capacitive coupling between interconnections for ensuring broadband model accuracy.

This section proposes a method to extract the capacitance and the conductance of cylindrical structures. The surface charge density distribution on a cylinder is approximated by the linear combination of global harmonic basis functions. In a manner similar to the *R-L* calculations, a scalar potential integral equation (SPIE) is converted to an equivalent circuit equation, which relates charge and potential at each node of the interconnection. Since the capacitances can be modified by surrounding media such as molding compound, the integral equations are re-considered to include the influence of homogeneous dielectric media.

2.4.1 *Cylindrical Accumulation Mode Basis Functions (AMBF)*

In a similar manner to the definition of cylindrical CMBFs for computing inductance and resistance, we can define another set of global basis functions that capture the charge density distribution on the surface of a cylindrical conductor. The basic idea is that any of the charge density distribution can be found from the electric scalar potential solution of Laplace's equation [Cheng, 1989]:

$$\nabla^2 \phi = 0 . \tag{2.20}$$

In cylindrical coordinates, the general solution of Laplace's equation is a function of all the coordinate variables (ρ, φ, and z). For example, a single cylinder has higher charge density distribution at the two edges of the cylinder [Scharstein, 2007] [Verolino, 1995]. However, the charges crowding at the edges cancel each other when cylinders are concatenated, as in the case of bonding wires. In addition, since the variation over the axial coordinate (z) can be described by axial discretization, we can assume the axial variation constant and find the general expression for potential as a function of ρ and φ. By applying the electric potential to the boundary condition of normal electric fields, the following expression for charge density distribution can be obtained, as explained in the Appendix A.3:

$$\sigma = \sigma_0 + \sum_{n=1}^{\infty} \left\{ \sigma_q \sin n\varphi + \sigma_d \cos n\varphi \right\}, \tag{2.21}$$

where σ_0, σ_d, and σ_q are undefined constants of surface charge density in C/m^2. Equation (2.21) indicates that the surface charge density distribution on a cylinder is the linear combination of harmonic functions, which is identical to a Fourier series expansion. Thus, the capacitance extraction approach in this chapter becomes equivalent to Method of Moments-based methods with harmonic basis functions [Faria et al., 2006], [Clements et al., 1975].

After finding normalization factors following a similar process as used with the cylindrical CMBFs, we can summarize the following surface modal basis functions:

Skin-effect (SE) mode ($n = 0$):

$$v_{l0} = \begin{cases} \dfrac{1}{a_{l0}} & \vec{r} \in S_l \\ 0 & \text{elsewhere} \end{cases}, \tag{2.22}$$

Proximity-effect, direct (PE-d) mode ($n > 0$):

$$v_{lnd} = \begin{cases} \dfrac{1}{a_{ln}} \cos(n\varphi_l) & \vec{r} \in S_l \\ 0 & \text{elsewhere} \end{cases}, \tag{2.23}$$

Proximity-effect, quadrature (PE-q) mode ($n > 0$):

$$v_{lnq} = \begin{cases} \dfrac{1}{a_{ln}} \sin(n\varphi_l) & \vec{r} \in S_l \\ 0 & \text{elsewhere} \end{cases}, \qquad (2.24)$$

where n is the order of the basis function and S_l is the lateral surface of the conductor l. The parameter a_{ln} is the effective area used to normalize the surface integral of the basis functions over the surface S_l given by:

$$a_{ln} = \begin{cases} 2\pi\rho_l l_l & n = 0 \\ 4\rho_l l_l & n > 0 \end{cases}, \qquad (2.25)$$

where ρ_l and l_l are the radius and the length of the conductor l, respectively. As in the classification of cylindrical CMBFs, direct and quadrature modes in the higher order bases are orthogonal to each other, and the linear combination of them describes arbitrary charge crowding caused by nearby conductors.

2.4.2 SPIE Formulation in Free Space

The AMBFs introduced in the previous section approximate the surface charge density distribution in SPIE, which represents the contribution of charge density distribution to the potential at a testing point $\vec{r}$. The SPIE can be written as:

$$\frac{1}{4\pi\varepsilon_0} \int_{V'} G(\vec{r},\vec{r}')q(\vec{r}',\omega)dV' = \Phi(\vec{r},\omega), \qquad (2.26)$$

where $q(\vec{r}',\omega)$ is volume charge density, $\Phi(\vec{r},\omega)$ is electric potential, and $G(\vec{r},\vec{r}')$ is the Green's function. The retardation term in the Green's function is assumed to be negligible. By inserting the approximation of charge density distribution $q = \sum_{n=0} Q_{knq} v_{knq}$ to (2.26) and applying the following inner product:

$$\left\langle v_{lmd}(\vec{r},\omega), x \right\rangle = \int_S v_{lmd}(\vec{r},\omega) x \, dS \qquad (2.27)$$

the following equation can be obtained that relates surface charges and conductor potentials:

$$\sum_{n,q} Q_{knq} P_{lmd,knq} = V_{lmd}^k , \qquad (2.28)$$

where k and l are conductor indices,

$$P_{lmd,knq} = \frac{1}{4\pi\varepsilon_0} \int_{S_l}\int_{S_k} v_{lmd}(\vec{r}_l) v_{knq}(\vec{r}_k) \frac{1}{|\vec{r}_l - \vec{r}_k|} dS_k \, dS_l$$

is the partial coefficient of potential (F^{-1}), and $V_{lmd}^k = \int_{S_l} \Phi_k(\vec{r}_l) v_{lmd}(\vec{r}_l) dS_l$ is the modal voltage on the conductor surface due to the coupling of the kth conductor. The modal voltage, which is the potential value relative to the ground at infinity, must be distinguished from the modal voltage difference (2.9) between two nodes on conductors.

The computation of the partial coefficient of potential $P_{lmd,knq}$ involves integrals over four variables (φ_l, φ_k, z_l, and z_k). Since the integrals have the same form as those of the partial inductances (2.9), we can use the techniques in Section 2.3.2.2, where analytic integrals over the axial variables (z_l and z_k), coordinate transformation of angular variables, and numerical quadrature approximations are discussed. Since cylindrical AMBFs are scalar surface functions, the partial coefficient of potential does not include the integrals over radial variables. The independency to the radial variable reduces the computational time for applying the numerical quadrature rule. In addition, cylindrical AMBFs do not depend on frequency, so frequency sweep is not necessary. Therefore, the required cost for computing integrals involving cylindrical AMBFs is much lower than the cost for computing CMBF based integrals.

The modal voltage V_{lmd}^k in (2.28) also shows similar properties to the modal voltage difference for fundamental and higher-order mode basis functions in (2.16). When an AMBF is in the fundamental mode ($m = 0$), the integral over the lateral surface becomes unity, so V_{lmd}

equals the actual nodal voltage Φ_l. When the AMBF is one of higher-order mode harmonic functions ($m > 0$), the integral becomes zero. Thus, the equivalent network of modal coefficients of potential is composed of the fundamental-order capacitive coupling between conductor nodes and higher-order capacitive coupling from the closed loops. The details of the construction of combined equivalent circuit will be discussed later in this chapter.

2.4.3 *SPIE Formulation Considering Homogeneous Media*

In the previous section, the surrounding medium of interconnections was assumed to be free space ($\varepsilon = \varepsilon_0$). The free space assumption is valid when interconnections are exposed for measurement, but the real environment for 3-D interconnections is usually filled with packaging or substrate material. For example, bonding wire interconnections are usually submerged in a molding compound, and vertical interconnections such as vias are fabricated in multilayered substrates using organic, ceramic, and silicon material.

Most of the packaged interconnection applications are based on inhomogeneous media, which includes the combination of various materials and free space. To consider inhomogeneous media in integral equations, we need to use a special type of Green's functions that can describe the specific media. A special but popular case of inhomogeneous media is a multilayered structure, which can be modeled by using the multilayered Green's function. In this chapter, only the homogeneous media is considered. For example, the molding compound surrounding bonding wires can be assumed to be a homogeneous media since the size of the molding region is large relative to the size of the entire wire structure.

2.4.3.1 *Vector and Scalar Potentials*

The permittivity ε_0 is shown in SPIE (2.26), but EFIE (2.6) does not contain any permittivity term (both equations neglect retardation). Thus, replacing the free space permittivity (ε_0) with a permittivity ($\varepsilon_0 \varepsilon_B$) where ε_B is the relative background permittivity in (2.26) seems to be sufficient for modeling interconnections in a homogeneous dielectric medium [Li et al., 2002]. However, the permittivity difference also influences the EFIE (2.6) if the background is not free space. To clarify the effect of the permittivity term in EFIE, we need to reconsider Maxwell's equation for appropriately modifying the formulation.

The displacement current term in Ampere's law can be rewritten as:

$$\nabla \times \vec{H} = \vec{J}^C + \varepsilon_r \varepsilon_0 j\omega \vec{E} = \vec{J}^C + \left[\varepsilon_0(\varepsilon_r - \varepsilon_B)\right]\varepsilon_r j\omega\vec{E} + \varepsilon_0 \varepsilon_B j\omega\vec{E} \quad (2.29)$$

where ε_r is the relative permittivity of the interconnection, and ε_B is the relative permittivity of the background media. Similar to the free-space media case [Ruehli et al., 1992], we can define the following total current, which is composed of conduction current due to free charge and equivalent polarization current due to dielectrics:

$$\vec{J} = \vec{J}^C + \left[\varepsilon_0(\varepsilon_r - \varepsilon_B)\right]\varepsilon_r j\omega\vec{E} \quad (2.30)$$

It is important to note that the polarization current is nonzero even in conductors having free-space permittivity (ε_0) since the background media may not be free space. In the equivalent circuit model, excess capacitance represents this effect. The total current density becomes the forcing term of the vector Helmholtz equation, the solution (vector potential) of which is:

$$\vec{A} = \frac{\mu}{4\pi} \int_{v'} G(\vec{r},\vec{r}')\vec{J}(\vec{r}')dv' \quad (2.31)$$

In the homogeneous dielectric media, Gauss' law can be written as:

$$\nabla \cdot \vec{E} = \frac{q^T}{\varepsilon_0 \varepsilon_B}, \quad (2.32)$$

where $q^T = q^F + q^B$ is total charge density, q^F is free charge density, and q^B is bound charge density. By using this total charge density as an excitation term of scalar Helmholtz equation, the solution (scalar potential) can be obtained as:

$$\Phi = \frac{1}{4\pi\varepsilon_0\varepsilon_B} \int_{v'} G(\vec{r},\vec{r}')q^T(\vec{r})dv' . \qquad (2.33)$$

If the exact form of Green's function $G(\vec{r},\vec{r}')$ is used, the non-free-space medium permittivity ε_B is also included in the retardation term ($e^{-jk|\vec{r}-\vec{r}'|}$) of (2.31) and (2.33). If the media is lossy, the attenuation in the retardation term may not be negligible even under the assumption that the interconnection structure is electrically small. However, when the media is lossless, we can neglect the retardation term and the effect of the medium permittivity on the Green's function.

Since the difference in scalar potential equation (2.33) and (2.26) is the addition of the non-free-space media permittivity, the capacitive coupling between interconnections in the background dielectric media can be calculated in the same way as discussed earlier. However, we need to modify the construction of the voltage equation from the vector potential integral equation (2.31), since the total current contains the polarization current.

2.4.3.2 *Equivalent Circuit Model of Conductor*

In a conductor, the total electric field is contributed by the conduction current $\vec{J}^C$ (Ohm's law) and inductive coupling. When the background media is not free space, total current $\vec{J}$ is not reduced to the conduction current only, but also has the polarization current term. Therefore, the integral equation has two coupling terms that originate from conduction current and polarization current. Approximating the conduction current with the cylindrical CMBFs and applying inner product based on Galerkin's method, we can obtain the following voltage equation:

$$I_{jnq}R_{imd,jnq} + j\omega\sum_{j,n,q} L_{imd,jnq}I_{jnq} + j\omega\sum_{j,n,q} L_{imd,jnq}\left[j\omega R_{jnq}C_{jnq}^{ex}I_{jnq}\right] = V_{imd} \qquad (2.34)$$

where

$$C_{jnq}^{ex} = \frac{\varepsilon_0(\varepsilon_r - \varepsilon_B)}{\int_{V_j} \vec{w}_{jnq}^* \cdot \vec{w}_{jnq}dV_j} ,$$

and R_{imd}, $L_{imd,jnq}$, and V_{imd} are identical to the expressions in (2.9). The added excess capacitance term C_{jnq}^{ex} represents the polarization current in the interconnection. The voltage equation (2.34) can be expressed by the equivalent circuit model shown in Figure 2.13, where the modal current I_{imd} flows through the partial resistance.

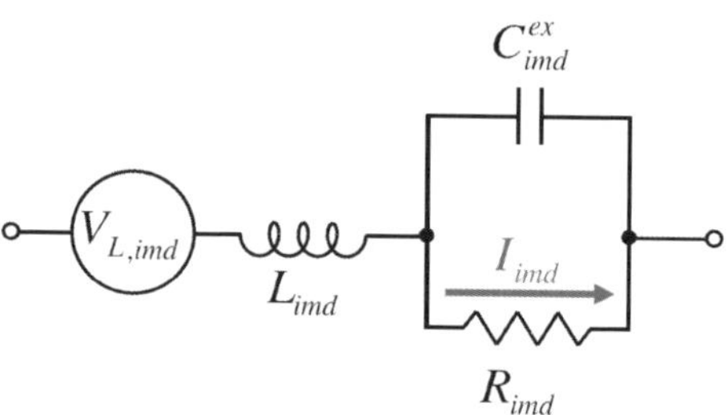

Figure 2.13: Equivalent circuit model including excess capacitance. $V_{L,imd}$ represents the voltage drop through inductive coupling.

To check the significance of the excess capacitance, the equivalent impedance of the parallel *R-C* network can be calculated as:

$$Z_{RC} = \frac{R}{1 + j\omega RC^{ex}} = \frac{R}{1 + j\omega \dfrac{1}{\sigma}\int_{V_i} \vec{w}_{imd}^{*} \cdot \vec{w}_{imd}\, dV_i \times \dfrac{\varepsilon_0(\varepsilon_r - \varepsilon_B)}{\int_{V_i} \vec{w}_{imd}^{*} \cdot \vec{w}_{imd}\, dV_i}}$$

$$= \frac{R}{1 + j\omega \dfrac{\varepsilon_0(\varepsilon_r - \varepsilon_B)}{\sigma}} \tag{2.35}$$

For good conductors such as copper and gold, the term $j\omega\varepsilon_0(\varepsilon_r - \varepsilon_B)/\sigma$ is very small over the typical frequency range of interest. For example, if gold bonding wires are used in a molding compound with $\varepsilon_B = 4.3$, the frequency where the magnitude of $j\omega\varepsilon_0(\varepsilon_r - \varepsilon_B)/\sigma$ becomes 0.1 is:

$$f = 0.1 \times \frac{1}{2\pi} \left| \frac{\sigma}{\varepsilon_0(\varepsilon_r - \varepsilon_B)} \right| = 0.1 \times \frac{1}{2\pi} \times \left| \frac{4.1 \times 10^7}{8.854 \times 10^{-12}(1 - 4.3)} \right|$$

$$= 2.23 \times 10^{16}\, \text{Hz}. \tag{2.36}$$

The effect of the excess capacitance therefore appears at very high frequencies, which means the charge relaxation time is negligible in the typical frequency range of interest, where package structures are currently being used. Therefore, we can omit the excess capacitance of the interconnection branch and use the original equivalent circuit constructed earlier.

Using (2.33) is still valid when the background media is a dispersive media, which is expressed by a complex-valued function of frequency. In this case, the resultant equivalent potential coefficients become frequency-dependent complex numbers, so the equivalent network is composed of capacitances and conductances.

2.5 Broadband Equivalent *RLC* Network

The shunt capacitance and the series *R-L* model together constitute a broadband interconnection model. The interconnection model is similar to an approximate transmission line model that consists of a cascaded ladder network of series impedances and shunt capacitances. The constructed equivalent model can cover a wide frequency band by subdividing interconnections along the axial direction. Therefore, the complexity, or the number of parasitic elements of the interconnection model, depends on the maximum modeling frequency. The constructed model provides broadband network parameters, examples of which will be shown later in this chapter. This section discusses the procedure for generating an equivalent *RLC* network to compute multi-port network parameters (or *RLGC* network if the surrounding material is lossy).

The initial step for extracting an equivalent model of an interconnection is to set the maximum modeling frequency, which determines the required number of inductive and capacitive cells along the axial direction. The initial number of cells is determined by the geometrical model input, but the length of the generated cell from the geometry can be electrically long at the maximum frequency. In this case, the initial cell is divided into additional subcells, the length of which is less than

$$lc_{\max} = \frac{\lambda_r c}{\sqrt{\varepsilon_B}\, f_{\max}}, \tag{2.37}$$

where $lc_{\max}$ is the maximum cell length, λ_r is the wavelength ratio, $c = 3 \times 10^8$ (m/s) is the velocity of light in free space, ε_B is the relative permittivity of the background media, and $f_{\max}$ is the maximum modeling frequency. The parameter λ_r is defined to ensure the cell length is sufficiently small at the modeling frequency. If $\lambda_r = 0.05$, the maximum cell length should be less than 1/20 of the wavelength at the maximum frequency. Since the maximum cell size $lc_{\max}$ is proportional to the inverse of the maximum frequency, the complexity of the interconnection problem increases with frequency. For example, the size of global impedance matrix becomes four-times larger if the maximum modeling frequency is doubled.

Figure 2.14 shows the method used to generate the inductive and capacitive cells. Each inductive cell defines the volume current flowing through the cross section of the interconnection, and each capacitive cell defines the surface charge on the lateral surface of the interconnection. Figure 2.14 also illustrates that the volume inductive cells and the surface capacitive cells overlap each other, so they establish an *RLC* network like

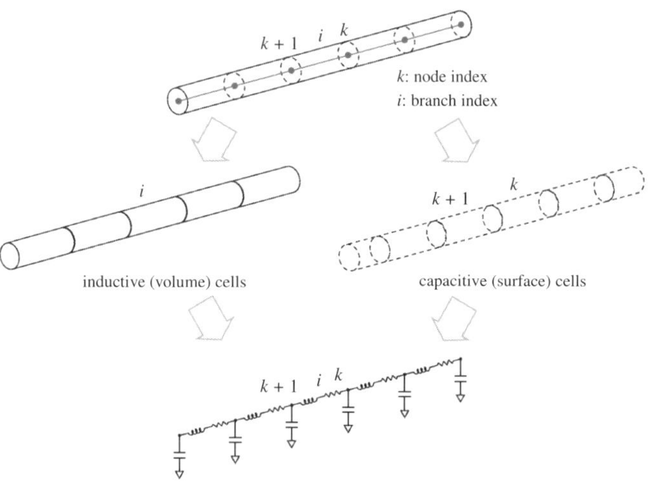

Figure 2.14: Axial cell discretization and the generation of *RLC* equivalent circuit model (Mutual coupling terms are omitted for simplicity).

the lumped-element approximation of transmission lines. Placing small capacitive cells at the ends of interconnections enables capturing charge density crowding at the ends of interconnections.

From the defined discretization, partial resistance, partial inductance, and partial coefficients of potential between modal basis functions are calculated by using (2.9) and (2.28), and voltage equations can be obtained as shown in the matrix form below:

$$\begin{pmatrix} \mathbf{Z}_{ss} & \mathbf{Z}_{sp} \\ \mathbf{Z}_{ps} & \mathbf{Z}_{pp} \end{pmatrix} \begin{pmatrix} \mathbf{I}_s \\ \mathbf{I}_p \end{pmatrix} = \begin{pmatrix} \Delta \mathbf{V}_i \\ 0 \end{pmatrix} \tag{2.38}$$

$$\begin{pmatrix} \mathbf{P}_{ss} & \mathbf{P}_{sp} \\ \mathbf{P}_{ps} & \mathbf{P}_{pp} \end{pmatrix} \begin{pmatrix} \mathbf{Q}_s \\ \mathbf{Q}_p \end{pmatrix} = \begin{pmatrix} \mathbf{V}_i \\ 0 \end{pmatrix} \tag{2.39}$$

Equation (2.38) is identical to (2.17). Equation (2.39) represents the relation of charge and modal scalar potential, which was discussed earlier. The above voltage equations are expressed in Figure 2.15 for a two-bonding wire example, where the *R-L* and the *C* circuitry are drawn separately for clarity. In a capacitance network, the self potential coefficient from the SE mode is attached at the physical node of the original interconnection, and the higher-order mode potential coefficients form grounded loops. Since the potential coefficients are independent of frequency, this capacitance equivalent model can be reduced to the capacitance matrix that does not show the higher-order mode loops explicitly.

The connection of the series *R-L*'s and the shunt *C*'s in Figure 2.15 is based on the following approximate matrix form of the continuity equation or Kirchhoff's Current Law (KCL) for each node:

$$\begin{pmatrix} -\mathbf{E} & \\ 0 & \Delta_{\mathbf{I}} \end{pmatrix} \begin{pmatrix} \mathbf{I}_t \\ \mathbf{I}_i \end{pmatrix} + j\omega \mathbf{Q}_s = 0, \tag{2.40}$$

where $\mathbf{I}_t$ is a terminal current vector, $\mathbf{I}_i$ is an internal current vector, $\mathbf{E}$ is an identity matrix, and $\mathbf{Q}_s$ is the SE-mode charge vector. The internal current vector $\mathbf{I}_i$ is identical to the SE-mode current vector $\mathbf{I}_s$. $\Delta_{\mathbf{I}}$

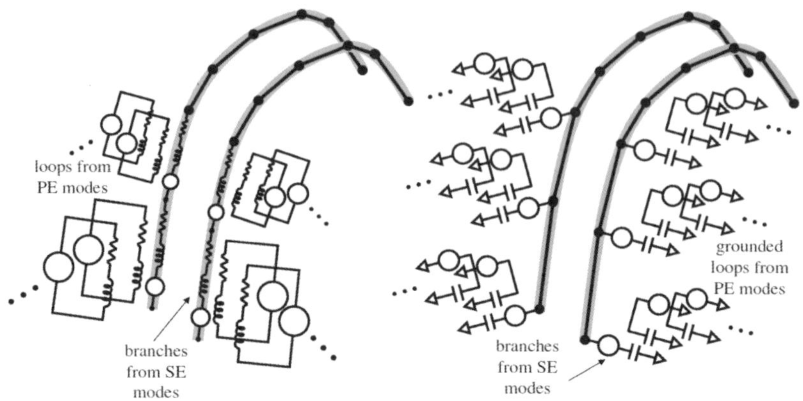

Figure 2.15: Modal *RLC* equivalent circuit model of two bonding wires. *RL* network and *C* network are separated for clarity.

contains information of incoming and outgoing currents for each node. The matrix form of the continuity equation and the voltage equations (2.38) and (2.39) constitute the entire matrix equation to compute the network parameters of the equivalent circuit. The terminal current $\mathbf{I}_t$, a known variable representing the excitation condition, appears on the right side of the following combined matrix equation:

$$
\begin{pmatrix}
\mathbf{Z}_{ss} & \mathbf{Z}_{sp} & -\boldsymbol{\Delta}_{\mathbf{V}} & 0 & 0 \\
\mathbf{Z}_{ps} & \mathbf{Z}_{pp} & 0 & 0 & 0 \\
\boldsymbol{\Delta}_{\mathbf{I}} & 0 & 0 & j\omega\mathbf{E} & 0 \\
0 & 0 & -\mathbf{E} & \mathbf{P}_{ss} & \mathbf{P}_{sp} \\
0 & 0 & 0 & \mathbf{P}_{ps} & \mathbf{P}_{pp}
\end{pmatrix}
\begin{pmatrix}
\mathbf{I}_s \\ \mathbf{I}_p \\ \boldsymbol{\Phi}_t \\ \boldsymbol{\Phi}_i \\ \mathbf{Q}_s \\ \mathbf{Q}_p
\end{pmatrix}
=
\begin{pmatrix}
\mathbf{0} \\ \mathbf{0} \\ \mathbf{I}_t \\ \mathbf{0} \\ \mathbf{0} \\ \mathbf{0}
\end{pmatrix}
\tag{2.41}
$$

In the above equation, $\boldsymbol{\Delta}_{\mathbf{V}}$ contains the information of voltage differences for each branch. $\begin{bmatrix}\boldsymbol{\Phi}_t & \boldsymbol{\Phi}_i\end{bmatrix}^T$ and $\mathbf{Q}_s$ are re-arranged to collect the terminal variables, so a permutation matrix multiplies $\boldsymbol{\Delta}_{\mathbf{V}}$, $\mathbf{P}_{ss}$, $\mathbf{P}_{sp}$, and $\mathbf{P}_{ps}$. The solutions $\boldsymbol{\Phi}_t$ of for each terminal current excitation provides the Z-parameter matrix $\mathbf{Z}$, which can be converted to an *S*-parameter matrix $\mathbf{S}$ using the following relation [Young, 2001]:

$$\mathbf{S} = (\mathbf{Z} + Z_0 \mathbf{E})^{-1} (\mathbf{Z} - Z_0 \mathbf{E}), \tag{2.42}$$

where Z_0 is a reference impedance. Examples of using the combined *RLC* network to obtain *S* parameters of bonding wires and TSV interconnections will be shown later, both in Chapter 2 and 3.

2.6 Inclusion of Planar Structures

The *RLC* model of cylindrical type wires can be constructed by using the approach discussed in the previous section. However, the wire *RLC* model captures parasitic elements for the cylindrical interconnections only. In realistic situations, the bonding wires and other cylindrical elements are also influenced by planar structures. Thus, the coupling effects of the planar structures should be considered to construct the complete bonding wire package model. This section discusses the usage of the *RLC* extraction method for generating models of bonding wires in stacked ICs, including the effect of planar coupling.

Figure 2.16 illustrates a typical bonding wire structure that is mounted on stacked dies and a package substrate [Kam et al., 2008], [Wane et al., 2007]. The bonding wires are mainly inductive components, and their partial inductance and resistance can be calculated by using the method described earlier. The capacitive coupling can be found by computing the partial coefficients of potentials, as discussed earlier as well. However, the complete electrical behavior of the bonding wires is also dominated by several other factors.

The first factor that influences bonding wire characteristic is the substrate ground plane. The actual inductance of a bonding wire in a package is the loop inductance formed by the signal line current and return current on the ground. Therefore, the location of the ground determines the total loop inductance. If the ground plane is finite or has irregular shapes with holes and slots, the loop inductance is also modified due to the increased return current path. Similarly, the capacitance of the bonding wire is also determined by the location of the ground, and the resultant characteristic impedance of the wire depends on the loop inductance and the capacitance between the wire and ground.

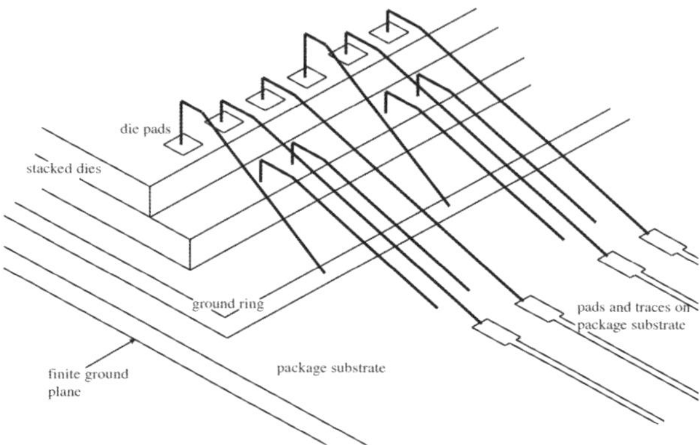

Figure 2.16: Typical bonding wires in stacked dies with planar structures.

The other factor is coupling from other planar structures, including bonding pads, power/ground rings, and planar-type interconnections such as microstrip or strip lines that can alter the impedance. The bonding pad, which is another discontinuity in the signal path, may increase the capacitive coupling to neighboring wires and ground [Chuang et al., 2004]. The power/ground rings are usually designed to ensure equipotential for all power/ground wires, but their high frequency effects on the bonding wires might be undesirable. The location of power/ground wires also influences the characteristics of signal propagation. In summary, the complete modeling of 3-D bonding wire structures requires not only the partial parasitic components of bonding wires but also other planar structures as well.

2.6.1 *Combining Cylindrical and Planar Structures*

This section presents two methods to model the coupling from planar structures. The first method is a general approach that combines conventional PEEC method [Ruehli, 1974] with the proposed modal equivalent circuit method. Since the PEEC method is able to extract the equivalent network of any finite planar structure, we can couple the effect of a finite ground plane, ground ring, and boding pads to bond wires. In case of modeling large solid ground planes, the image method can be used as an alternative. By using a simple simulation example, the two methods are validated with analytic and 3-D EM simulation results.

2.6.1.1 *Conventional PEEC Method*

For considering the coupling from planar structures, the conventional PEEC method can be combined with the proposed method. In the context of solving an integral equation, this combination is completed by computing additional partial elements involving the interaction between piecewise planar basis functions and cylindrical modal basis functions.

Inductive and capacitive couplings among planes can be found from the conventional PEEC method [Ruehli, 1974]. For inductive cells, each splanar structure is discretized so that X-directional and Y-directional cells form a grid model as shown in Figure 2.17 [Xie et al., 1999]. For the self inductance calculation of inductive cells, several analytical formulae can be used as described in [Ruehli, 1972, Zhong et al., 2003]. For the mutual inductance between two parallel bricks, use of virtual self inductance provides good analytical results [Zhong et al., 2003]. Capacitive cells are defined to cover a node, where X- and Y-directional conduction currents and displacement current through the capacitor should satisfy KCL. The analytical formulae of mutual coefficients of potential between two conducting panels can be used [Ruehli et al., 1973]. All the closed-form formulae of partial elements used are summarized in the Appendix A.4.

In general, the capacitive cells should cover the entire surface of a planar structure, which is composed of six faces. However, if we can assume the plane is sufficiently thin, the capacitive cells on the side surfaces can be omitted. In addition, the charge distribution on the upper and the lower surfaces can be considered at the same time under the assumption that the mutual coefficients of potential between the upper and lower plates are the same as those of the self coefficients of potential. In this case, simply doubling the coefficient of potential on the upper plane approximates the total planar capacitive coupling.

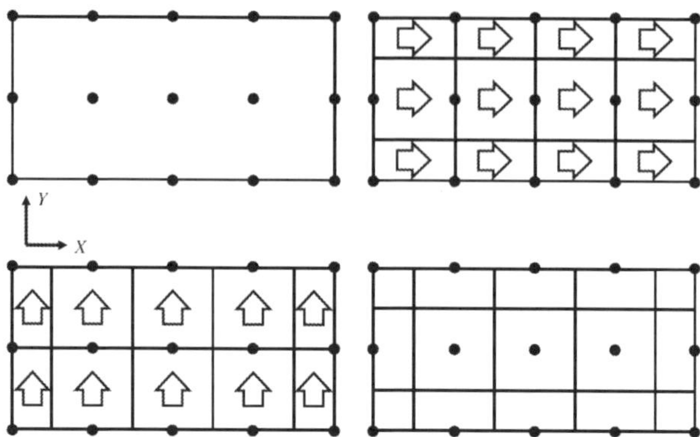

Figure 2.17: Planar structure cell generation (upper left: node definition. upper right: inductive cells in X direction. lower left: inductive cells in Y direction. lower right: capacitive cells).

Computing the electrical coupling between a plane segment and a cylindrical conductor requires an additional integral involving piecewise constant basis function and cylindrical CMBF. Figure 2.18 shows the configuration of a cylinder segment and a planar conductor segment with their geometrical parameters. Similar to the integral calculations used in earlier sections, the following partial inductances and coefficients of potentials can be found by using a combination of analytical and numerical integrals:

$$L_{i,jnq} = \frac{\mu}{4\pi} \int_{V_i} \int_{V_j} \hat{l}_i \cdot \vec{w}_{jnq}(\vec{r}_j, \omega) \frac{1}{|\vec{r}_i - \vec{r}_j|} dV_j dV_i \tag{2.43}$$

$$P_{i,jnq} = \frac{1}{4\pi\varepsilon} \int_{S_i} \int_{S_j} v_{jnq}(\vec{r}_j) \frac{1}{|\vec{r}_i - \vec{r}_j|} dS_j dS_i \tag{2.44}$$

For example, the integral corresponding to inductive coupling (2.43) is reduced to the following form after the analytical integration over the axial variable of the cylinder (z_j):

$$L_{i,jnq} = \frac{\mu}{4\pi} \int_{x_i,y_i} \int_{\rho_j,\phi_j} \left(\hat{l}_i \cdot \vec{w}_{jnq}(\vec{r}_j, \omega) \right) I_z(x_i, y_i, \rho_j, \varphi_j) \rho_j d\rho_j d\varphi_j dx_i dy_i \tag{2.45}$$

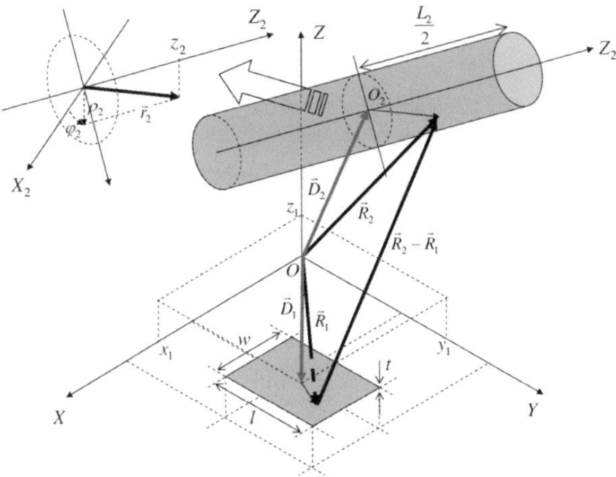

Figure 2.18: Definition of the orientation parameters for a plane and a cylinder segment.

In the above expression, we simplify the integral over z_i under the assumption that the thickness of the planar structure is negligibly small. Another expression similar to (2.45) can be obtained from (2.44) for the capacitive coupling. (The details of deriving the analytical integral over the axial variable (I_z) are discussed in the Appendix A.5). After computing all the partial element matrices of wires and planes, continuity equations (KCL) are applied for each wire and plane, and the following global matrix can be generated:

$$
\begin{pmatrix}
\mathbf{Z}^{\text{pl}} & \mathbf{Z}^{\text{pl,w}}_{\text{s}} & \mathbf{Z}^{\text{pl,w}}_{\text{p}} & -\boldsymbol{\Delta}^{\text{pl}}_{\text{V}} & & & & \\
\mathbf{Z}^{\text{w,pl}}_{\text{s}} & \mathbf{Z}_{\text{ss}} & \mathbf{Z}_{\text{sp}} & & -\boldsymbol{\Delta}^{\text{w}}_{\text{V}} & & & \\
\mathbf{Z}^{\text{w,pl}}_{\text{p}} & \mathbf{Z}_{\text{ps}} & \mathbf{Z}_{\text{pp}} & & & & & \\
\boldsymbol{\Delta}^{\text{pl}}_{\text{I}} & & & & & j\omega\mathbf{E} & & \\
& \boldsymbol{\Delta}^{\text{w}}_{\text{I}} & & & & & j\omega\mathbf{E} & \\
& & & -\mathbf{E} & & \mathbf{P}^{\text{pl}} & \mathbf{P}^{\text{pl,w}}_{\text{s}} & \mathbf{P}^{\text{pl,w}}_{\text{p}} \\
& & & & -\mathbf{E} & \mathbf{P}^{\text{w,pl}}_{\text{s}} & \mathbf{P}_{\text{ss}} & \mathbf{P}_{\text{sp}} \\
& & & & & \mathbf{P}^{\text{w,pl}}_{\text{p}} & \mathbf{P}_{\text{ps}} & \mathbf{P}_{\text{pp}}
\end{pmatrix}
\begin{pmatrix}
\mathbf{I}^{\text{pl}} \\
\mathbf{I}^{\text{w}}_{\text{s}} \\
\mathbf{I}^{\text{w}}_{\text{p}} \\
\boldsymbol{\Phi}^{\text{pl}} \\
\boldsymbol{\Phi}^{\text{w}} \\
\mathbf{Q}_{\text{pl}} \\
\mathbf{Q}^{\text{w}}_{\text{s}} \\
\mathbf{Q}^{\text{w}}_{\text{p}}
\end{pmatrix}
=
\begin{pmatrix}
\mathbf{I}^{\text{pl}}_{\text{t}} \\
\mathbf{I}^{\text{w}}_{\text{t}}
\end{pmatrix}
$$

$$(2.46)$$

where the terminal currents $\mathbf{I}_t^{pl}$ and $\mathbf{I}_t^{w}$ are from planes and wires, respectively. With the defined terminal excitation currents, network parameters are obtained in the same way as discussed in the earlier sections. The ports on planes and wires can be connected to each other to obtain the network parameter of the entire interconnection signal path.

2.6.2 *Infinite Ground Plane: Image Method for Modeling Infinite Ground*

Although the reference or ground plane in a package is a finite and irregularly shaped structure including discontinuities such as holes and slots, approximating the ground plane as a perfect and infinite one is sometimes useful for analyzing bonding wires or vias in a package. Thus, together with the classical PEEC method to model small finite planar structures, the image method can be employed to capture the effect of relatively large ground planes [Schuster et al., 2000]. Compared to the rigorous use of the conventional PEEC method, the image method reduces computational time.

As shown in Figure 2.19, symmetric image conductors are defined on the opposite side of the ideal ground plane. Since the image conductor has the identical shape of the original conductor, we can use the same resistance, inductance, and coefficients of potential matrices of the original conductor. However, the inductive and capacitive coupling between the original and the image conductors need to be calculated.

The simulation setup with the image conductors provides a network parameter with a doubled port number, which can be converted to single-ended S parameter with ideal ground by using the following modal conversion matrix [Bockelman et al., 1995]:

$$\begin{pmatrix} a_d \\ a_c \end{pmatrix} = \frac{1}{\sqrt{2}} \begin{pmatrix} 1 & -1 \\ 1 & 1 \end{pmatrix} \begin{pmatrix} a_r \\ a_i \end{pmatrix} \tag{2.47}$$

where a_r and a_i are incoming or reflected power waves of the actual port and the image port, and a_d and a_c are differential and common mode power waves, respectively. The differential power wave can be

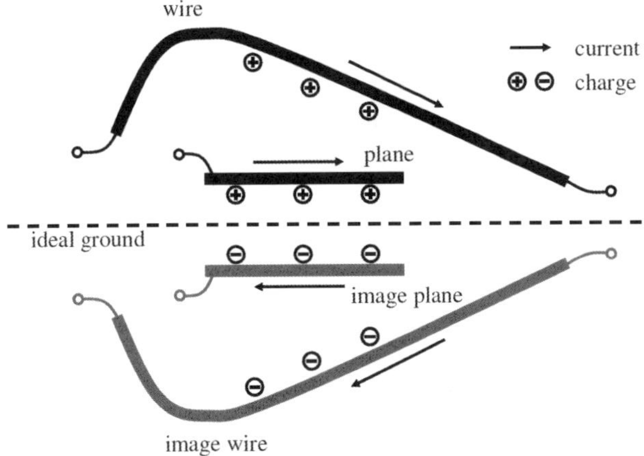

Figure 2.19: Example of image conductors and port definition.

used as the resultant incoming wave. By stamping (2.47) into a global conversion matrix $\mathbf{M}$, the following transform leads to the total scattering (S) parameter:

$$\begin{pmatrix} \mathbf{S}_{dd} & \mathbf{S}_{dc} \\ \mathbf{S}_{cd} & \mathbf{S}_{cc} \end{pmatrix} = \mathbf{M}\mathbf{S}\mathbf{M}^{-1}, \tag{2.48}$$

where $\mathbf{S}_{dd}$ is the resultant S parameter that provides response of the original interconnections.

2.7 Examples with Bonding Wires

This section presents three bonding wire modeling examples to illustrate various coupling effects. Frequency-domain S parameters are extracted from *RLC* models of JEDEC4 type bonding wires, bonding wires in a plastic ball grid array package, and wires in stacked ICs. Through these examples, we will show how electrical coupling from planar structures modifies the expected characteristics of bonding wires. In addition, the effects of signaling and grounding design schemes can be observed, and the complicated contribution of coupling from horizontal and vertical wires can be quantified. This section validates the results with a 3-D full-wave EM simulator.

2.7.1 *Three JEDEC4 Type Bonding Wires*

The bonding wire structure shown in Figure 2.20(a) is used in low-frequency package applications such as thin quad flat pack (TQFP) and lead-frame type packaging. The length of wires is about 5mm, which is rather long as compared to current technologies. The shape of each wire is constructed from the simplified model of EIA/JEDEC standard (JEDEC4) [JEDEC, 1997]. The coordinate values of each wire segmentation point are also shown in Figure 2.20(a). The diameter of the wires is 25um. Wire material is copper, with conductivity 5.8×10^7 S/m.

The three bonding wires were simulated with an ideal ground. As discussed in the previous section, additional image wires were defined to describe the ideal ground effect, and then the resultant single-ended S parameter matrix was converted to differential- and common-mode S parameter matrix using (2.47) and (2.48). Figure 2.20(b) shows the magnitudes of all the S-parameters. The comparison of the proposed method described in this chapter, and from CST MWS, a commercially available 3D full wave solver, shows good agreement, except for small errors in coupling.

Equivalent circuit of the three bonding wires can be extracted as shown in Figure 2.21. The equivalent circuit model includes all the inductive and capacitive coupling terms between the three wires and three image wires, but some of the small couplings can be neglected for practical simulations. Negative mutual inductances are observed because

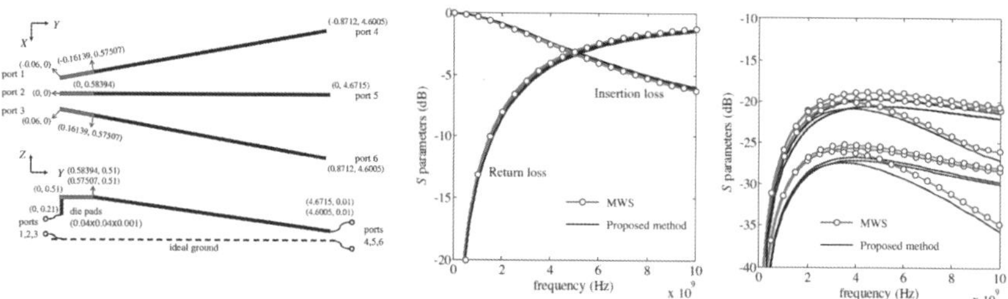

Figure 2.20: Three JEDEC4 bonding wires above ground plane (a) Structure showing dimensions and (b) Return Loss, Insertion Loss (left) and Coupling (right) showing comparison between proposed method and CST MWS.

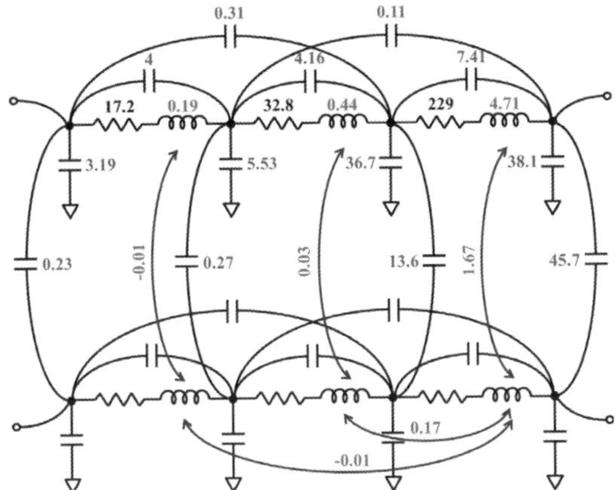

Figure 2.21: Part of equivalent circuit for the three JEDEC4 type bonding wires above ideal ground at 100MHz. Upper RLC network is from the left wire connecting Port 1 and Port 4, and lower RLC network comes from the image wire of the left wire. Resistance (mohm), inductance (nH), and capacitance (fF).

of the convention of current direction assumed when computing mutual inductances. For simplicity, PE-mode *RL* loops are omitted, and the higher-order capacitance loops are reduced to the finalized capacitance expressions in Figure 2.21.

2.7.2 *Bonding Wires in a Plastic Ball Grid Array (PBGA) Package*

In this subsection, bonding wire structures in wire-bonded PBGA [Kam et al., 2008] are simulated. By using the proposed method, the characteristics of different signaling structures are validated and compared with CST MWS. Simulation results show that a proper choice of signaling scheme as well as the wire structure design determines interconnection bandwidth.

Figure 2.22(a) shows the configuration of wires and grounds to be simulated. All the structures are based on differential signaling, and ground-signal-signal-ground (GSSG) structures have additional ground wires. In case of GSSG signaling, the ground wires can be assigned on a separate region of the package or can be tied to a common ground ring as shown in the GSSG-2 structure. For considering the effect of plastic molding around the wires, the dielectric constant of molding compound is defined as 4.3 [Chen et al., 2003]. An ideal ground is assumed at a location of 100um below the package surface. To capture wire shapes more accurately, we used analytical expressions that describe wedge-ball bonding wire geometry [Tay et al., 1995]. The curves ensure both zero slope at the wedge bonding on the package and an infinite slope at the ball bonding on the die. After obtaining the curves with pad distances and die/wire heights, approximate conductor segments were generated. The lengths of the long and the short wires are 1.75mm and 1.12mm, respectively.

Figure 2.22(b) shows the magnitude of scattering parameters for the three structures in Figure 2.22(a). As expected, the GSSG structure (especially GSSG-1) is better for signal transmission than the SS structure. However, the *S* parameters of the GSSG-2 structure show that using grounding bars does not ensure improvement in signal transmission. For GSSG design to work effectively, ground wires should follow the signal wires in parallel. Comparison with CST MWS results shows good accuracy.

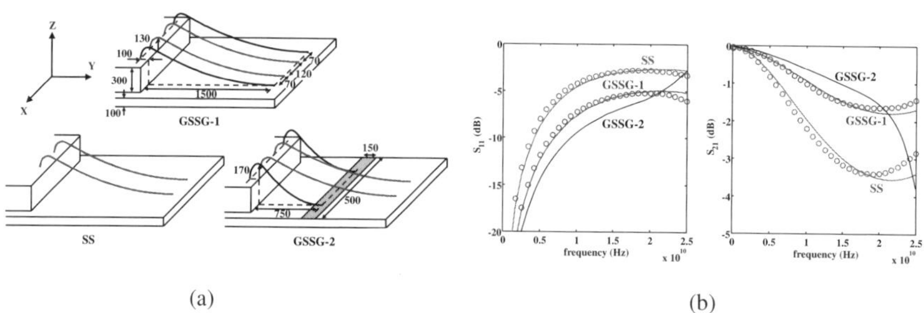

Figure 2.22: PBGA (a) Bond wire structure and assignments and (b) Return Loss (left) and Insertion Loss (right) solid line (proposed) and circles (CST MWS).

2.7.3 *Bonding Wires in Three Stacked ICs: The Effect of Vertical Coupling*

In this section, parasitic elements of bonding wires on three stacked ICs are extracted. The bonding wire structure in Figure 2.24 has been

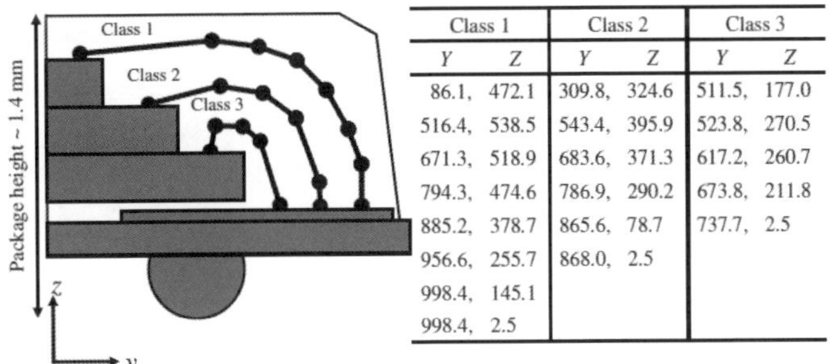

Class 1		Class 2		Class 3	
Y	Z	Y	Z	Y	Z
86.1,	472.1	309.8,	324.6	511.5,	177.0
516.4,	538.5	543.4,	395.9	523.8,	270.5
671.3,	518.9	683.6,	371.3	617.2,	260.7
794.3,	474.6	786.9,	290.2	673.8,	211.8
885.2,	378.7	865.6,	78.7	737.7,	2.5
956.6,	255.7	868.0,	2.5		
998.4,	145.1				
998.4,	2.5				

Figure 2.23: Side view and segmentation geometry (in um) of gold bonding wires on three stacked ICs.

obtained from the model of a triple-chip stacked chip-scale package (CSP) in Figure 2.23 [Fukui et al., 2000]. Wire diameter is 25um, and the pitch between adjacent wires is 60 or 80um [Milke et al., 2007]. All wires are in parallel. The approximate lengths of wires are 1.26mm (class 1), 0.82mm (class 2), and 0.48mm (class 3), respectively.

For validation, inductance and resistance values of six bonding wire structures as shown in Figure 2.24(a) were extracted, and S parameters from the R-L equivalent circuit model were compared with values from CST MWS.

In the simulation setup, six ports were defined as shown in Figure 2.24(a), which assumes each wire as a signal or a ground wire. Simulation results in Figure 2.24(b) indicate good correlation, except for small error at high-frequencies. This high-frequency error can be attributed to the quasi-static assumption used and the non-inclusion of capacitive coupling in this example. Since bonding wires are dominated by inductive coupling, the absence of capacitive coupling is expected to have a small effect on the response. The other source of error is the approximation of conductor segments used. As discussed earlier, the 1-D current assumption used may be inaccurate, especially at the sharp edges of adjoining conductor segments. For example, the sharp edge shown in class 3 wire of Figure 2.24(a) will generate some high-frequency error. The two combined error effects translate to a maximum error of 15.1% for $|S_{33}|$ in Figure 2.24(b), but this maximum error is acceptable since it occurs at small return loss values.

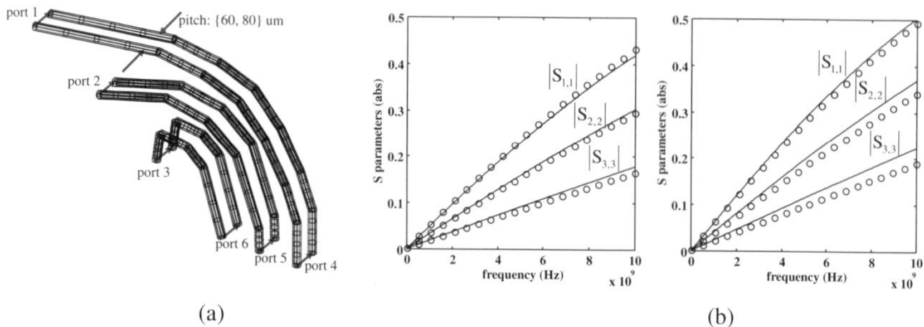

Figure 2.24: Stacked IC Bonding Wires (a) Configuration for six wires modeled and (b) S-parameters.

2.8 Examples with Vias

In this section we will address vias used in glass interposers, similar to the structures described in Chapter 4. Glass interposer is an alternative to the silicon interposer where vias can be created in glass with dimensions similar to through silicon vias (described in Chapter 3 and 4). Borosilicate glass is considered as the dielectric medium with relative permittivity 4.8 and loss tangent 0.003. The two examples considered include the effect of pitch on cross talk and repercussions of creating via chains in the interposer on insertion loss. Both examples assume a homogeneous medium. The 3D Path Finder [3DPF, 2012], a 3D extraction tool based on the methods described in this chapter, was used for the extraction.

2.8.1 *Glass Interposer Vias*

Consider a glass interposer with 3×3 array of vias. The vias have a diameter of 35µm and length 200µm, as shown in Figure 2.25. The via pitch is varied with a 70µm in Figure 2.25(a) and 250µm in Figure 2.25(b). Therefore, in Figure 2.25(b), the vias are depopulated to create a coarser pitch. As shown in Figure 2.25, signal ports (P1, P2, P3 and P4) are shown with ground returns, similar to a Ground-Signal-Ground ratio with ports placed at the top and bottom of the vias in 3D Path Finder [3DPF, 2012]. The objective here is to evaluate the effect of pitch on both near end and far end cross talk.

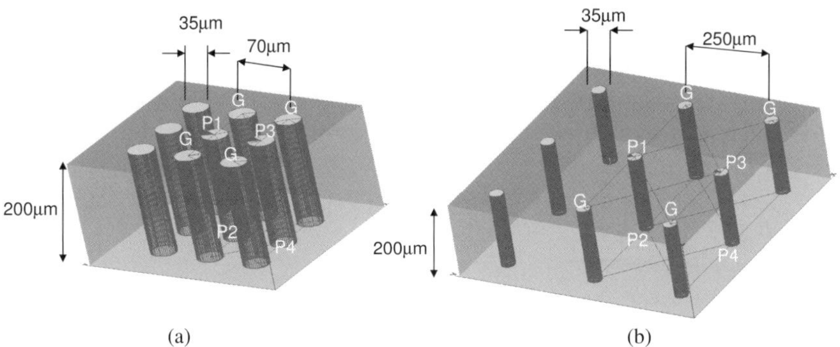

Figure 2.25: Glass interposer vias (a) 70um pitch and (b) 250um pitch.

The near end cross talk is shown in Figure 2.26 where the via pitch has a small effect on the coupling between ports 1 and 3. The far end coupling between ports 1 and 4 is affected much more by the via pitch, as can be seen in Figure 2.26, where the difference is ~2.5dB at 10GHz.

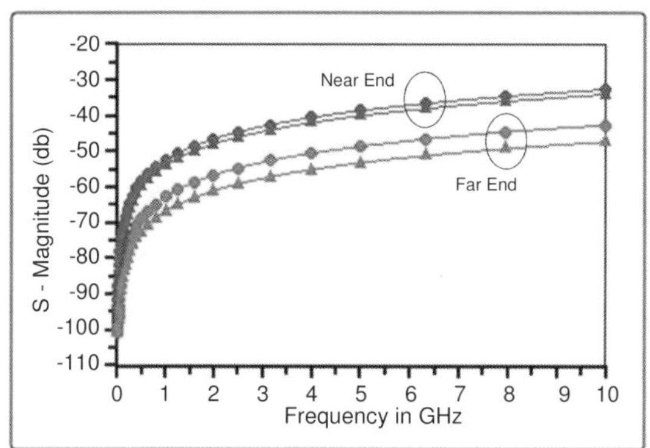

Figure 2.26: Near end and far end cross talk comparison for 70μm and 250μm pitch (circle: 70μm pitch; triangle: 250μm pitch).

2.8.2 *Via Chains*

With high density vias in the glass interposer, often times situations may arise where jogs may be necessary to complete a signal path. The general assumption is that since the interposer thickness is small, such jogs may not be detrimental to signal integrity. Here, we consider an example of signal jogs in a glass interposer substrate. The thickness of the interposer is 200μm with via radius of 35μm on a 70μm pitch. Metal straps with width of 35μm and thickness 10μm are used to jog from the top to the bottom of the interposer, as shown in Figure 2.27(a). Two neighboring vias are strapped together at the top and bottom, which represent the ground vias. The objective is to compute the insertion loss between port 1 and port 2 where the reference used is the ground vias. Another objective to check the effect of via pitch on the insertion loss for the same structure, where the via pitch is now increased to 100μm. In Figure 2.27(a), with a 70μm pitch, the total length of the signal line including the jogs is 1150μm (350μm length and 800μm combined via length). When the pitch is increased to 100μm, the total length increases to 1300μm (500μm length and 800μm via length).

The insertion loss between port 1 and 2 is shown in Figure 2.27(b). For the 70μm pitch, the insertion loss is ~0.15dB at 10GHz which increases to ~0.27dB for the 100μm pitch. Clearly, the effect of signal jogs is noticeable, since the lengths of these signal lines are 1.15mm to 1.3mm (very short lines).

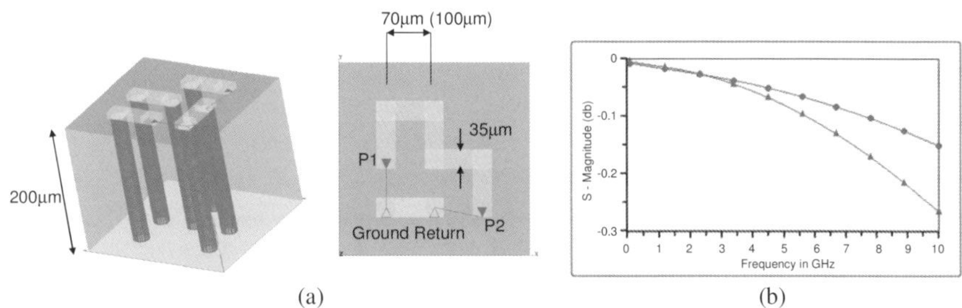

(a)   (b)

Figure 2.27: Signal jog (a) Structure showing the signal jog and position of ports and (b) Insertion loss (circles: 70um pitch; triangles: 100um pitch).

2.9 Example of Package on Package

A popular System in Package (SIP) technology is Package on Package (POP) being used by LPDDR3 and other mobile applications. Here the dies are packaged and the packages are mounted using BGA on top of each other as depicted in Figure 2.28(a). Signals propagating through the stack will be affected by the physical structure causing signal degradation. In this section, we evaluate the signal insertion loss in a 3 die stack package on package structure. The detailed structure with dimensions is shown in Figure 2.28(b) (3D view) and Figure 2.28(c) (cross section). The structure is assumed to be in a homogenous medium with relative permittivity 4.3 and loss tangent 0.01. The structure can be sectioned to account for an inhomogeneous medium as well (for example using air as the dielectric for the BGA), which has been done in the present analysis. The metal pads and lines have a thickness of 10μm. The objective is to calculate the insertion loss from port 1 (P1) to port 2 (P2) which represents the signal propagation from the bottom of the stack to die 1. A common reference or ground return is used for both ports, as shown in Figure 2.28(c).

The insertion loss and return loss are shown in Figure 2.29 where the signal degrades by ~2dB at 10GHz. Measuring the distance from port 1 to port 2, a distance of roughly 2mm results in an insertion loss of ~2dB

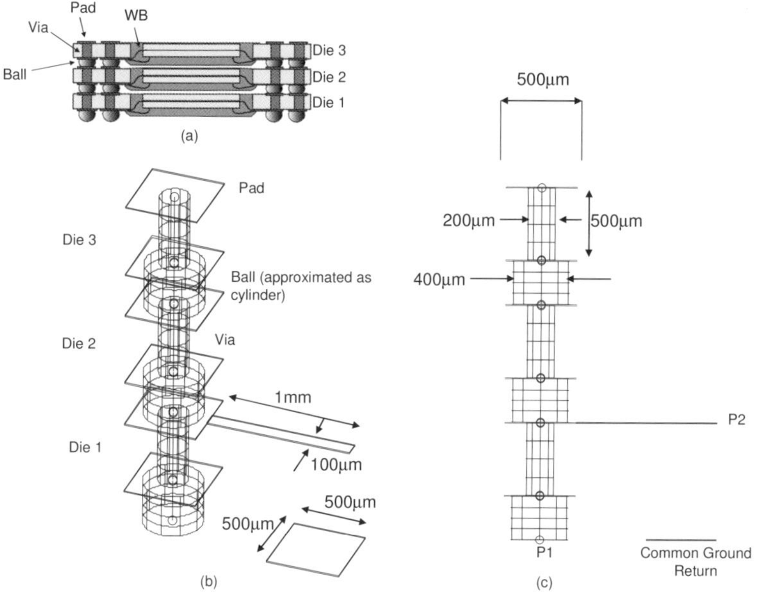

Figure 2.28: Package on Package (a) 3 die stack, (b) 3D view and (c) Cross section.

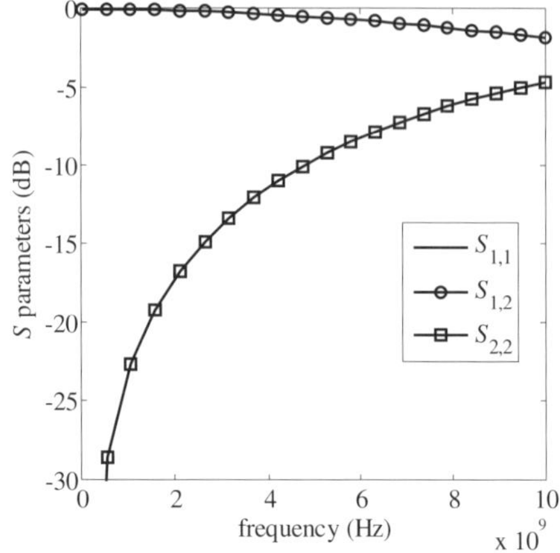

Figure 2.29: Insertion Loss (circles) and Return Loss (squares) for POP structure.

which is quite large. This can be attributed to many discontinuities shown in the stack (vias, pads and balls) that cause signal reflection. For low power applications such as in mobile phones, any increase in insertion loss represents wastage of power. This is therefore becoming the driver for creating newer technologies using through silicon vias (TSV) that can reduce the interconnect length and therefore reduce the insertion loss of the signal lines.

2.10 Summary

In summary, this chapter presented the development of a modeling method that enables the electrical analysis of interconnections used in 3D integration. By using specialized basis functions that are applicable for cylindrical structures such as bonding wires and via interconnections, the proposed method can extract parasitic elements and coupling of a large number of strongly coupled interconnections arising in 3-D integration at considerably reduced computational time.

The modeling method in this chapter is composed of three parts. First part is the extraction of frequency-dependent series impedance. To capture inductive coupling cylindrical CMBFs with EFIE was used. The use of cylindrical CMBF to compute mutual inductances can be further improved by controlling the required number of PE-mode basis functions. Second part is the extraction of capacitances and conductances, which play a role at higher frequencies. Similar to cylindrical CMBF for impedance calculations, cylindrical AMBFs were used. In addition, the integral equation formulation was revisited to consider the case when the background material was a (non-free-space) dielectric. Without compromising on the efficiency of *R-L* computations, the method discussed was used to obtain broadband *RLGC* model of large number of 3-D interconnections. Third part is the inclusion of the effects of various planar structures including strip-type interconnections, bonding pads, and finite ground structures into the analysis of cylindrical structures. To consider the coupling from planar structures, this chapter discussed the combined integral-equation-based model by adding staircase (piecewise constant) basis functions used in the conventional PEEC method. Combining the *RL* extraction, *CG* extraction, and planar coupling, this chapter discussed examples for modeling various bonding wires, vias and BGAs. The method discussed in this chapter has been extended for analyzing through silicon vias (TSVs) in the next chapter.

References

1. [Tummala, 2001] R. R. Tummala, Fundamentals of Microsystems Packaging. McGraw-Hill, 2001.
2. [Milke et al., 2007] E. Milke and T. Mueller and A. Bischoff, "New Bonding Wire for Fine Pitch Applications," in Proc. IEEE Electronics Packaging Technology Conference, 2007, pp. 750–754.
3. [Kam et al., 2008] D. G. Kam and J. Kim, "40-Gb/s Package Design Using Wire-Bonded Plastic Ball Grid Array," IEEE Trans. Advanced Packaging, Vol. 31, No. 2, pp. 258–266, May 2008.
4. [Kim et al., 2009] J.-H. Kim, R. Schmitt, D. Oh, W. Beyene, M. Li, A. Vaidyanath, J. Feng, and C. Yuan, "Feasibility Study of a 3.2Bb/s Memory Interface in Ultra Low-Cost LQFP Packages," in IEC DesignCon 2009, Feb. 2009.
5. [Hu et al., 2008] Y. Hu, J. Chen, M. Lamson, and R. Bashirullah, "An Active Crosstalk Reduction Technique for Parallel High-Speed Links in Low Cost Wirebond BGA Packages," in Proc. IEEE Conference on Electrical Performance of Electronic Packaging, Oct. 2008, pp. 37–40.

6. [Fukui et al., 2000] Y. Fukui, Y. Yano, H. Juso, Y. Matsune, K. Miyata, A. Narai, Y. Sota, Y. Takeda, K. Fujita, M. Kada, "Triple-Chip Stacked CSP," in Proc. IEEE Electronic Components and Technology Conference, May 2000, pp. 385–389.

7. [Tummala, 2005] R. R. Tummala, "Packaging: Past, Present and Future," in Proc. IEEE 6th International Conference on Electronic Packaging Technology, Aug.–Sep. 2005, pp. 3–7.

8. [Dreiza et al., 2005] M. Dreiza, A. Yoshida, J. Micksch, and L. Smith, "Stacked Package-on-Package Design Guide," Chip Scale Review, Jul. 2005.

9. [Han et al., 2013] K. J. Han, M. Swaminathan, "Modeling electrical interconnections in three-dimensional structures", United States Patent: 8,352,232, 1/08/2013.

10. [Ruehli, 1974] A. E. Ruehli, "Equivalent Circuit Models for Three-Dimensional Multiconductor Systems," IEEE Trans. Microwave Theory and Techniques, Vol. 22, No. 3, pp. 216–221, Mar. 1974.

11. [Daniel et al., 2001] L. Daniel, A. Sangiovanni-Vincentelli, and J. White, "Using Conduction Mode Basis Functions for Efficient Electromagnetic Analysis of On-Chip and Off-Chip Interconnect," in Proc. ACM/IEEE 38th Design Automation Conference, Jun. 2001, pp. 563–566.

12. [Silvester, 1966] P. Silvester, "Modal Network Theory of Skin Effect in Flat Conductors," Proc. IEEE, Vol. 54, No. 9, pp. 1147–1151, Sep. 1966.

13. [Han et al., 2008a] K. J. Han, M. Swaminathan, and E. Engin, "Electric Field Integral Equation with Cylindrical Conduction Mode Basis Functions for Electrical Modeling of Three-dimensional Interconnects," in Proc. ACM/IEEE 45th Design Automation Conference, Jun. 2008.

14. [Abramowitz et al., 1965] M. Abramowitz and A. Stegun, Handbook of Mathematical Functions: with Formulas, Graphs, and Mathematical Tables. New York: Dover Publications, 1965.

15. [Han et al., 2009] K. J. Han, M. Swaminathan, "Inductance and Resistance Calculations in Three-Dimensional Packaging Using Cylindrical Conduction Mode Basis Functions," IEEE Transactions on Computer-Aided Design of Integrated Circuits and Systems, Vol. 28, No. 6, pp. 846–859, Jun. 2009.

16. [Daniel et al., 2003] L. Daniel, A. Sangiovanni-Vincentelli, and J. White, "Proximity Templates for Modeling of Skin and Proximity Effects on Packages and High Frequency Interconnect," in Proc. IEEE/ACM International Conference on Computer Aided Design, Nov 2003, pp. 326–333.

17. [Ruehli et al., 2001] A. E. Ruehli and A. C. Cangellaris, "Progress in the Methodologies for the Electrical Modeling of Interconnects and Electronic Packages," Proceedings of The IEEE, Vol. 89, No. 5, pp. 740–771, May 2001.

18. [Paul, 1994] C. R. Paul, Analysis of Multiconductor Transmission Lines. New York: Wiley, 1994.

19. [Ortiz et al., 2007] S. Ortiz and R. Suaya, "Efficient Implementation of Conduction Modes for Modelling Skin Effect," in Proc. IEEE Computer Society Annual Symposium on VLSI, Mar. 2007, pp. 500–505.

20. [Press et al., 1995] W. H. Press, S. A. Teukolsky, W. T. Vetterling and B. P. Flannery, Numerical Recipes in C - The Art of Scientific Computing, 2nd ed. Cambridge,
U. K.: Cambridge Univ. Press, 1995.

21. [Gander et al., 2000] W. Gander and W. Gautschi, "Adaptive Quadrature-Revisited," BIT Numerical Mathematics, Vol. 40, No. 1, pp. 84{101, Mar. 2000.

22. [Han et al., 2008b] K. J. Han, M. Swaminathan, and E. Engin, "Wideband Electrical Modeling of Large Three-Dimensional Interconnects using Accelerated Generation of Partial Impedances with Cylindrical Conduction Mode Basis Functions," in Proc. IEEE MTT-S International Microwave Symposium, Jun. 2008.

23. [Ray et al., 1999] S. Ray, A. K. Gogoi, and R. K. Mallik, "Closed Form Expression for Mutual Partial Inductance between Two Skewed Linear Current Segments," in Proc. Antennas and Propagation Society International Symposium, July 1999, pp. 1278–1281.

24. [Antonini et al., 2003] G. Antonini and A. E. Ruehli, "Fast Multipole and Multifunction PEEC Methods," IEEE Trans. Mobile Computing, Vol. 2, No. 4, pp. 288–298, Oct.–Dec. 2003.

25. [Kamon et al., 1994] M. Kamon, M. J. Tsuk, and J. K. White, "FastHenry: A multipoleaccelerated 3D inductance extraction program," IEEE Trans. Microw. Theory Tech., Vol. 42, No. 9, pp. 1750–1758, Sep. 1994.

26. [Cheng, 1989] D. K. Cheng, Field and Wave Electromagnetics. Addison-Wesley, 1989.

27. [Scharstein, 2007] R. W. Scharstein, "Capacitance of a Tube," Journal of Electrostatics, Vol. 65, pp. 21–29, 2007.

28. [Verolino, 1995] L. Verolino, "Capacitance of a Hollow Cylinder," Electrical Engineering, Vol. 78, pp. 201–207, 1995.

29. [Faria et al., 2006] J. B. Faria and M. G. das Neves, "Accurate Evaluation of Indoor Triplex Cable Capacitances Taking Conductor Proximity Effect Into Account," IEEE Trans. Power Delivery, Vol. 21, No. 3, pp. 1238–1244, Jul. 2006.

30. [Clements et al., 1975] J. C. Clements, C. R. Paul, and A. T. Adams, "Computation of the Capacitance Matrix for Systems of Dielectric-Coated Cylindrical Conductors," IEEE Trans. Electromagnetic Compatibility, Vol. 17, No. 4, pp. 238–248, Nov. 1975.

31. [Li et al., 2002] G.-L. Li, Z.-H. Feng, "Consider the Losses of Dielectric in PEEC," in Proc. International Conference on Microwave and Millimeter Wave Technology. IEEE, 2002, pp. 793–796.

32. [Ruehli et al., 1992] A. E. Ruehli and H. Heeb, "Circuit Models for Three-Dimensional Geometrices Including Dielectrics," IEEE Trans. Microwave Theory and Techniques, Vol. 40, No. 7, pp. 1507–1516, Jul. 1992.

33. [Young, 2001] B. Young, Digital Signal Integrity: Modeling and Simulation with Interconnects and Packages. Upper Saddle River, NJ: Prentice Hall PTR, 2001.

34. [Wane et al., 2007] S. Wane and O. Tesson, "Compact Equivalent Circuit Derivation of Bond Wire Arrays for Power and Signal Integrity Analysis," in Proc. European Microwave Conference, Oct. 2007, pp. 524–527.

35. [Chuang et al., 2004] J. Y. Chuang, S. P. Tseng, and J. A. Yeh, "Radio Frequency Characterization of Bonding Wire Interconnections in a Modeled Chip," in Proc. IEEE 54th Electronic Components and Technology Conference, Vol. 1, Jun. 2004, pp. 392–399.

36. [Xie et al., 1999] X. Xie and J. L. Prince, "Frequency Response Characteristics of Reference Plane Effective Inductive and Resistance," IEEE Trans. Advanced Packaging, Vol. 22, No. 2, pp. 221–229, May 1999.

37. [Ruehli, 1972] A. E. Ruehli, "Inductance Calculations in a Complex Integrated Circuit Environment," IBM J. Res. Develop., pp. 470–481, Sep. 1972.

38. [Zhong et al., 2003] G. Zhong, C. K. Koh, "Exact Closed-Form Formula for Partial Mutual Inductances of Rectangular Conductors," IEEE Trans. Circuits and Systems I: Fundamental Theory and Applications, Vol. 50, No. 10, pp. 1349–1353, Oct. 2003.

39. [Ruehli et al., 1973] A. E. Ruehli, P. A. Brennan, "Efficient Capacitance Calculations for Three-Dimensional Multiconductor Systems," IEEE Trans. Microwave Theory and Techniques, Vol. 21, No. 2, pp. 76–82, Feb. 1973.

40. [Schuster et al., 2000] C. Schuster, G. Leonhardt, W. Fichtner, "Electromagnetic Simulation of Bonding Wires and Comparison with Wide Band Measurements," IEEE Trans. Advanced Packaging, Vol. 23, No. 1, pp. 69–79, Feb. 2000.

41. [Bockelman et al., 1995] D. E. Bockelman and W. R. Eisenstadt, "Combined Differential and Common-Mode Scattering Parameters: Theory and Simulation," IEEE Trans. Microwave Theory and Techniques, Vol. 43, No. 7, pp. 1530–1539, Jul. 1995.

42. [JEDEC, 1997] "Bond Wire Modeling Standard," EIA/JEDEC Standard, No. 59, Jun. 1997.

43. [Chen et al., 2003] N. Chen, K. Chiang, T. Her, Y.-L. Lai, and C. Chen, "Electrical Characterization and Structure Investigation of Quad Flat Non-Lead Package for RFIC Applications," Solid-State Electronics, Vol. 47, pp. 315–322, Feb. 2003.

44. [Tay et al., 1995] A. A. O. Tay, K. S. Yeo, and J. H. Wu, "The Effect of Wirebond Geometry and Die Setting on Wire Sweep," IEEE Trans. Components, Packaging, and Manufacturing Technology - Part B, Vol. 18, No. 1, pp. 201–209, Feb. 1995.

45. [3DPF, 2012] 3D Path Finder, www.e-systemdesign.com.

Chapter 3

Electrical Modeling of Through Silicon Vias

Through silicon via (TSV) is a new technology that provides short electrical connections between the top and bottom surface of a silicon substrate. When used in silicon stacking, TSVs provide short connections between transistors that are vertically separated from each other. The manner of these connections depends on whether via first, middle or last technology is used. In the area of packaging, TSVs used in silicon interposers also provide a short electrical path to the printed circuit board. What makes the modeling of TSVs interesting and its design challenging is the material properties of the medium surrounding these vertical interconnections. Due to the semi-conducting properties of the silicon medium; losses, capacitance effects and coupling behavior of TSVs are unique and are quite different from similar structures in a perfectly insulating medium. Hence, the electrical modeling of TSVs is discussed in detail in this chapter. This chapter is a follow-on to Chapter 2 where the electromagnetic modeling of cylindrical cross section interconnections such as wirebonds and vias or Column Grid Arrays (CGA) have been discussed. Since 3D integration can be enabled by using a combination of all three technologies (wirebonds, CGA and TSV), some of these have been considered in Chapter 4 to analyze multiple scenarios for stacking and packaging ICs.

3.1 Benefits of Through Silicon Vias

As is well known, TSVs provide short interconnection lengths as opposed to wirebond technology for stacking of ICs. Recent studies have shown that 3D DDR3 DRAM [Kang et al., 2010] can be enabled by

115

using TSVs whereby 50% reduction in standby power and 25% reduction in active power are possible as compared to quad-die package with an increase in I/O speed from 1066Mbps to 1600Mbps. An emerging application is in the area of wide I/O memory for mobile applications where logic and memory are being stacked on top of each other using TSV technology. In an interesting plenary talk given by Oh Hyun Kwon [Kwon, 2011], he compared a conventional 3D package using Flip Chip Package on Package (FC-POP) with LPDDR2 memory (low power DDR2) to an equivalent System in Package (SIP) with wide IO memory, as shown in Figure 3.1. Dramatic improvements in package size (35% reduction) and power consumption (50% reduction) were seen as shown in Figure 3.1(c). A very interesting aspect is the increase in bandwidth by 8X by supporting 512 I/Os transmitting at a data rate of 12.8Gbps as compared to 3.2Gbps in LPDDR2 memory.

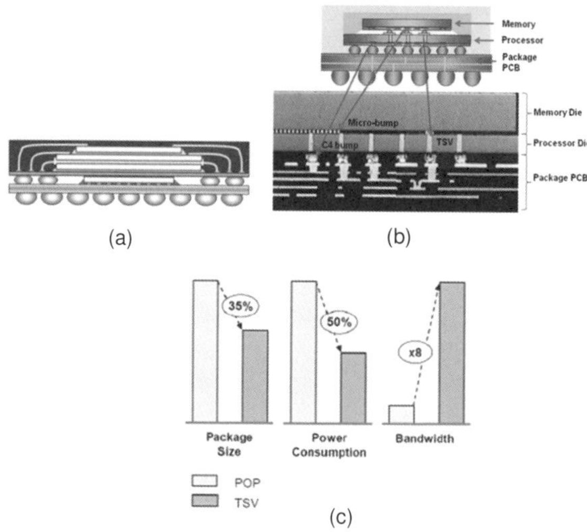

Figure 3.1: (a) Flip Chip POP (FC-POP), (b) TSV-System in Package (SIP) with wide I/O DRAM and (c) Performance benefits.

The reduction in package size is obvious since the wirebonds on the top tier in the memory stack in Figure 3.1(a) is around 1mm long, thereby using a large amount of area to package the memory stack as opposed to the flip chip processor in the bottom package. By replacing the wirebonds using TSVs in Figure 3.1(b), the interconnection length and therefore the package area can be reduced. The reduction in power can be attributed to the reduction in the capacitance of the TSV-SiP that needs to be charged and discharged as compared to the wirebond and routing capacitance in the FC-POP. Finally, the higher bandwidth for TSV-SIP is due to the fine pitch of the TSVs that provides more interconnections per unit area as compared to FC-POP, leading to 512 signal I/O connections (total connections is around 1200 including power and ground) between the processor and memory. Due to the shorter delay of TSVs because of the smaller interconnection length (1mm long wirebond as compared to 60µm long TSV), the speed of the processor-memory interface has increased from 3.2Gbps to 12.8Gbps. Clearly, the parasitics of the interconnections dictate to a large extent the electrical performance of either the FC-POP or TSV-SIP. In this chapter, the extraction of the electrical parasitics of TSVs that include its resistance (R), inductance (L), capacitance (C) and conductance (G) are discussed along with the computation of its scattering parameters.

3.2 Challenges in Modeling Through Silicon Vias

Consider a silicon stack as shown in Figure 3.2(a) containing multiple tiers, where each tier represents a die. The tiers are bonded to each other through microbumps or pads. Without loss of generality, the bottom most tier could be considered as the silicon interposer. The physical geometry associated with one of the tiers is enlarged in Figure 3.2(b) showing the copper connections at the center of the TSV, surrounded by an oxide liner in a silicon medium. Depending on the process used, copper can be substituted with tungsten which provides a better coefficient of thermal expansion (CTE) match to the silicon substrate but at the expense of higher resistance. The structure shown in Figure 3.2(b) is also used in silicon interposers to package the stacked ICs. Two of the TSVs are shown in Figure 3.2(c) to illustrate the physical geometry and material properties.

In Figure 3.2(c), each TSV consists of a metal center conductor of radius R. The metal can either be copper (conductivity $\sigma = 5.8 \times 10^7 \text{S/m}$) or tungsten (conductivity $\sigma = 1.9 \times 10^7 \text{S/m}$). The center conductor is surrounded by an oxide (typically SiO_2) with a relative permittivity of 3.9. The metal conductor with oxide liner is embedded in a silicon based semi-conductor medium with relative permittivity of 11.9. A very important electrical parameter of the silicon medium is its conductivity which depends on the doping used. Standard CMOS grade silicon has a conductivity of 10S/m with high resistivity silicon having conductivity in the range of 0.01S/m. In Figure 3.2(c), the ports (points of excitation and measurement) connected to the TSVs are referenced to 50ohms for computing the scattering parameters (explained later). Typical dimensions for the TSVs are: diameter (2R) in the range of 1.6–50μm, pitch D in the range of 2.4–80μm and length L in the range of 6–250μm. The oxide thickness d_{ox} is typically around 100–200nm or larger depending on the process.

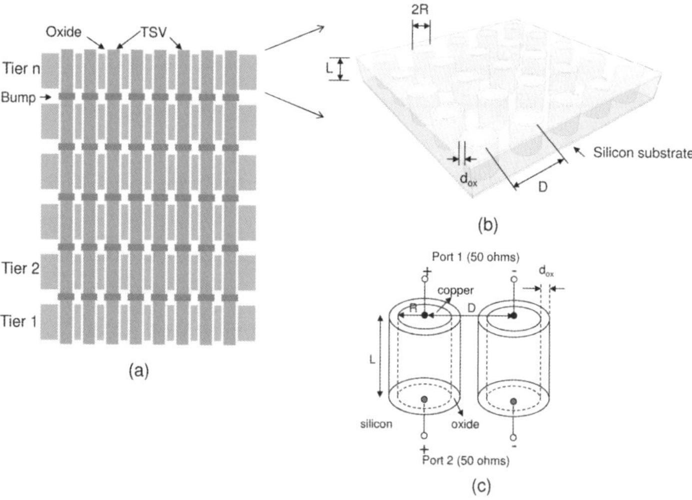

Figure 3.2: Through silicon vias (a) TSVs in 3D Stack with multiple dies (tiers), (b) TSV array in a tier and (c) two TSVs.

The extraction of the electrical parasitics of TSVs can be a challenging task for the following reasons: 1) the dimensions are multi-scale with an aspect ratio of 20:1 (L/2R) and oxide thickness of 100–200nm (d_{ox}), 2) it is embedded in a semiconducting medium which is lossy (CMOS grade silicon) making the electromagnetic wave propagation effects complex (explained later), 3) with TSV densities being greater than $10^5/cm^2$, the conductance and capacitance of the silicon substrate can cause significant leakage and coupling between TSVs, 4) biasing of the silicon substrate can change the TSV capacitance due to Metal-Oxide-Semiconductor (MOS) capacitance behavior, 5) with temperature dependent conductivity, the resistance and conductance of TSVs become temperature dependent (explained in a later chapter) and 6) the electrical parameters are strongly frequency dependent. Due to these effects, with some being local and others global, arrays of TSVs have to be modeled in parallel making the problem a very complex one from a computational standpoint. Since, most electromagnetic solvers use a meshing scheme to discretize Maxwell's equations, it severely limits the number of TSVs that can be analyzed due to the multi-scale dimensions of the structures and the large number of TSVs that need to be analyzed. To approach this problem from a practical standpoint, two mesh-free methods for computing the electrical parasitics of TSVs are described. These are 1) a physical model based approach using analytical equations and 2) a rigorous electromagnetic analysis based approach by solving Maxwell's equations using specialized basis functions that approximate the current and charge in TSVs without requiring a mesh.

3.3 Propagating Modes in Through Silicon Vias – An Electromagnetic Perspective

A microstrip line on a semiconductor substrate such as Si (silicon) separated by SiO_2 (silicon-di-oxide) supports three fundamental modes of propagation namely, slow-wave, quasi-TEM and skin-effect modes. These modes have been discussed elegantly in [Hasegawa et al., 1971]. The three modes are separated based on the frequency, thickness of the SiO_2 layer, thickness of the Si substrate and silicon conductivity. Since, the TSV structure is similar to the microstrip line described by [Hasegawa et al., 1971] with a metal-SiO_2-Si interface, a similar set of propagating modes can be expected, which is described in this section. A detailed analysis of the mode propagation in TSVs is available in [Ndip et al., 2011]. In this section we use a circuit representation to explain the modal behavior.

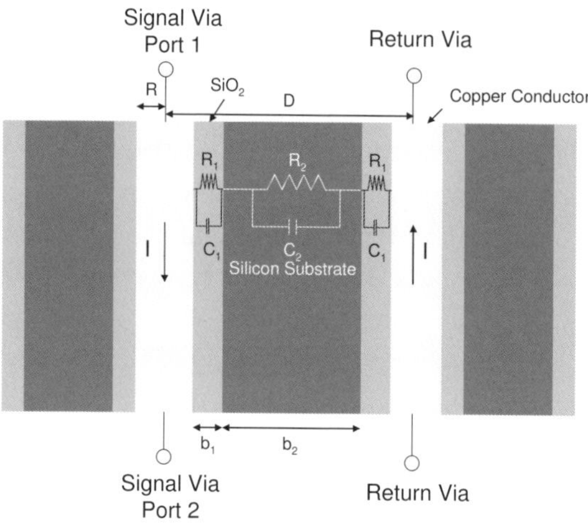

Figure 3.3: Signal and Ground TSV.

Consider a signal and ground (return) TSV shown in Figure 3.3. The cylindrical copper conductor is surrounded by SiO$_2$ liner of thickness b$_1$. The region between the two oxide liners contains the silicon substrate of thickness b$_2$. Based on [Hasegawa et al., 1971], when the product of frequency and resistivity of the silicon substrate is large enough to produce a small dielectric loss angle, then the silicon substrate acts like a low-loss dielectric. In such a case the wave propagates in the presence of two dielectrics (SiO$_2$ and Si) between the signal and ground TSV and the fundamental mode is a quasi-TEM mode where the velocity of the wave is governed by the permittivity of the silicon substrate. The skin-effect mode occurs when the product of substrate conductivity and frequency is large enough where the electric and magnetic fields have a small depth of penetration into silicon. In such a scenario, the silicon substrate acts as a conductor wall and appears as a lossy ground plane to the signal conductor. The minimum frequency at which this occurs is when the skin depth $\delta = b_2$ (since $b_1 \ll b_2$). Since $\delta = \dfrac{1}{\sqrt{\pi f \mu \sigma_{Si}}}$ where $\mu = \mu_0$ is the permeability of free space and σ_{Si} is the conductivity of silicon, the frequency at which skin-effect mode begins is $f = \dfrac{1}{\pi\mu\sigma_{Si}b_2^2}$. With typical silicon substrate conductivity of 10S/m and $b_2 < 50\mu m$, the onset of skin-effect mode occurs at high frequency when $f > 10^{12}$Hz and is therefore not considered here. In addition to the dielectric and skin-effect modes, a third mode called the "slow wave mode" can exist when the frequency is not high and the conductivity of the silicon substrate is moderate. This mode is a surface wave that occurs due to the strong interfacial polarization across the SiO_2 liner and has a velocity of propagation much slower than the silicon substrate due to the Maxwell-Wagner effect that increases the effective permittivity at lower frequencies [Hasegawa et al., 1971]. Given the frequency range of interest for most applications from DC to 100GHz and the typical dimensions of TSVs, the modes to consider for signal propagation are the slow wave and dielectric modes, which are discussed further in this section.

The behavior of the slow wave and dielectric modes can be explained by the Maxwell-Wagner effect which is an interfacial relaxation process that occurs in all systems where the electric current must pass an interface between two dielectrics [Barlea et al., 2008]. The dispersion of the dielectric occurs in TSVs due to the series connection of the dielectric slabs formed by SiO_2 and Si. When the SiO_2 and Si dielectric layers can each be represented as a circuit with conductance and capacitance in parallel, connected to each other in series, then the interface can be charged by the conductance. This equivalent circuit representation and connectivity of the interfaces is shown in Figure 3.3, where R_1, C_1 and R_2, C_2 are the resistance and capacitance of the SiO_2 and Si substrate, respectively. The admittance of the three dielectric layers in series can be written as [Barlea et al., 2008]:

$$Y = \frac{\dfrac{R_1}{1+\omega^2\tau_1^2} + \dfrac{R_2}{1+\omega^2\tau_2^2} + \dfrac{R_1}{1+\omega^2\tau_1^2} + j\omega\left[\dfrac{\tau_1 R_1}{1+\omega^2\tau_1^2} + \dfrac{\tau_2 R_2}{1+\omega^2\tau_2^2} + \dfrac{\tau_1 R_1}{1+\omega^2\tau_1^2}\right]}{\left[\dfrac{R_1}{1+\omega^2\tau_1^2} + \dfrac{R_2}{1+\omega^2\tau_2^2} + \dfrac{R_1}{1+\omega^2\tau_1^2}\right]^2 + \omega^2\left[\dfrac{\tau_1 R_1}{1+\omega^2\tau_1^2} + \dfrac{\tau_2 R_2}{1+\omega^2\tau_2^2} + \dfrac{\tau_1 R_1}{1+\omega^2\tau_1^2}\right]^2} \qquad (3.1)$$

where 'ω' is the angular frequency and $\tau_i = R_i C_i$ $i = 1,2$ is the time constant. The equivalent capacitance 'C' and conductance 'G' can be calculated as:

$$C = \frac{\text{Im}(Y)}{\omega}; G = \text{Re}(Y). \tag{3.2}$$

The capacitance C_1 of the oxide can be calculated approximately assuming a coaxial cable representation with inner radius 'R' and outer radius 'R + b1' as:

$$C_1 = \frac{2\pi\varepsilon_{SiO_2} L}{\ln((R + b_1)/R)} \tag{3.3}$$

where 'L' is the length and ε_{SiO_2} is the permittivity of the oxide. For a TSV with $L = 100\mu m$, $R = 15\mu m$, $b_1 = 0.1\mu m$ and using a relative permittivity of 3.9 for SiO_2, the capacitance C_1 can be calculated as 3.26pF. The capacitance C_2 can be calculated approximately assuming a two wire transmission line as:

$$C_2 = \frac{\pi\varepsilon_{Si} L}{\ln\left(\dfrac{D}{2R} + \sqrt{\dfrac{D^2}{4R^2} - 1}\right)} \tag{3.4}$$

where 'D' is the pitch and ε_{Si} is the permittivity of the silicon substrate (since $b_1 \ll b_2$). For a TSV with $L = 100\mu m$, $R = 15\mu m$, $D = 100\mu m$ and using a relative permittivity of 11.9 for silicon, the capacitance C_2 can be calculated as 0.0176pF. The conductance of SiO_2 is typically small since it is a good insulator and therefore its resistance R_1 is large, which is assumed to be 1MΩ. However, since the conductivity of the silicon substrate is large (10S/m), its resistance R_2 is much smaller than R_1. For a parallel RC circuit in the silicon substrate, the dissipation factor or loss tangent can be defined as $\tan d = \dfrac{G}{\omega C}$. In a dielectric, the loss tangent can also be computed as $\tan d = \dfrac{\sigma}{\omega\varepsilon}$, where σ is the conductivity and ε is the real part of permittivity. Therefore, the conductance of the silicon substrate can be computed as:

$$G_2 = \frac{1}{R_2} = \frac{\pi \sigma_{Si} L}{\ln\left(\frac{D}{2R} + \sqrt{\frac{D^2}{4R^2} - 1}\right)} .$$

(3.5)

Using (3.5), for the dimensions described, the resistance R_2 can be computed as 596Ω. With the defined parameters, the effective conductance and capacitance of the signal TSV with respect to the ground TSV has been plotted in Figure 3.4 using equations (3.1) and (3.2).

Both the conductance and capacitance versus frequency curves in Figure 3.4 exhibit the classic Debye dispersion behavior. At low frequencies, $\omega \to 0$ and the time constant $\omega \tau_i \ll 1$. Therefore, the conductance G and capacitance C from equation (3.1) and (3.2) can be approximated as:

$$G = \frac{1}{2R_1 + R_2} = 0.5 x 10^{-3} mS$$

$$C = \frac{2R_1^2 C_1 + R_2^2 C_2}{(2R_1 + R_2)^2} = 1.63 pF$$

(3.6)

Hence, at low frequencies, the capacitance is large and the conductance small. Comparing with the capacitance of the two wire line from equation (3.4) which is 0.0176pF where the wave propagates in the silicon substrate with relative permittivity of 11.9, the capacitance at low frequencies is 92 times larger, indicating an effective permittivity of 1094 for wave propagation. This is the reason why at low frequencies, the TSV supports the propagation of a slow wave. It is important to note that the conductance (or leakage) of the slow wave is small, indicating that the wave attenuation is minimum at low frequencies (f < 0.1MHz). At high frequencies, $\omega \to \infty$ and $\omega \tau_i \gg 1$. In such a scenario, the conductance and capacitance from equations (3.1) and (3.2) can be approximated as:

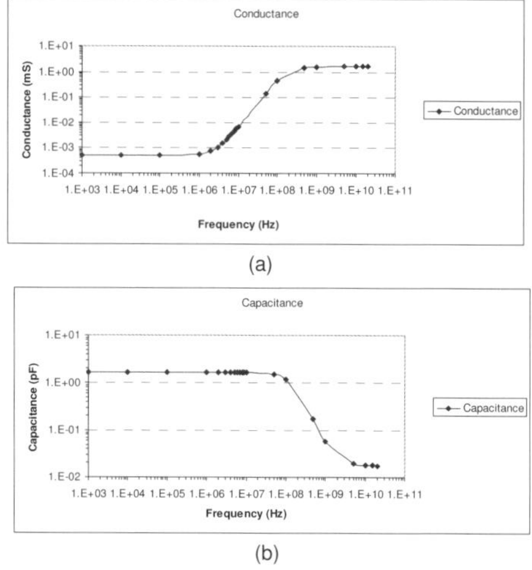

(a)

(b)

Figure 3.4: Frequency behavior of (a) Effective Conductance and (b) Effective Capacitance.

$$G = \frac{2R_2C_2^2 + R_1C_1^2}{R_1R_2(2C_2 + C_1)^2} = 1.64\,mS$$

$$C = \frac{C_1C_2}{C_1 + 2C_2} = 0.0174\,pF$$

$$(3.7)$$

Here, the conductance and capacitance are dictated by the material properties of the silicon substrate indicating that the silicon is acting as a lossy dielectric material. Here, the wave propagation is in the dielectric region between the two TSVs and is a quasi-TEM mode where the loss arises due to the displacement current. A question that often arises is, when does the wave transition from a slow wave to a quasi-TEM mode. This can be explained using the discussion in [Hasegawa et al., 1971] by plotting the loss tangent given by $\tan d = \omega C / G$ for an equivalent series RC circuit.

In Figure 3.5, the variation of the loss tangent with frequency is shown where the loss tangent is small at low frequencies, reaches a maximum at around 3MHz and then decreases to a low value beyond 0.5GHz. The frequency at which the maxima occurs for the loss tangent is the transition frequency where the slow wave mode transitions into a

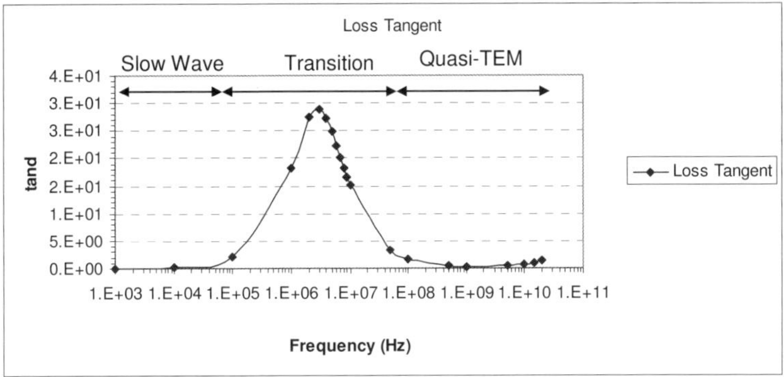

Figure 3.5: Loss tangent Vs Frequency.

diffusion type TEM mode followed by a quasi-TEM mode. During the transition phase between the slow wave to the quasi-TEM mode (0.1MHz - .5GHz), the attenuation increases significantly per wavelength, followed by a reduction in the loss per wavelength at frequencies beyond 0.5GHz. Such a behavior will not be seen if a good insulator is used instead of Si (conductivity of $\sim 1 \times 10^{-7}$S/m) since the insulator will behave like a dielectric supporting only the quasi-TEM mode.

The effects described can be confirmed using any full wave electromagnetic solver by monitoring the intrinsic impedance $\eta = E_\rho / H_\phi$, where E_ρ is the radial component of the electric field and H_ϕ is the axial component of the magnetic field of the TSV in the silicon substrate (cylindrical coordinates used here). At a slow wave frequency, this ratio will be smaller than the intrinsic impedance of the silicon substrate and will converge to the intrinsic impedance of the silicon substrate (109Ω) as the frequency increases into the quasi-TEM region. The electric and magnetic fields for a TSV pair at a frequency close to the quasi-TEM region (0.5GHz) are shown in Figure 3.6 using a full wave electromagnetic solver [CST] where the intrinsic impedance is 93.8Ω.

Though this section covered the slow wave and quasi-TEM mode propagation effects in TSVs, it did not consider the resistance and inductance variations of the conductors with frequency. Moreover,

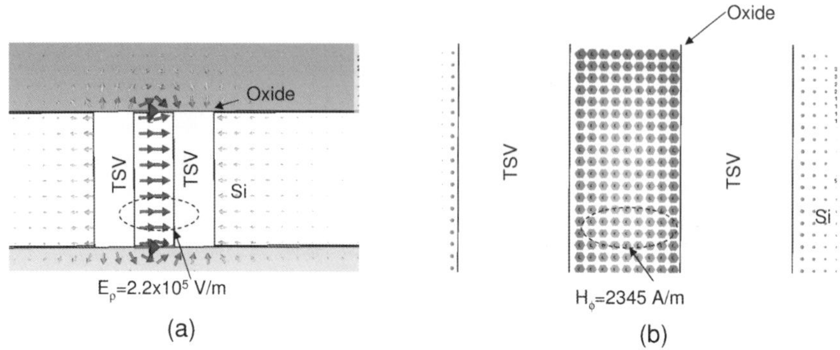

Figure 3.6: (a) Electric Field and (b) Magnetic Field for a TSV pair.

simple analytical equations that have been used do not consider the proximity-effects between conductors carrying current. However, this section provides the basis for more rigorous analysis and explanations provided later in this chapter.

3.4 Physics Based Modeling of Through Silicon Vias

In the previous section approximate equations were derived for the oxide capacitance, substrate capacitance and substrate conductance between a signal and ground TSV. This captures the insulating and semiconductor behavior of the dielectric material used. These equations were derived based on the physics associated with the wave propagation in a coaxial transmission line and a two wire line. A similar approach can be used to compute the inductance and the resistance of the conductors due to current flowing through them. A two wire model can be used to compute the loop inductance L_{ind} of two TSVs as [Kim et al., 2011]:

$$L_{ind} = \frac{\mu_0 \mu_r}{2\pi} \ln(\frac{D}{R})L \tag{3.8}$$

where μ_0 is the permeability of free space and $\mu_r = 1$ is the relative permeability of the substrate. From Figure 3.3, setting D = 100μm, L = 100μm and R = 15μm, the loop inductance can be computed as 37.9pH. Two effects that are not captured in (3.8) is the frequency dependence of inductance due to skin-effect and proximity-effect due to current flowing on neighboring conductors. For TSVs, the frequency dependent variation of inductance is small (discussed later) and can be neglected. However, the proximity-effect which is due to the non-uniform distribution of current in the conductor due to neighboring conductors can have a large effect on the inductance, which is not captured in (3.8). In [Kim et al., 2011], a proximity factor has been used to modify (3.8) based on the ratio of pitch (D) to diameter (2R) of the TSVs. When the number of conductors is more than two, it becomes difficult to capture the proximity-effect using analytical equations. Later in this chapter, the proximity-effect is discussed in more detail based on rigorous electromagnetic analysis. For now, let's use (3.8) to calculate the loop inductance of the TSV pair. Based on [Kim et al., 2011], the resistance variation with frequency can be computed using:

$$R = \sqrt{R_{dc}^2 + R_{ac}^2}$$

$$R_{dc} = \frac{1}{\sigma_{Cu}} \times \frac{L}{\pi R^2}; R_{ac} = \frac{1}{\sigma_{Cu}} \times \frac{L}{\pi\left[R^2 - (R-\delta)^2\right]} \qquad (3.9)$$

$$\delta = \frac{1}{\sqrt{\pi f \mu_0 \sigma_{Cu}}}$$

where δ is the skin depth, σ_{Cu} is the conductivity of copper (5.8×10^7 S/m), μ_0 is the permeability of free space, R_{dc} is the DC resistance, R_{ac} is the ac resistance and the other physical parameters are defined as shown in Figure 3.2(c). The variation of resistance with frequency for $L = 100\mu m$ and $R = 15\mu m$ is shown in Figure 3.7 where the resistance increases from 2.44 mohms at low frequency to 39.84 mohms at 20GHz. It is important to note that the resistance shown in Figure 3.7 is for a single TSV. For a pair of TSVs (signal and ground as in Figure 3.3), the resistance doubles.

3.4.1 *Creating an Equivalent Circuit*

Using the computed R, L, G, C parameters, an equivalent circuit for a differential TSV pair (signal TSV with reference supporting current in

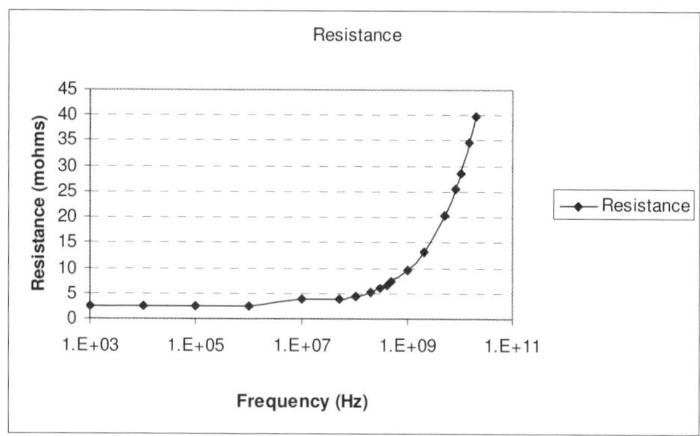

Figure 3.7: Resistance variation with frequency.

opposite directions) can be constructed as shown in Figure 3.8(a). This equivalent circuit is a lumped T-element circuit that is symmetric, where R, L are series elements and G, C are shunt elements. The parameters R, G, C are frequency dependent parameters as described earlier with L being frequency independent. The T-element circuit in Figure 3.8(a) can be further simplified to the circuit shown in Figure 3.8(b) where the impedance z and admittance y, which are frequency dependent parameters, can be computed as:

$$z(f) = 2R + j\omega L; \qquad y(f) = G + j\omega C \qquad (3.10)$$

where G, C are the equivalent conductance and capacitance from (3.2) and ω is the angular frequency in rad/s. This representation is useful in deriving the scattering parameters for a TSV pair, as described in the next sub-section.

3.4.2 *Computing Scattering Parameters*

Scattering (S) parameters provide insight into the performance of an interconnect structure in the frequency domain especially at high frequencies where wave propagation effects become important. These parameters can be measured using a vector network analyzer (VNA). A very useful parameter is the insertion loss of a TSV pair since this

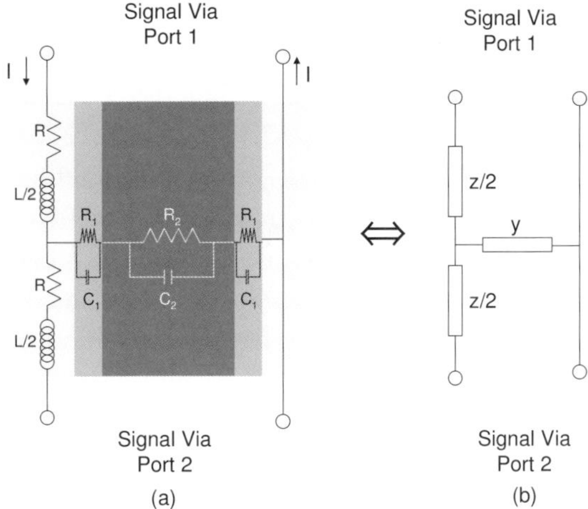

Figure 3.8: T-element equivalent circuit showing (a) R,L,G,C parameters and (b) z,y parameters.

provides information on the signal loss in the TSV. In this section, the equivalent circuit in Figure 3.8(b) will be used to derive the S-parameters.

The 2 port impedance (Z) parameters between port 1 and 2 from Figure 3.8(b) can be derived as:

$$Z(f) = \begin{bmatrix} \dfrac{z}{2} + \dfrac{1}{y} & \dfrac{1}{y} \\ \dfrac{1}{y} & \dfrac{z}{2} + \dfrac{1}{y} \end{bmatrix} \tag{3.11}$$

The S-parameters based on 50ohm reference impedance can be computed from (3.11) as:

$$S(f) = \left[Z(f) + 50I\right]^{-1} \left[Z(f) - 50I\right] \tag{3.12}$$

where 'I' is the identity matrix. Note that for a 2-port structure, the S-parameters are a 2×2 matrix where the important response is S(1,2), which when expressed on a dB scale, represents the insertion loss of the TSV pair.

Based on the physical dimensions and material properties used to compute the R,L,G,C parameters (D $= 100\mu$m, R $= 15\mu$m, L $= 100\mu$m, $d_{ox} = 0.1\mu$m, $\varepsilon_{SiO2} = 3.9$ and $\varepsilon_{Si} = 11.9$), using the T-element equivalent circuit in Figure 3.8(b) and using (3.11) and (3.12), the computed insertion loss S(1,2) (decibels dB(S12)) for the differential TSV pair is shown in Figure 3.9 as 'o' (Physics) from 1KHz to 20GHz (return loss S(1,1) not shown). The correlation of the physics based model to electromagnetic simulations (shown as line and explained later) is quite good with a small deviation at higher frequencies. From the figure, the sharp slope associated with the insertion loss up to ~0.5GHz can be seen due to the transition from the slow wave to the quasi-TEM mode described earlier. After around 0.5GHz, the slope of the curve decreases indicating that the displacement currents in the silicon substrate begin to contribute towards the loss. Such a sharp increase in insertion loss up to ~0.5GHz will never be seen for vias passing through good insulators and is therefore unique to through silicon vias. From Figure 3.9, an insertion loss of ~0.37dB at 20GHz is quite large considering that the length of the TSVs is just 100µm. Later on in this chapter, the effect of varying the physical and material parameters on the insertion loss will be discussed in detail.

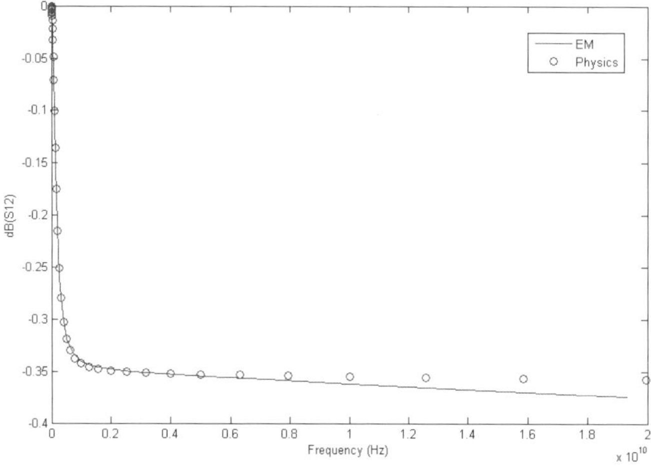

Figure 3.9: Insertion Loss for differential TSV.

3.5 Rigorous Electromagnetic Modeling

In this section a full-wave 3D electromagnetic method is used to compute the response of through silicon via (TSV) interconnections. An important observation about TSV interconnections is that they have a circular cross section and a cylindrical structure. This property can be used to derive specialized basis functions which approximate the current and charge in the TSVs. The specialized basis functions can be used to extract the electrical response of the structure by solving Maxwell's equations [Han et al., 2013]. The approach for electromagnetic modeling of TSVs is therefore an extension of the methods described in Chapter 2 for wirebonds and via arrays with the difference that TSVs have an oxide liner. A TSV pair is shown in Figure 3.10, where the conduction current flows through the center conductor, charge is generated on the conductor and dielectric surfaces and polarization current flows through the oxide between the conductor and silicon substrate. By using specialized basis functions to approximate the current and charge, solving the appropriate Maxwell's equations, and calculating equivalent circuit parameters, an electrical circuit can be derived for the TSV pair, as shown in Figure 3.10. This electrical circuit looks similar to Figure 3.8 with the difference that the conduction current, charge and polarization current are

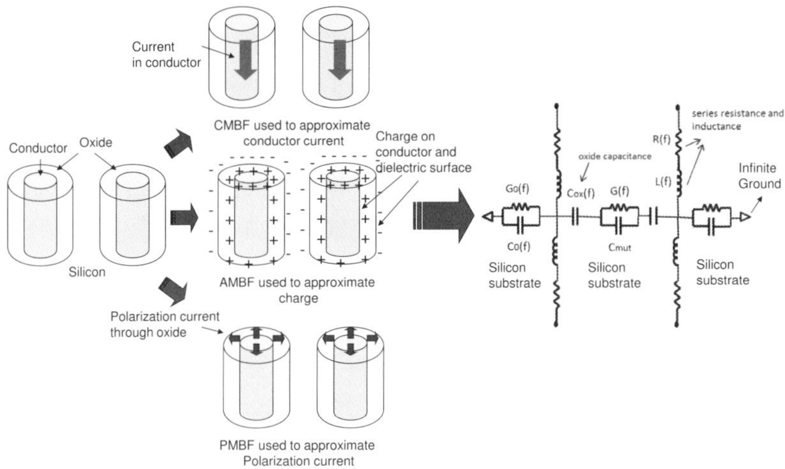

Figure 3.10: Electromagnetic modeling of TSVs using cylindrical basis functions and equivalent circuit.

approximated by accounting for the non-uniformity of the current and charge (proximity-effect) and by accounting for all the modal variations expected in an electromagnetic response. It is important to note that circuit elements shown in Figure 3.10 are partial elements and can be used to compute the loop elements by defining a ground return path. In Figure 3.10, ground at infinity is assumed. The frequency dependent RLGC parameters of TSVs through electromagnetic modeling can therefore be computed as follows [Han et al., 2010]:

a) *Conductor series resistance and inductance*: This represents the loss and inductive coupling in copper conductors, which are due to the volume current density distribution. The conductor series impedance (R and L) in Figure 3.10 can be extracted by solving the electric field integral equation (EFIE) with cylindrical conduction mode basis functions (CMBF) [Han et al., 2010], as in Chapter 2.

b) *Substrate parallel conductance and capacitance*: This represents the conductance and capacitance between conductor and infinite ground, between dielectric and infinite ground and between dielectrics in the substrate (Figure 3.10) produced by the surface charge density distribution on the conductor and dielectric surfaces. The parallel admittance can be extracted by solving scalar potential integral equation (SPIE) with cylindrical accumulation mode basis functions (AMBF) as in Chapter 2. The conductance terms can be computed by using the complex permittivity for silicon defined in (3.13).

c) *Excess capacitance in oxide liner*: This represents the effect of the insulator between conductor and silicon substrate which originates from polarization current in the insulator. Unlike the conduction current which flows along a longitudinal direction, the polarization current flows radially, between the conductor and silicon substrate. To capture the polarization current density distribution, new basis functions are required in addition to CMBF and AMBF. These basis functions are called polarization mode basis functions (PMBF) and are derived in this chapter.

By solving for the RLGC elements that capture effects related to non-uniform charge and current density distribution in the TSVs due to proximity-effect, the parasitics of TSVs can be extracted more accurately. As illustrated in Figure 3.10, the extracted individual elements are combined to generate the complete equivalent circuit model. As mentioned in Section 3.2, a major challenge arising with the modeling of TSVs are the multi-scale dimensions involved due to the thin oxide thickness, aspect ratio and the need for modeling multiple TSVs to extract coupling effects. This is due to the need for meshing the structure where many mesh elements may be required due to the multi-scale dimensions involved. Using specialized basis functions as described in this section eliminates the need for meshing and therefore the solution described is both memory efficient and computationally less expensive.

3.5.1 *Defining the Complex Permittivity of Silicon Substrate*

An important parameter to use during electromagnetic modeling is the complex permittivity of the silicon substrate which is affected by the doping level. The complex permittivity of the silicon substrate can be written as:

$$\varepsilon_{Si} = \varepsilon_0 \varepsilon_{Si,i} (1 - j \tan \delta - j \frac{\sigma_{Si}}{\omega \varepsilon_0 \varepsilon_{Si,i}}) \tag{3.13}$$

where $\varepsilon_{Si,i} = 11.9$ is the relative permittivity of silicon, σ_{Si} is the conductivity of silicon based on the doping and $\tan \delta$ is the intrinsic loss tangent. The intrinsic loss tangent originates from the intrinsic loss of the silicon substrate without doping and therefore represents the loss characteristics of silicon viewed as a dielectric. By comparing (3.4) and (3.5) with (3.13), it is important to note that the intrinsic loss tangent of undoped silicon is not included in equations (3.4) and (3.5).

3.5.2 *Modeling the Conductor Currents*

To extract conductor loss and inductive coupling, electric field integral equation (EFIE) combined with cylindrical conduction mode basis functions (CMBF) has been used, similar to Chapter 2. As explained in the previous chapter, cylindrical CMBFs are obtained from the solution of the diffusion equation for current density in a cylindrical conductor. The combination of skin- and proximity-effects mode basis functions can describe current density distribution influenced by nearby conductors.

The main benefit of using EFIE with cylindrical CMBF is that it can address a large number of cylindrical conductors to capture skin- and proximity-effects with less computational cost. Since the details of the extraction procedure of modal resistances and inductances have been discussed in Chapter 2, a summary of the equations used are described here. The electric field integral equation (EFIE) to be solved to extract the resistances and inductances is:

$$\frac{\overrightarrow{J^C}(\vec{r},w)}{\sigma} + j\frac{\omega u}{4\pi}\int_{V'} G(\vec{r},\vec{r}')\overrightarrow{J^C}(\vec{r}',\omega)dV' = -\nabla\Phi(\vec{r},\omega) \tag{3.14}$$

where $\vec{J}^C$ is the conductor current density, σ is the conductivity of the conductor, $G(\vec{r},\vec{r}')$ is the Green's function defined in Chapter 2 and $\Phi(\vec{r},\omega)$ is the electric scalar potential. Since the conductor is surrounded by an oxide, (3.14) should also include the polarization current. However, as described in Chapter 2, the frequency at which the polarization current term becomes dominant is of the order of 10^{16} Hz and therefore has been neglected in (3.14). The conduction current can now be approximated using the CMBF as:

$$\vec{J}_j^C(\vec{r},\omega) \cong \sum_{n,q} I_{jnq}\, \vec{w}_{jnq}(\vec{r},\omega) \tag{3.15}$$

where, $\vec{w}_{jnq}(\vec{r},\omega)$ is the CMBF in conductor segment j with order n and orientation q, as described in Chapter 2. By using the inner product based on Galerkin's method described in Chapter 2, the modal resistance and inductance can be computed as:

$$\sum_{n,q} I_{jnq} R_{imd,jnq} + j\omega \sum_{n,q} I_{jnq} L_{imd,jnq} = \Delta V_{imd} \tag{3.16}$$

where

$$R_{imd,jnq} = \frac{1}{\sigma}\int_{V_i} \vec{w}_{imd}^{\,*}(\vec{r}_i,\omega)\cdot\vec{w}_{jnq}(\vec{r}_j,\omega) \tag{3.17}$$

$$L_{imd,jnq} = \frac{\mu}{4\pi}\int_{V_j}\int_{V_i} \vec{w}_{imd}^{\,*}(\vec{r}_i,\omega)\cdot\vec{w}_{jnq}(\vec{r}_j,\omega)\frac{1}{\left|\vec{r}_i-\vec{r}_j\right|}dV_j dV_i \tag{3.18}$$

$$\Delta V_{imd} = - \int_{S_i} \Phi(\vec{r}_i) \vec{w}_{imd}^{\,*}(\vec{r}_i, \omega) \cdot d\vec{S}_i \, . \tag{3.19}$$

In the above equations, i, m, d represent the index of inductive cell, the order of CMBF and the orientation of CMBF, respectively. $\vec{w}_{imd}(\vec{r}_i, \omega)$ is the cylindrical CMBF with the *(i, m, d)-th* order consisting of the skin-effect and proximity-effect (d and q modes) described in Chapter 2. In the above equations, ω is the angular frequency, $\vec{r}$ is the radial distance, V is the volume and S is the surface area of the conductor. From the above equations, the retardation term has been ignored due to the small length and fine pitch of TSVs and therefore this is a quasi-static solution. Using the vector identity described in Chapter 2, applying the divergence theorem, and forcing the conductor currents to be longitudinal, (3.19) can be simplified to:

$$\Delta V_{imd} = - \int_{S_i^+} \phi(\vec{r}^{\,+}) \vec{w}_{imd}^{\,*} \bullet dS_i^+ - \int_{S_i^-} \phi(\vec{r}^{\,-}) \vec{w}_{imd}^{\,*} \bullet dS_i^- , \tag{3.20}$$

where the potential difference is evaluated at the two end point surfaces of the conductor segment. Since the potential on a conductor is constant, only the skin-effect mode contributes to the potential difference in (3.20) with the proximity-effect modes resulting in zero potential due to the harmonic behavior of the functions.

Unlike (3.8) and (3.9) which accounts for the frequency independent inductance and skin-effect resistance respectively, (3.16) accounts for the non-uniform current density through the higher order modes introduced (d and q modes) and therefore accounts for the proximity-effects as well. This leads to frequency dependent inductance and resistance terms in the matrix equation generated from (3.15) and (3.16). Hence, the conductor series resistance and inductance can be extracted by solving (3.16).

3.5.3 Modeling the Charge Density

Three charge densities need to be addressed for TSVs namely, the free charge density on the surface of the conductor, the bound charge density at the conductor-oxide interface and the bound charge density on the silicon substrate. This is shown in Figure 3.11. Therefore the total charge on the conductor and silicon surfaces can be expanded as:

$$q^C = \sum_{l,n,q} Q^C_{l,n,q} v^C_{l,n,q}$$

$$q^D = \sum_{l,n,q} Q^D_{l,n,q} v^D_{l,n,q}$$

(3.21)

where q^C is the total charge density on the conductor surface with surface area S_C, q^D is the total charge density on the silicon insulator surface with surface area S_D, $v^C_{l,n,q}$ and $v^D_{l,n,q}$ are specialized accumulation mode basis functions (AMBF) as described in Chapter 2.

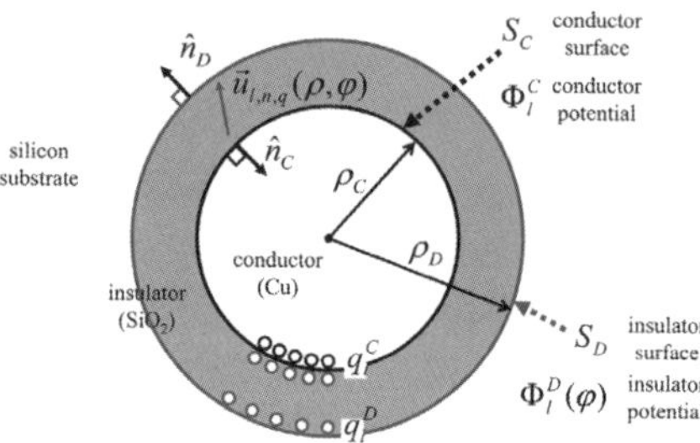

Figure 3.11: Cross section of the TSV showing free charge on the conductor surface, bound charge on the conductor-oxide interface and bound charge on the oxide-silicon interface.

In (3.21), the subscript (l,n,q) represent the lth conductor, n^{th} mode and q^{th} orientation, similar to the CMBF. From Figure 3.11, since the bound charge on the conductor-oxide interface has opposite polarity as compared to the bound charge q^D on the silicon insulator, the total charge $q^C = q^F - q^D$, resulting in the free charge density on the conductor surface $q^F = q^C + q^D$. In Figure 3.11, the potentials on the conductor surface (ϕ^C_l) and insulator surface (ϕ^D_l) are also shown. Since, the conductor is an equipotential surface while the insulator surface is not, Dirichlet and Neumann boundary conditions for the potential are enforced at the conductor and insulator surfaces, respectively. The Dirichlet boundary condition on the conductor surface requires that the higher order terms of the modal potential become zero.

The potential on the insulator surface can therefore be expressed as a periodic function of the angular variable φ as [Han et al., 2010]:

$$\phi_k^D(\varphi) = \phi_{k0}^D + \frac{4}{\pi}\sum_{m=1}[\phi_{kmd}^D\cos m\varphi + \phi_{kmq}^D\sin m\varphi] \qquad (3.22)$$

where the subscript refers to the kth conductor, m^{th} mode and d^{th} orientation.

The relationship between charge and potential can now be computed by solving the Scalar Potential Integral Equation (SPIE) shown below:

$$\frac{1}{4\pi\varepsilon_{Si}}\int_{V'}G(\vec{r},\vec{r'})q(\vec{r'},\omega)dV' = \Phi(\vec{r},\omega) \qquad (3.23)$$

where q is the charge density consisting of q^C, q^D, ϕ is the potential on the conductor and insulator surface, G is the Green's function as defined in Chapter 2, V' is the volume, r is the radial distance and prime/unprimed coordinates represent the source and field points, respectively. Equation (3.23) can now be solved using the expansion functions defined in (3.21), boundary conditions defined in (3.22) and the inner product based on Galerkin's method defined in Chapter 2. This results in the matrix equation:

$$\sum_{\ln q}P_{kmd,\ln q}^{C,D}Q_{\ln q} = \phi_{kmd}^{C,D} \qquad (3.24)$$

where

$$P_{kmd,\ln q}^{C,D} = \frac{1}{4\pi\varepsilon_{si}}\int_{S_l}\int_{S_k}v_{kmd}(\vec{r_k})v_{\ln q}(\vec{r_l})\frac{1}{|\vec{r_i}-\vec{r_j}|}$$

A few items to take away from equation (3.24): As with the EFIE, the Green's function is quasi-static and does not include the retardation term. The permittivity used is that of the silicon medium indicating that the substrate conductance and capacitance are being extracted here. From (3.13), ε_{Si} is a complex number that makes the coefficient of potential P complex as well. The coefficient of potential (units Farad^{-1}) contains the self and interaction terms between the modal behavior of charge on the conductor and insulator surfaces. Inversion of the P matrix provides the conductance and capacitance between the TSVs in the silicon substrate. In (3.23), though the charge is a function of the angular frequency ω, the AMBFs are independent of frequency and hence the coefficients of potential are frequency independent. The modal potential $\phi_{kmd}^{C,D}$ is defined with respect to the ground at infinity. Details of (3.24) can be better understood through Figure 3.12 which shows the global coordinates for an N-TSV system, where cylindrical AMBF $v_{kmd}(\vec{r})$ and $v_{\ln q}(\vec{r}')$ are defined on conductors k and l, respectively. As compared to (3.4) and (3.5), (3.24) allows for the non-uniform distribution of charge density based on the proximity of conductors.

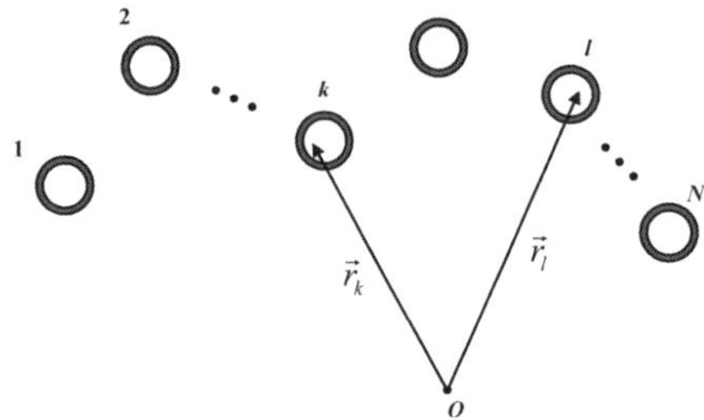

Figure 3.12: TSV coordinate details for solving SPIE.

3.5.4 *Modeling the Polarization Current*

To model the polarization current, the EFIE similar to (3.14) needs to be solved in the oxide region. Equation (3.14) is the EFIE in free space. Similar to the discussion in [Ruehli et al., 1992], the equivalent current in the oxide region can be written as:

$$\vec{J} = \vec{J}^C + j\omega\varepsilon_0(\varepsilon_{SiO_2} - \varepsilon_{Si})\vec{E} \tag{3.25}$$

where $\vec{J}^C$ is the conduction current in the oxide (which is zero) and the second term is the polarization current $\vec{J}^D$ in the oxide. The EFIE from (3.14) can now be rewritten by accounting for the contribution from the polarization current as:

$$\vec{E}(\vec{r}) + j\omega\frac{\mu}{4\pi}\int_{V'}G(\vec{r},\vec{r}')\vec{J}^{C}(\vec{r}',\omega)dv' + j\omega\frac{\mu}{4\pi}\int_{V'}G(\vec{r},\vec{r}')\varepsilon_0(\varepsilon_{SiO_2}-\varepsilon_{Si})j\omega\vec{E}dv' = -\nabla\phi$$

$$(3.26)$$

where $\vec{E}(\vec{r})$ is the electric field in the oxide region, ϕ is the scalar potential, the second term is the electric field generated by the conduction current on other conductors (mutual coupling) and the third term is the contribution due to the polarization current. As expected, the conduction current in the oxide region is zero since it is an insulator. Since, the oxide thickness in TSVs is small, the electric field is assumed to have radial (ρ) and angular (φ) dependences only. Since $\vec{J}^{C}$ is longitudinal, the contribution due to the conductor current $\vec{J}^{C}$ from other conductors can be neglected. Since the polarization current $\vec{J}^{P} = j\omega\varepsilon_0(\varepsilon_{SiO_2}-\varepsilon_{Si})\vec{E}(\vec{r})$, (3.26) can be rewritten as:

$$\frac{\vec{J}^{P}(\vec{r},\omega)}{j\omega\varepsilon_0(\varepsilon_{SiO_2}-\varepsilon_{si})} + j\omega\frac{\mu}{4\pi}\int_{V'}G(\vec{r},\vec{r}')\vec{J}^{P}(\vec{r}',\omega)dV' = -\nabla\Phi(\vec{r},\omega) \qquad (3.27)$$

where ε_{ox} is the relative permittivity of the oxide.

Using a philosophy similar to CMBF and AMBF, the polarization current density can be expanded as:

$$\vec{J}^{P}(\vec{r},\omega) = \sum_{l,n,q}I_{l,n,q}u_{l,n,q} \qquad (3.28)$$

where $u_{l,n,q}$ are the polarization mode basis functions (PMBF). As before subscript (l, n, q) corresponds to the lth conductor, n^{th} mode and q^{th} orientation. By substituting (3.28) into (3.27) and using the inner product based on Galerkin's method described in Chapter 2, the following equations can be derived:

$$\sum_{l,n,q}I_{\ln q}\frac{1}{j\omega C^{ex}_{kmd,\ln q}} + \sum_{l,n,q}j\omega L_{kmd,\ln q}I_{\ln q} = \int_{V_k}\vec{u}_{kmd}\cdot(-\nabla\Phi)dV_k \qquad (3.29)$$

where

$$C^{ex}_{kmd,\ln q} = \frac{\varepsilon_0(\varepsilon_{SiO_2}-\varepsilon_{si})}{\int\vec{u}_{kmd}\cdot\vec{u}_{\ln q}dV_k}$$

$$L_{kmd,\ln q} = \frac{\mu}{4\pi} \int_{V_k} \int_{V_l} G(\vec{r_k}, \vec{r_l}) \vec{u}_{kmd}(\vec{r_k}) \cdot \vec{u}_{\ln q}(\vec{r_l}) \, dV_l dV_k .$$

A few things to note from (3.28) and (3.29) namely, 1) the polarization currents have both radial and angular dependencies and 2) both capacitive $C_{kmd,\ln q}^{ex}$ and inductive $L_{kmd,\ln q}$ terms occur. Due to the small ratio of the oxide thickness to the surface area in TSVs, the inductance term contribution is small and therefore can be neglected [Han et al., 2010].

The PMBF can be derived by solving Laplace's equation in the oxide region with the Dirichlet boundary condition on the conductor surface S^C and Neumann boundary condition on the dielectric surface S^D. This results in the polarization mode basis functions which can be defined as [Han et al., 2010]:

$$\vec{u}_{kmd}(\rho,\varphi) = \begin{cases} \dfrac{1}{A_{k0}} \dfrac{\rho_C}{\rho} \hat{\rho} & m = 0 \\[2ex] \dfrac{1}{A_{km}\rho} \{ [P_m(\dfrac{\rho}{\rho_C})\cos m\varphi]\hat{\rho} - [Q_m(\dfrac{\rho}{\rho_C})\sin m\varphi]\hat{\varphi}\} & m > 0; d-\mathrm{mode} \\[2ex] \dfrac{1}{A_{km}\rho} \{ [P_m(\dfrac{\rho}{\rho_C})\sin m\varphi]\hat{\rho} + [Q_m(\dfrac{\rho}{\rho_C})\cos m\varphi]\hat{\varphi}\} & m > 0; q-\mathrm{mode} \end{cases}$$

$$(3.30)$$

where

$$P_m(\frac{\rho}{\rho_C}) = m \left\{ \frac{\rho^m}{\rho_C^m} + \frac{\rho^{-m}}{\rho_C^{-m}} \right\}$$

$$Q_m(\frac{\rho}{\rho_C}) = m \left\{ \frac{\rho^m}{\rho_C^m} - \frac{\rho^{-m}}{\rho_C^{-m}} \right\}$$

and the normalization factors are:

$$A_{k0} = 2\pi\rho_C l_k$$

$$A_{km} = 4l_k m \left\{ \frac{\rho_D^m}{\rho_C^m} + \frac{\rho_D^{-m}}{\rho_C^{-m}} \right\}$$

where l_k is the length of the kth conductor. From (3.30) the PMBF has $1/\rho$ dependence and is radially directed for the fundamental mode while it has both radial and angular dependence for the higher order mode. A plot of the PMBF basis functions are shown in Figure 3.13 for the fundamental and two higher order modes. From (3.28), since the PMBFs are orthogonal to each other, only the self terms are non-zero and therefore the excess capacitance becomes:

$$C_{kmd,kmd}^{ex} = \frac{\varepsilon_0(\varepsilon_{ox} - \varepsilon_{si})}{\int_{V_k} \vec{u}_{kmd} \cdot \vec{u}_{kmd} \, dV_k} = \begin{cases} \dfrac{2\pi l_k \varepsilon_0(\varepsilon_{SiO_2} - \varepsilon_{Si})}{\ln(\dfrac{\rho_D}{\rho_C})} & m = 0 \\[4ex] \pi \dfrac{16 m l_k \varepsilon_0(\varepsilon_{SiO_2} - \varepsilon_{Si})}{Q_m(\dfrac{\rho_D}{\rho_C})}{P_m(\dfrac{\rho_D}{\rho_C})} & m > 0 \end{cases} \tag{3.31}$$

Comparing (3.3) and (3.31), the capacitance for the fundamental mode is similar to that of a coaxial cylinder having an inner and outer radius of ρ_C and ρ_D, respectively. However, it is important to note that fundamental mode capacitance in (3.31) is the excess capacitance and not the total capacitance as in (3.3). This is because a single homogenous medium consisting of the silicon substrate is used to calculate the contribution due to the charges.

Let's now focus on the potential in (3.29). Using the vector identity $\nabla \bullet \psi \vec{A} = \psi \nabla \bullet \vec{A} + \vec{A} \bullet \nabla \psi$, the right hand side of (3.29) can be written as:

$$\int_{V_k} \vec{u}_{kmd} \cdot (-\nabla \Phi) dV_k = -\int_{V_k} \nabla \bullet (\Phi \vec{u}_{kmd}) dV_k + \int_{V_k} \Phi \nabla \bullet \vec{u}_{kmd} dV_k \tag{3.32}$$

Due to the nature of the PMBF, from (3.30), $\nabla \bullet \vec{u}_{kmd} = 0$. Therefore, (3.32) can be simplified after applying the divergence theorem to:

$$\int_{V_k} \vec{u}_{kmd} \cdot (-\nabla \Phi) dV_k = -\int_{V_k} \nabla \bullet (\Phi \vec{u}_{kmd}) dV_k = -\oint_{S_k} \Phi \vec{u}_{kmd} \bullet dS_k \tag{3.33}$$

where S_k is the surface area that completely encloses the volume V_k. The polarization current density $\vec{J}^P$ does not have any longitudinal

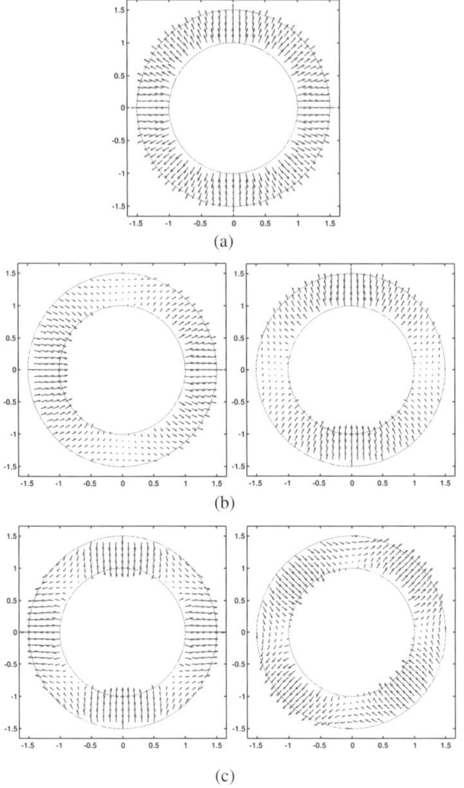

Figure 3.13: Example plot of PMBFs using conductor radius of 1m, insulator radius of 1.5m and length 1m (a) fundamental mode (b) $m = 1$ d and q mode (c) $m = 2$ d and q mode.

component. Therefore, using Figure 3.11, (3.33) can be simplified further as:

$$-\oint_{S_k} \Phi \vec{u}_{kmd} \bullet dS_k = \Delta V_{kmd} = -\int_{S_c} \Phi_k^C \vec{u}_{kmd} \bullet \hat{n}_C \rho_C d\phi dz - \int_{S_D} \Phi_k^D(\varphi) \vec{u}_{kmd} \bullet \hat{n}_D \rho_D d\varphi dz$$

$$(3.34)$$

where ΔV_{kmd} is the potential difference between the (k^{th}) conductor and dielectric for the mth mode, the potential Φ_k^C on the conductor surface is a constant while the potential $\Phi_k^D(\varphi)$ on the dielectric surface varies with angle and is given by (3.22). For the $m = 0$ mode and $m > 0$ modes, the potential difference can be written as:

$$\Delta V_{kmd} = \begin{cases} \Phi_k^C - \Phi_{k0}^D & m = 0 \\ -\Phi_{kmd}^D & m > 0 \end{cases} \tag{3.35}$$

where, the potential contribution from the higher order modes are zero on the surface of the conductor while they are non-zero on the surface of the dielectric.

3.5.5 *Combining the Current and Charge Contributions*

The current flowing through the conductor causes a voltage drop across the conductor due to the resistance and inductance given by the matrix equation:

$$\begin{bmatrix} Z_{ss} & Z_{sp} \\ Z_{ps} & Z_{pp} \end{bmatrix} \begin{bmatrix} I_s \\ I_p \end{bmatrix} = \begin{bmatrix} \Delta\Phi^C \\ 0 \end{bmatrix} \tag{3.36}$$

where subscript 's' and 'p' represent the skin- and proximity-effect modes, I is the current, Z is the complex impedance and $\Delta\Phi^C$ is the potential difference across the two ends of the conductor. Notice that the potential difference across the conductor due to the proximity-effect modes is zero in (3.36).

The coefficient of potential which captures the capacitive coupling between the conductor and dielectric boundaries from (3.24) can be written as:

$$\begin{bmatrix} P_{ss}^C & P_{sp}^C & P_{ss}^{CD} & P_{sp}^{CD} \\ P_{ps}^C & P_{pp}^C & P_{ps}^{CD} & P_{pp}^{CD} \\ P_{ss}^{DC} & P_{sp}^{DC} & P_{ss}^D & P_{sp}^D \\ P_{ps}^{DC} & P_{pp}^{DC} & P_{ps}^D & P_{pp}^D \end{bmatrix} \begin{bmatrix} Q_s^C \\ Q_p^C \\ Q_s^D \\ Q_p^D \end{bmatrix} = \begin{bmatrix} \Phi_s^C \\ 0 \\ \Phi_s^D \\ \Phi_p^D \end{bmatrix} \tag{3.37}$$

where the contributions from the skin (s) and proximity (p) effect modes have been separated for the charge distribution Q with superscript representing the charge on the conductor (C) or dielectric (D) surface, the coefficient of potential P is due to the interaction between the skin-skin (ss), skin-proximity (sp), proximity-skin (ps), proximity-proximity (pp) modes with the superscript representing the interaction between the two modes on the conductor (C), on the dielectric (D), on the conductor-dielectric (CD) and dielectric-conductor (DC) surfaces. The potential on the right side of (3.36) is the potential due to the skin- and proximity-effect modes on the conductor (C) and dielectric (D) surfaces. Note that the potential due to the proximity-effect modes on the conductor is zero due to its harmonic behavior, as explained earlier.

From (3.31) and (3.34), the polarization current equation across the oxide can be written as:

$$\begin{bmatrix} Z_{ss}^{ex} & Z_{sp}^{ex} \\ Z_{ps}^{ex} & Z_{pp}^{ex} \end{bmatrix} \begin{bmatrix} I_s^{pol} \\ I_p^{pol} \end{bmatrix} = \begin{bmatrix} \Phi^C - \Phi_s^D \\ -\Phi_p^D \end{bmatrix} \tag{3.38}$$

where the impedance Z due to the excess (ex) capacitance is $Z^{ex} = 1/(j\omega C^{ex})$, I is the polarization current (pol) due to the skin (s) and proximity (p) effect modes with the potential due to the conductor (C) and dielectric (D) from the skin (s) and proximity (p) effect modes shown on the right side of (3.38). It is important to note that (3.38) is a diagonal matrix.

The KCL or continuity equation on the conductor can be used to relate the terminal currents I_t (forcing or excitation current) and internal currents I_i (same as I_s) with the free charge on the conductor surface q^F as:

$$\begin{bmatrix} -E \\ 0 \end{bmatrix} \Delta_I \begin{bmatrix} I_t \\ I_s \end{bmatrix} + j\omega(Q_s^C + Q_s^D) = 0 \tag{3.39}$$

where E is an identity matrix and Δ_I relates the current at the internal nodes of the conductor segments. Similarly, applying the continuity equation at the dielectric surface:

$$\begin{bmatrix} I_s^{pol} \\ I_p^{pol} \end{bmatrix} + j\omega \begin{bmatrix} Q_s^D \\ Q_p^D \end{bmatrix} = 0 \tag{3.40}$$

where the polarization current equals the time variation of bound charge.

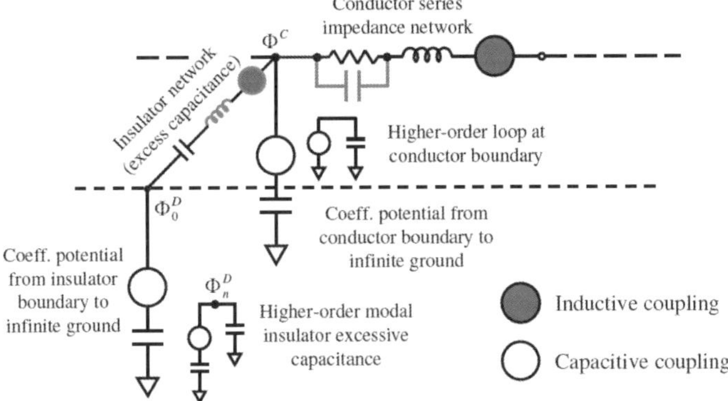

Figure 3.14: Equivalent circuit network for TSV interconnections (high order loops for inductive proximity-effect not included) [Han, 2009].

Equations (3.36) to (3.40) can be represented using an equivalent circuit network for a single TSV, as shown in Figure 3.14 [Han, 2009] where the nodes are connected together based on the current flow and potential differences. In the figure, the light colored components can be eliminated as per the description provided earlier due to approximations based on the dimensions of TSVs. In Figure 3.14, the potential on the conductor-dielectric (C) and dielectric-insulator (D) interfaces are shown with the excess capacitance between them. The conductors are represented using self resistance and inductance elements along with inductive coupling to neighboring conductors. The self capacitance from conductor and dielectric surfaces to infinite ground is also shown along with capacitive coupling and higher order capacitance loops. Though the network in Figure 3.14 can be solved in a circuit simulator, it is not conducive to interpretation and needs to be modified to a representation as shown in Figure 3.15.

3.5.6 *Equivalent Circuit Simplification*

By eliminating the internal current and charge vectors in (3.36)–(3.40) the reduced matrix equation that relates the terminal currents and

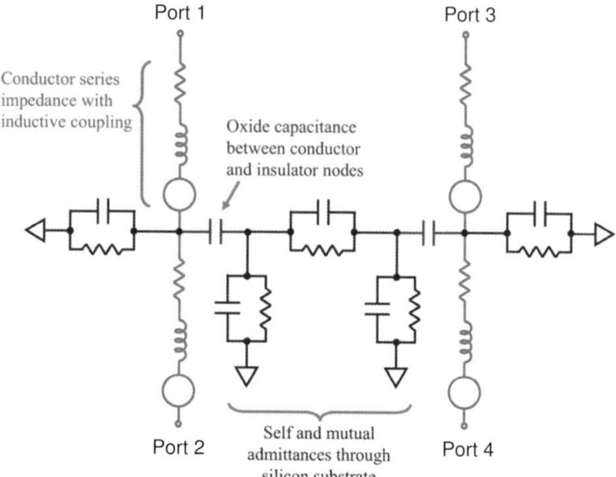

Figure 3.15: Equivalent circuit model for two TSVs (higher order loops not shown).

conductor node voltages can be derived as [Han et al., 2010]:

$$\left[Y_C + j\omega C^{eq} \right] \Phi^C = \begin{bmatrix} I_t \\ 0 \end{bmatrix} \tag{3.41}$$

where

$$Y_C = \Delta_{I,T} Z_C^{-1} \Delta_{V,T}$$

$$\Delta_{I,T} = \begin{bmatrix} \Delta_I & 0 \end{bmatrix}; \Delta_{V,T} = \begin{bmatrix} \Delta_V \\ 0 \end{bmatrix}$$

$$Z_C = \begin{bmatrix} Z_{ss} & Z_{sp} \\ Z_{ps} & Z_{pp} \end{bmatrix}; C^{eq} = \begin{bmatrix} E & 0 \end{bmatrix} \begin{bmatrix} C_C^{eq} - C_{CD}^{eq} C_D^{eq-1} C_{DC}^{eq} \end{bmatrix} \begin{bmatrix} E \\ 0 \end{bmatrix}$$

$$\begin{bmatrix} C_C^{eq} & C_{CD}^{eq} \\ C_{DC}^{eq} & C_D^{eq} \end{bmatrix} = \begin{bmatrix} C_p^C & C_p^{CD} \\ C_p^{DC} & C_p^D \end{bmatrix} + \begin{bmatrix} C^{ex} & -C^{ex} \\ -C^{ex} & C^{ex} \end{bmatrix}$$

$$\begin{bmatrix} C_p^C & C_p^{CD} \\ C_p^{DC} & C_p^D \end{bmatrix} = \begin{bmatrix} P_{ss}^C & P_{sp}^C & P_{ss}^{CD} & P_{sp}^{CD} \\ P_{ps}^C & P_{pp}^C & P_{ps}^{CD} & P_{pp}^{CD} \\ P_{ss}^{DC} & P_{sp}^{DC} & P_{ss}^D & P_{sp}^D \\ P_{ps}^{DC} & P_{pp}^{DC} & P_{ps}^D & P_{pp}^D \end{bmatrix}^{-1}$$

The equivalent circuit for a pair of TSVs resulting from (3.41) is shown in Figure 3.15, where the network is more representative of the physical structure, with all of the modal behavior of the current and voltage included in the calculation of the components. The model in Figure 3.15 is a partial element equivalent circuit (PEEC) model consisting of four ports implying that to compare with the equivalent circuit in Figure 3.8 for a differential via (two ports), the matrix has to be further reduced by making the currents equal and opposite in the two vias. This can be done by using 25Ω as the reference impedance while calculating the four port S-parameters for the PEEC model in Figure 3.15 and converting the results into differential mode parameters (two ports), with a reference impedance of 50Ω, to compare the results from the physics based model. This is shown in Figure 3.9 for the insertion loss of the TSV ($D = 100\mu m$, $R = 15\mu m$, $L = 100\mu$, $d_{ox} = 0.1\mu m$, $\varepsilon_{SiO2} = 3.9$ and $\varepsilon_{Si} = 11.9$) as a line (EM) from 1KHz to 20GHz.

An important point to note is that the method presented applies to homogeneous silicon media and therefore does not capture the finite thickness of oxide–silicon–oxide substrate. In TSV structures, all the faces of conductors are surrounded by oxide or air, which does not permit conduction currents to leak out to silicon substrate. However, the assumption of homogeneous silicon media causes fictitious leakage current to flow, resulting in offset in the insertion loss. In the equivalent circuit model in Figure 3.15, the offset in the insertion loss is generated by the self conductance of the conductor. This offset in the insertion loss can be eliminated by removing the self admittance terms, which are generated by charge on the edge of conductors. The removal of charge on the edge of conductors does not affect the accuracy of the model for TSVs due to their aspect ratios.

A fair question that needs to be answered is the following: If the simple physics based model provides a reasonably good accuracy as compared to the rigorous electromagnetic solution as shown in Figure 3.9, why then bother to develop such a sophisticated solution for analyzing TSVs. This is because the distribution of charge on the TSVs becomes non-uniform as the density and number of TSVs increases. This effect is compounded when non-uniform spacing between TSVs due to the presence of keep out zones (explained in Chapter 4) further complicates the distribution of charge making it difficult to use simple analytical models. This effect, called as the proximity-effect, is an important effect to capture during the modeling of TSVs to compute both insertion loss and coupling, which requires rigorous electromagnetic analysis. Since the electromagnetic analysis described in this chapter is customized to solving TSVs with cylindrical cross section, it provides an advantage over other generic electromagnetic solvers in terms of accuracy, speed and memory utilization.

3.5.7 *Extraction of RLGC and S-parameters*

The PEEC model shown in Figure 3.15 has several advantages from a design stand point namely, 1) the RLGC parameter variation with frequency can be computed for all of the TSVs without having to assign either a signal, power or ground connection. This is useful in understanding skin- and proximity-effect for the TSVs as a function of their placement since the strong coupling between TSVs can alter their behavior; 2) The impact of the TSV dimensions on the RLGC parameters can be evaluated to better design the structures; 3) The S-parameters obtained from the PEEC model can be converted to a SPICE netlist using commercially available solvers [Idem, 2009]; 4) The effect of differential via or ground via placement on S-parameters can be assessed directly by reducing the computed S-parameters and 5) The SPICE netlist can be used to connect to other linear (package) or non-linear (transistor) models to assess performance either in time or frequency domain. This section analyzes the RLGC and S-parameters for a differential TSV with details related to parameter tuning and coupling provided in Chapter 4.

Consider the differential TSV shown in Figure 3.2(c) with length (L), diameter (2R), and pitch (D) fixed at 100μm, 30μm and 60μm, respectively. Four cases are simulated for oxide thicknesses (d_{ox}) of 0.1μm and 0.5μm with silicon conductivities of $\sigma_{Si} = 10S/m$ and 100S/m. The relative dielectric constants of silicon and silicon dioxide are 11.9 and 3.9, respectively. The equivalent circuit in Figure 3.16 shows the frequency-dependent RLGC parameters that have been simplified from the original equivalent circuit shown in Figure 3.15. The TSV pair in

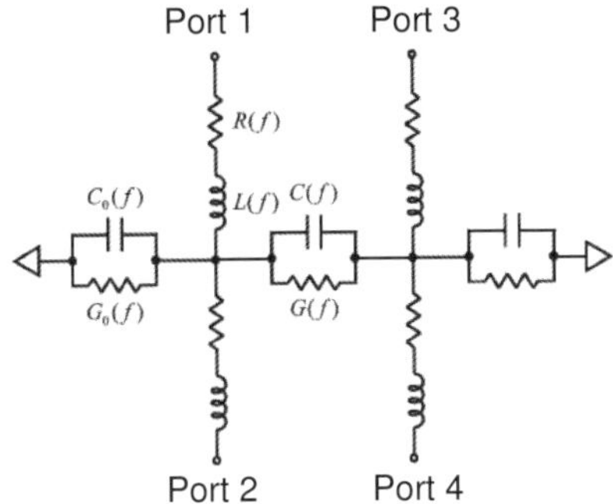

Figure 3.16: Equivalent circuit model for two TSVs (higher order loops not shown).

Figure 3.16 is therefore represented using a four port equivalent circuit. The oxide and substrate capacitance between the TSVs have been combined in Figure 3.16 into a single capacitance $C(f)$. For a differential via, the capacitance and conductance for each TSV is $C_0(f) + 2C(f)$ and $G_0(f) + 2G(f)$, respectively. The resistance $R(f)$, self inductance $L(f)$, mutual inductance (not shown), conductance $G_0(f) + 2G(f)$ and capacitance $C_0(f) + 2C(f)$ in Figure 3.16 extracted using the electromagnetic method with cylindrical modal basis functions are shown in Figure 3.17 for a frequency range of 10KHz–10GHz.

From Figure 3.17, the resistance shows strong frequency dependence due to skin-effect while the inductance shows small variation with frequency. The equivalent conductance and capacitance variation in the figure is similar to Figure 3.4 due to the Maxwell-Wagner effect described earlier. The change in silicon conductivity and oxide thickness only affects the conductance and capacitance parameters, as expected. For fixed silicon conductivity, decreasing the oxide thickness from 0.5µm to 0.1µm increases the separation between the conductances beyond ~100MHz while the separation for the capacitances is larger at lower frequencies. A similar trend can be seen for fixed oxide thickness as the silicon conductivity is varied from 10S/m to 100S/m. It is important to note that in both cases, in the transition frequency region, an increase in conductance is followed by a decrease in capacitance.

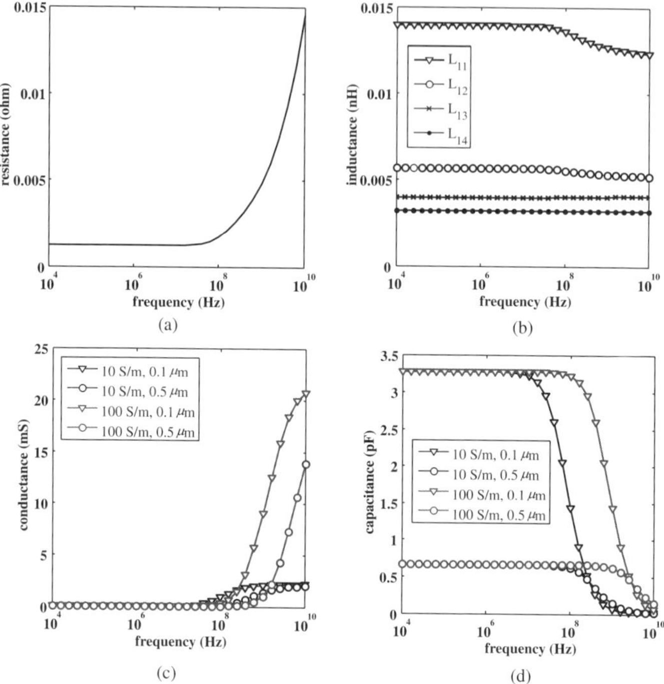

Figure 3.17: (a) Resistance, (b) Self and Mutual inductance, (c) Conductance G0 + 2G and (d) Capacitance C0 + 2C for a TSV pair [Han et al., 2010].

The insertion loss for the differential via based on a 50Ω reference impedance is shown in Figure 3.18 for silicon conductivities of 10S/m and 100S/m and oxide thicknesses of 0.5µm to 0.1µm. Comparing Figures 3.18(a) and (b), increase in silicon conductivity from 10S/m to 100S/m for a fixed oxide thickness increases the insertion loss while a decrease in oxide thickness from 0.5µm to 0.1µm for a fixed silicon conductivity increases the insertion loss as well. In the figure, the line represents the results from the electromagnetic modeling using cylindrical basis functions while the circles are from CST, an electromagnetic solver [CST]. In both Figures 3.18(a) and (b), a dashed line is shown indicating the results without the oxide capacitance. As expected the insertion loss shows an offset at low frequencies due to the contact between the conductor and semi-conductor, requiring the oxide capacitance as an electrical barrier.

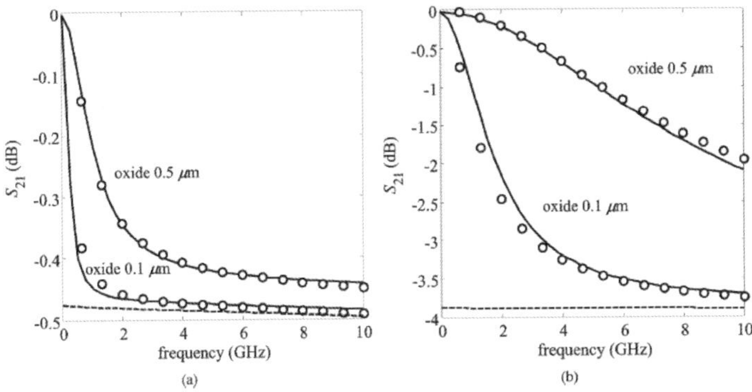

Figure 3.18: Insertion loss for the TSV pair (a) 10S/m and (b) 100S/m [Han et al., 2010].

Table 3.1: Comparison of CPU time.

Example	CST MWS	Proposed Method	Number of Frequency Points
	CPU Time (sec)	CPU Time (sec)	
Two TSV, 0.1μm, 10S/m	1081	467.82	40
Two TSV, 0.5μm, 10S/m	3533	414.20	40
Two TSV, 0.1μm, 100S/m	1428	382.12	40
Two TSV, 0.5μm, 100S/m	5568	378.95	40

As a comparison, Table 3.1 shows the computational time required to simulate the TSV structures using the proposed electromagnetic method and CST MWS [CST]. All the simulations were performed using an Intel Pentium IV 3.2GHz CPU with 2GB RAM. For the four two-TSV examples, the proposed method is 2–14 times faster than MWS.

3.6 Modeling of Conical Through Silicon Via

Due to process related effects, the TSVs designed as structures with cylindrical cross sections may have a tapered profile, as shown in Figure 3.19. The conical TSV structures can have electrical parameters that are quite different as compared to the cylindrical structures. The electromagnetic modeling method using cylindrical basis functions described earlier can be modified to include conical TSV structures [Xie et al., 2012]. In order to model the conical TSV structures, this section uses an approach that extends the method presented in the previous

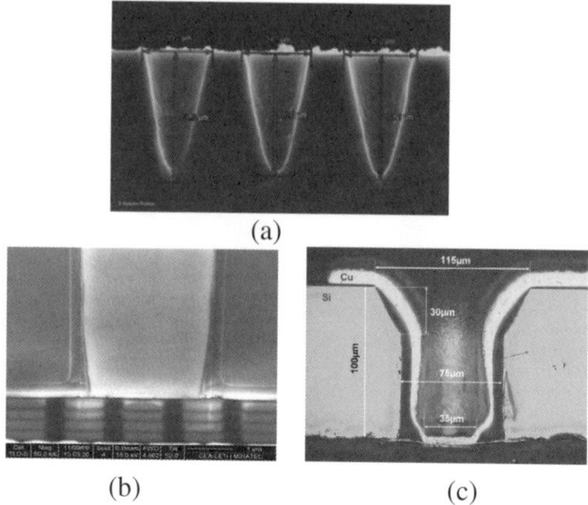

Figure 3.19: Cross-sectional image of TSV structures (a) SEM cross-section view of laser drilled blind vias in silicon [Industrial Laser Application Note] (b) and (c) SEM.

section. The modified approach uses conical modal basis functions to extract electrical parasitic elements. A hybrid modeling approach that combines the conical TSV modeling method with the original cylindrical TSV modeling method is also used to model complicated TSV structures. Examples of several complex TSV structures are simulated and the results are compared with commercial full-wave EM solvers.

3.6.1 *Conical TSV Modeling*

The procedure for modeling conical TSVs is similar to the procedure presented in Section 3.5, where the method can be divided into three parts, as discussed in Sections 3.5.2, 3.5.3, and 3.5.4. The only required modification for modeling conical TSVs is that the radius of each TSV is not a constant but a varying parameter.

To calculate the series resistance and inductances of conical conductors, EFIE (3.14) can be used, considering the integration volume V' as a conical shape. By using the same procedure discussed in Chapter 2 (Section 2.3.2), a circuit equation similar to (3.16) can be obtained. The conical CMBFs are obtained by modifying the normalization factor of cylindrical CMBFs by letting the cross-sectional radius be a variable.

Thus, the radius of the conical CMBFs varies according to the cross-sectional radius of the conical structure as plotted in Figure 3.20. By using the modified basis functions, the equation to calculate partial resistance and inductance can be obtained as (3.42) and (3.43), respectively.

$$R_{imd,jnq} = \frac{1}{\sigma} \int_0^{2\pi} \cos(m\varphi_i - \varphi_d)\cos(n\varphi_j - \varphi_q)d\varphi \int_{\rho_1 \frac{l}{\rho_2-\rho_1}}^{\rho_2 \frac{l}{\rho_2-\rho_1}} \frac{1}{A_{in}A_{jm}^*}dz \int_0^{z\frac{\rho_2-\rho_1}{l}} \rho J_m(\alpha^*\rho)J_n(\alpha\rho)d\rho$$

$$(3.42)$$

$$L_{imd,jnq} = \frac{\mu}{8\pi} \int_{\rho_1 \frac{l}{\rho_2-\rho_1}}^{\rho_2 \frac{l}{\rho_2-\rho_1}} \int_{\rho_1 \frac{l}{\rho_2-\rho_1}}^{\rho_2 \frac{l}{\rho_2-\rho_1}} \frac{1}{A_{im}^*A_{jn}} \frac{1}{|\vec{r}_i - \vec{r}_j|}dz \int_0^{z\frac{\rho_2-\rho_1}{l}} \int_0^{z\frac{\rho_2-\rho_1}{l}} \rho J_m(\alpha^*\rho_i)J_n(\alpha\rho_j)I_\varphi d\rho_i d\rho_j$$

$$(3.43)$$

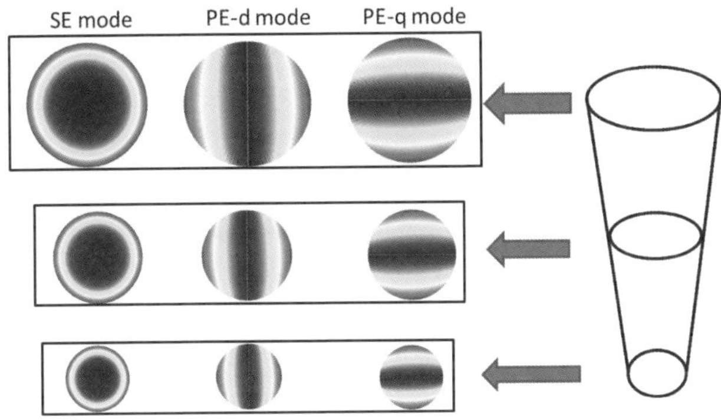

Figure 3.20: Conical CMBFs at different cross-section of the conical structure at 0.1GHz.

To calculate the integral for resistances and inductances, the coordinate system as defined in Figure 3.21 can be used.

Substrate parallel conductances and capacitances can be extracted by using SPIE (3.23) combined with conical AMBFs. The conical AMBFs are obtained by modifying the normalization factors of cylindrical AMBFs shown in (2.22), (2.23), and (2.24). A similar extraction procedure results in the equivalent circuit equation as in (2.28). The

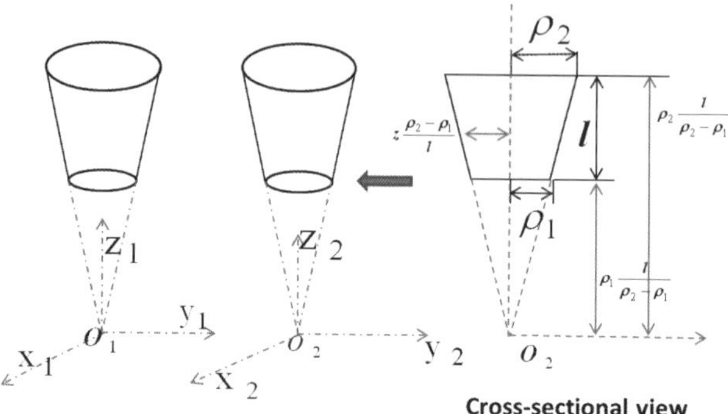

Figure 3.21: Definition of coordinate used to calculate numerical integral.

normalization factor of conical AMBF is obtained by equating the integration of the basis function over the lateral conical surface to unity. In order to calculate the surface integral for potential coefficients, we first transform the Cartesian coordinates to circular conical coordinates. Then the potential coefficients can be expressed as shown in (3.44).

$$P_{lmd,kmq} = \frac{1}{4\pi\varepsilon_0} \int_{\rho_1\frac{l}{\rho_2-\rho_1}}^{\rho_2\frac{l}{\rho_2-\rho_1}} \int_{\rho_1\frac{l}{\rho_2-\rho_1}}^{\rho_2\frac{l}{\rho_2-\rho_1}} dz dz'$$

$$\times \int_0^{2\pi}\int_0^{2\pi} v_{lmd}(\vec{r_l})v_{knq}(\vec{r_k})\frac{1}{|\vec{r_l}-\vec{r_k}|}\left\|\frac{\partial\Phi}{\partial\varphi}(\varphi,z)\times\frac{\partial\Phi}{\partial t}(\varphi,z)\right\|\cdot\left\|\frac{\partial\Phi'}{\partial\varphi}(\varphi,z)\times\frac{\partial\Phi'}{\partial\varphi}(\varphi,z)\right\| d\varphi d\varphi$$

$$(3.44)$$

To calculate the oxide liner excess capacitance, the EFIE combined with conical PMBFs can be used. The conical PMBFs are obtained by modifying the normalization factor of the cylindrical PMBFs (3.30). The oxide capacitance extraction procedure is extended from the procedure discussed in Sec. 3.5.4 by changing the integral volume to a conical annulus. In the resultant oxide capacitance calculation formula, the integration volume v_k is the conical oxide region as shown in Figure 3.22. The oxide capacitance formula can be expressed as equation (3.45).

$$C_{kmd,\ln q}^{ex} = \frac{\varepsilon_0 (\varepsilon_{ox} - \varepsilon_{si})}{\displaystyle\int_{\rho_1^C \frac{l}{\rho_2^C - \rho_1^C}}^{\rho_2^C \frac{l}{\rho_2^C - \rho_1^C}} dz \int_{z\frac{\rho_2^C - \rho_1^C}{l}}^{z\frac{\rho_2^C - \rho_1^C}{l} + \rho_2^D - \rho_2^C} \vec{u}_{kmd} \cdot \vec{u}_{\ln q} d\rho \int_0^{2\pi} d\varphi} \tag{3.45}$$

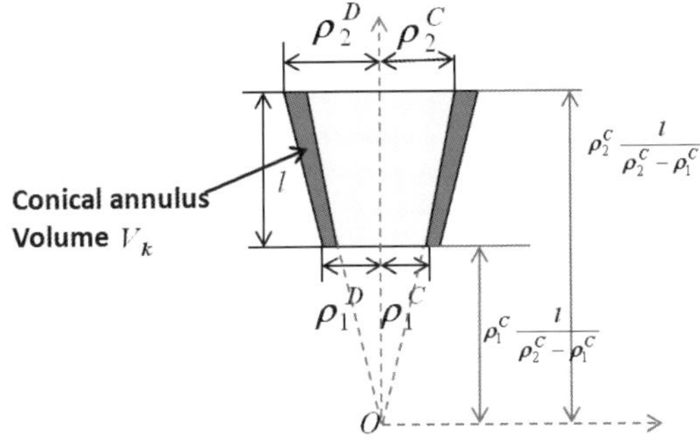

Figure 3.22: Parameters used to calculate oxide capacitance.

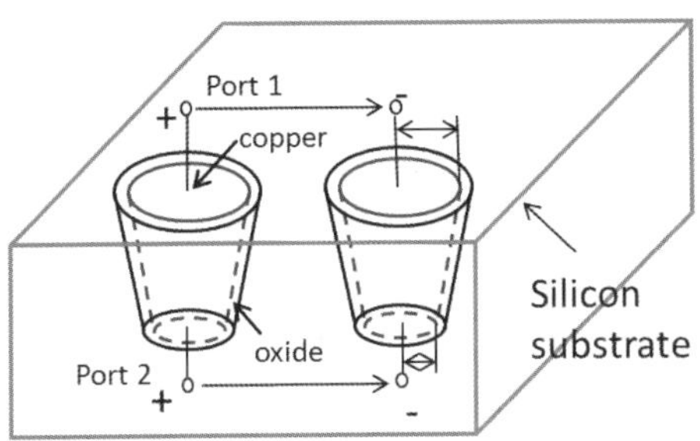

Figure 3.23: Geometry and ports setup for two conical TSVs.

3.6.2 *Simulation Results of Conical TSVs*

The modeling method is applied here to two conical TSVs. The ports setup of the conical TSV pair is shown in Figure 3.23. The length, pitch, oxide thickness, top radius, bottom radius are 200um, 50um, 0.1um, 10um, and 5um, respectively. The conductivity of the silicon substrate is set as 10S/m. The dielectric constants of silicon dioxide and silicon are set to be 3.9 and 11.9, respectively. For comparison, two cases of cylindrical TSV pairs are simulated using the TSV modeling method described earlier. The radii of the cylindrical TSVs in these two cases are fixed at 10um and 5um, respectively. All other parameters are the same as the conical TSV pair.

The equivalent circuit of the two conical TSVs can be constructed similar to Figure 3.16. The frequency-dependent values of RLCG parameters of the conical TSVs obtained using the method described are shown in Figure 3.24, with the parameters of two cylindrical TSVs obtained using the method in Section 3.5. As can be seen from Figure 3.24, the RLCG values of conical TSVs show similar frequency-dependent trend as cylindrical TSVs, and RLCG values of conical TSVs are between the two cylindrical TSV pairs.

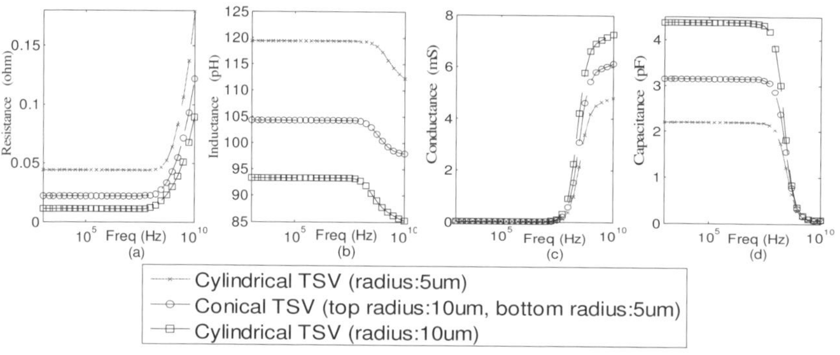

Figure 3.24: Equivalent parasitic elements of conical TSVs with 10um top radius and 5um bottom radius, cylindrical TSVs with 5um radius, cylindrical TSVs with 10um radius. (a) Series resistance of the conductor. (b) Self-inductance of the conductor. (c) Equivalent conductance. (d) Equivalent capacitance.

The insertion loss of the conical TSVs using the proposed modeling approach and CST Microwave Studio [CST] are shown in Figure 3.25, where the insertion loss obtained using the two different methods show good correlation up to 10GHz.

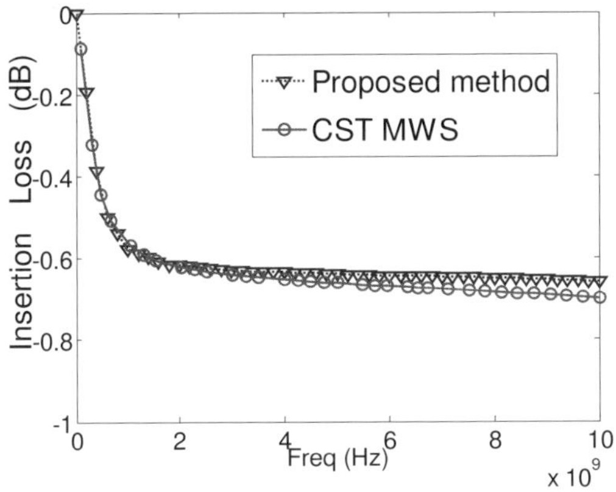

Figure 3.25: Insertion loss for two conical TSVs.

3.6.3 *Modeling of Complex TSV Structures*

In this subsection, a hybrid modeling method is proposed to model complex TSV structures. This modeling method can be divided into two steps. In the first step, the tapered TSVs are segmented into conical or cylindrical segments. Each segment can be modeled using cylindrical TSV modeling method or conical TSV modeling method based on the shape of the segment. In the second step, under the assumption that the maximum size of the problem space is much smaller than the wavelength corresponding to the maximum modeling frequency, the S-parameters of the segments can be cascaded to obtain the insertion loss of the tapered TSV pair.

The proposed hybrid modeling method is applied to model three different complex tapered TSV pairs. The geometry of the three cases such as the length, oxide thickness, pitch, radius is defined as in Figure 3.26. The conductivity of the silicon substrate is set as 10S/m. The dielectric constants of silicon dioxide and silicon are set to be 3.9 and 11.9, respectively. To simulate the complex TSV structure, the proposed method consumes only about 30MB memory, which is 33 times less than CST which consumes about 1GB memory. The insertion loss of the three tapered TSV pairs using the hybrid modeling method and CST are compared in Figure 3.27. For comparison, the three cases are also modeled using only one cylindrical segment, the radius of the cylindrical

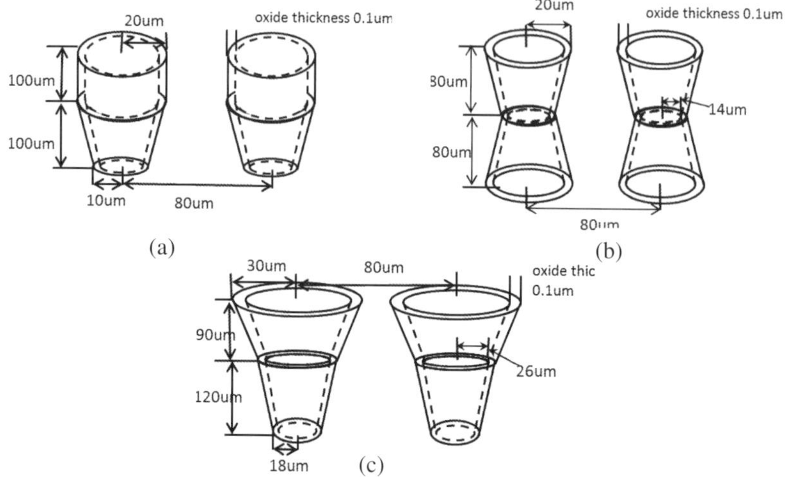

Figure 3.26: Complex TSV structures.

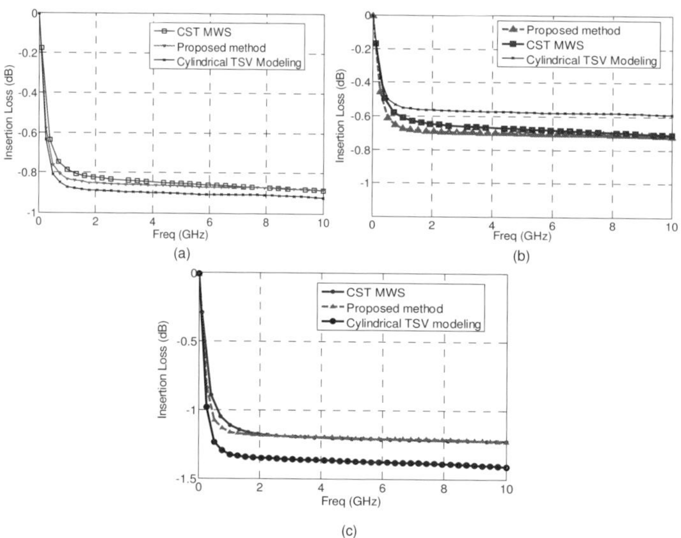

Figure 3.27: Insertion loss of Complex TSV structures (a) Case a, (b) Case b and (c) Case c.

segment is set at 20um, 14um and 30um, respectively. As can be seen from Figure 3.27, the insertion loss obtained using the hybrid method and CST show good correlation over a bandwidth of 10GHz. However, the insertion loss show larger deviation for the three cases using only cylindrical TSV modeling method.

3.7 MOS Capacitance Effect

In all of the structures considered, the silicon substrate has been considered as a lossy material and its semiconductor properties have been ignored. These models ignore the voltage-dependent MOS capacitance of the TSVs. In this section, two methods are used to extract this capacitance based on a full depletion approximation (FDA) [Band et al., 2009; Katti et al., 2010] and rigorous numerical modeling [Band et al., 2011; Band, 2011]. The voltage-dependent MOS capacitance is then used to extract the frequency response of the TSVs.

Consider the TSV structure shown in Figure 3.28(a) which consists of a cylindrical conductor surrounded by an oxide liner embedded in a silicon substrate with silicon dioxide (SiO_2) inter-layer dielectric (ILD) on either side. Such a TSV under bias conditions exhibits capacitance behavior similar to a planar MOS capacitor, as shown in Figure 3.28(b), where V_{FB} is the flat band voltage, V_T is the threshold voltage and V_g is the potential difference between the conductor and the silicon substrate.

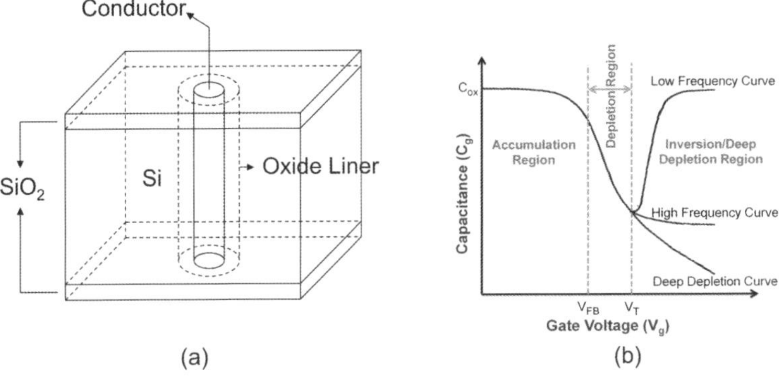

Figure 3.28: (a) TSV Structure and (b) Capacitance-voltage plot for planar MOS capacitor on p-type silicon.

A behavior similar to Figure 3.28(b) is expected for a TSV as well where the capacitance between the conductor and silicon substrate equals the oxide capacitance in the accumulation region but decreases in the depletion, inversion and deep depletion regions and is based on the bias voltage (shown as gate voltage for a MOS capacitor in Figure 3.28(b)). Beyond the threshold voltage V_T, three curves are shown which can be used to differentiate a signal TSV from power/ground TSVs, as explained in [Band et al., 2011]. When the DC component of the gate voltage changes very fast (~5V/ns), the generation of minority carriers cannot keep up with the rate of change of gate voltage. Hence no inversion region is formed around the oxide and any increase in the gate voltage is matched by an increase in the width of the depletion region. This occurs in signal TSVs where high frequency signals propagate and follows the deep depletion curve shown in Figure 3.28(b). When the DC component of the gate voltage changes slowly, an inversion region is formed for high gate voltage. The MOS capacitance in this scenario is dictated by the small signal AC component of the gate voltage. When the AC signal contains high frequency components (~>1MHz), the minority carrier generation rate is unable to keep up with it, resulting in an increase in the depletion region as the gate voltage increases. This corresponds to the high frequency curve in Figure 3.28(b) and occurs in power and ground TSVs containing high frequency noise signals. When the gate voltage has a low-frequency small-signal AC component, the minority carrier generation rate matches any change in the gate voltage, resulting in the inversion region width being proportional to the gate voltage. In such a situation, the low frequency curve in Figure 3.28(b) results. The low-frequency characteristic occurs in the power and ground TSVs containing low frequency noise transients.

3.7.1 *Extraction of MOS Capacitance*

Consider a TSV with circular cross section as in Figure 3.28(a) in a p-type silicon substrate. The cross section of the TSV biased in the depletion region is shown in Figure 3.29. Here, the silicon substrate is at

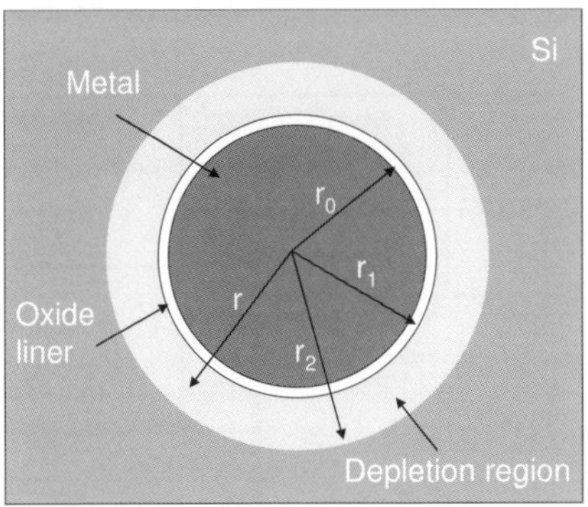

Figure 3.29: TSV biased in the depletion region.

ground potential and the voltage on the TSV can be expressed as:

$$V_{TSV} = \frac{Q_{TSV}}{C_{ox}} + V_{FB} + \varphi_s \tag{3.46}$$

where Q_{TSV} is the charge on the TSV, C_{ox} is the oxide liner capacitance, V_{FB} is the flat band voltage and φ_s is the surface potential of the Si-SiO$_2$ interface. The oxide capacitance from (3.3) is:

$$C_{ox} = \frac{2\pi\varepsilon_{ox}L}{\ln(\frac{r_1}{r_0})} \tag{3.47}$$

where r_1 is the metal radius, r_0 is the oxide liner radius, L is the TSV length and ε_{ox} is the permittivity of the oxide. The flat band voltage is the voltage that needs to be applied to the TSV to obtain a flat band energy diagram shown in Figure 3.30. In this condition there are no charges in the semiconductor. From Figure 3.30, the flat band voltage can be calculated as:

$$V_{FB} = \varphi_M - X - \frac{E_g}{2q} - \frac{kT}{q}\ln(\frac{N_a}{n_i}) \tag{3.48}$$

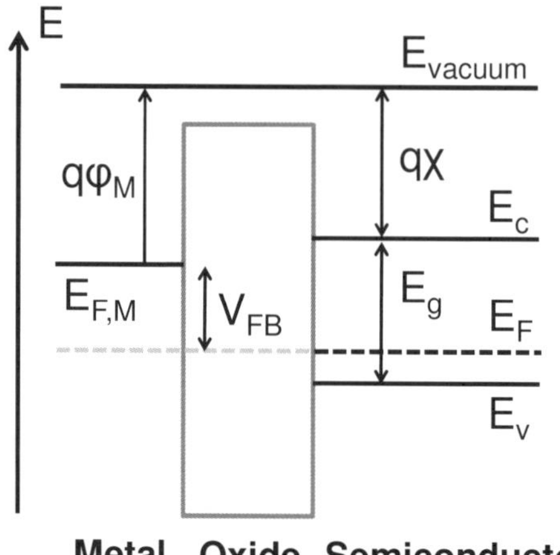

Figure 3.30: Energy band diagram of Metal Oxide Semiconductor junction at flat band condition for p-type substrate [Band et al., 2011].

where φ_M is the metal work function, X is the electron affinity (4.05V for silicon), E_g is the bandgap energy (1.12eV for silicon at room temperature), q is the electron charge (1.6022×10^{-19}C), k is the Boltzman constant (1.3807×10^{-23}m^{2}kg/s^{2}K), T is the absolute temperature in K, n_i is the intrinsic carrier concentration of silicon 1.18×10^{10}/cm^3 at 300K and N_a is the doping concentration of the acceptor ions.

To calculate the surface potential φ_s, the charge distribution in the depletion region can be represented by Poisson's equation in cylindrical coordinates as:

$$\frac{d^2\varphi}{dr^2} + \frac{1}{r}\frac{d\varphi}{dr} = -\frac{\rho}{\varepsilon_{si}} \qquad (3.49)$$

where φ and ρ are the potential and charge density at distance r in the depletion region and ε_{si} is the permittivity of silicon. The charge distribution is assumed to be invariant with length and is assumed to be axisymmetric.

3.7.2 *Full Depletion Approximation*

One method for solving (3.49) is by using the full depletion approximation (FDA) where the depletion region charge is only due to the ionized doping atoms. Hence, (3.49) can be rewritten as:

$$\frac{1}{r}\frac{d}{dr}(r\frac{d\varphi}{dr}) = \frac{qN_a}{\varepsilon_{Si}} \tag{3.50}$$

with the boundary condition that the potential φ and electric field $d\varphi/dr$ go to zero at $r = r_2$, which is the edge of the depletion region. Integrating twice in the depletion region, the surface potential φ_s at the Si-SiO$_2$ interface can be obtained as:

$$\int_r^{r_2} d(r\frac{d\varphi}{dr}) = \int_r^{r_2}\frac{qN_a}{\varepsilon_{Si}}rdr \Rightarrow -r\frac{d\varphi}{dr} = \frac{qN_a}{2\varepsilon_{Si}}(r_2^2 - r^2)$$

$$\int_{r_1}^{r_2} -d\varphi = \int_{r_1}^{r_2}\frac{qN_a}{2\varepsilon_{Si}}(\frac{r_2^2}{r} - r)dr \Rightarrow \varphi_s = \frac{qN_a}{2\varepsilon_{Si}}[r_2^2 \ln(\frac{r_2}{r_1}) - \frac{r_2^2 - r_1^2}{2}] \tag{3.51}$$

The charge on the metal equals the charge in the depletion region due to the ionized doping atoms. Therefore,

$$Q_{TSV} = qN_a\pi(r_2^2 - r_1^2)L \tag{3.52}$$

Combining (3.46), (3.47), (3.51) and (3.52) results in:

$$V_{TSV} = \frac{qN_a(r_2^2 - r_1^2)}{2\varepsilon_{ox}}\ln(\frac{r_1}{r_0}) + V_{FB} + \frac{qN_a}{2\varepsilon_{Si}}[r_2^2 \ln(\frac{r_2}{r_1}) - \frac{r_2^2 - r_1^2}{2}] \tag{3.53}$$

From (3.53), for a given bias voltage V$_{TSV}$, the radius r$_2$ to the end of the depletion region can be found. This radial distance can be used to calculate the depletion capacitance C$_{dep}$ as:

$$C_{dep} = \frac{2\pi\varepsilon_{Si}L}{\ln(\frac{r_2}{r_1})} \tag{3.54}$$

with the total capacitance C_{tot} from the metal to the silicon substrate equaling:

$$C_{tot} = \left[\frac{1}{C_{ox}} + \frac{1}{C_{dep}} \right]^{-1}$$
(3.55)

At threshold, the surface potential φ_s equals twice the bulk φ_B. The outer radius of the deletion region r_T at threshold can be calculated from (3.51) using:

$$\varphi_s = \frac{qN_a}{2\varepsilon_{Si}} [r_T^2 \ln(\frac{r_T}{r_1}) - \frac{r_T^2 - r_1^2}{2}] = 2\varphi_B = 2\frac{kT}{q} \ln(\frac{N_a}{n_i})$$
(3.56)

with the resulting threshold voltage V_T calculated as:

$$V_T = \frac{qN_a(r_T^2 - r_1^2)}{2\varepsilon_{ox}} \ln(\frac{r_1}{r_0}) + V_{FB} + 2\frac{kT}{q} \ln(\frac{N_a}{n_i})$$
(3.57)

For the deep depletion curve in Figure 3.28(b), (3.53) can be used to calculate the radius to the edge of the depletion region (r_2) for a given bias voltage V_{TSV}. Using the calculated r_2, the total capacitance can be calculated from (3.55). For the high frequency curve in Figure 3.28(b), the capacitance beyond the threshold voltage is the same as the capacitance at the threshold voltage. For the low frequency curve, beyond the threshold voltage, the capacitance reverts back to the oxide capacitance value.

3.7.3 *Numerical Analysis*

The FDA described is an approximation which can be avoided if (3.49) can be solved numerically where the charge density ρ consists of the fixed doping ions (N_a), holes (p) and electrons (n) in the semiconductor and is given by:

$$\rho = q(-N_a + p - n) = qN_a[-1 + e^{-\frac{q\varphi}{kT}} - e^{\frac{q(\varphi - 2\varphi_B)}{kT}}]$$
(3.58)

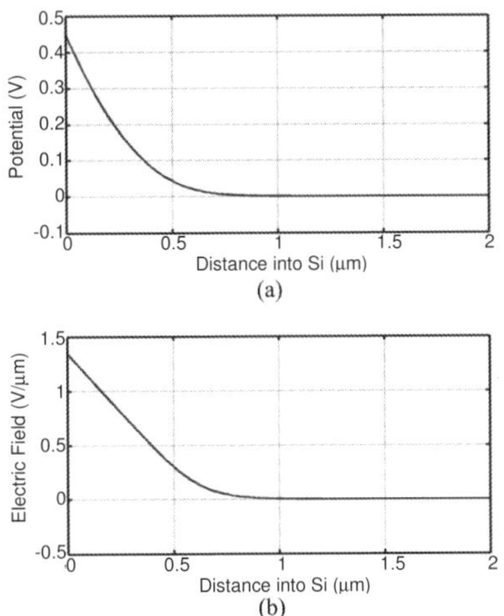

Figure 3.31: (a) Potential and (b) Electric field from Si-SiO$_2$ interface; oxide thickness = 0.1um.

The initial conditions are the surface potential φ and electric field ($-d\varphi/dr$) at a distance $r = r_1$ at the Si-SiO$_2$ interface. Equation (3.58) can be solved iteratively by using the 4th order Runge-Kutta method. For a given surface potential, the electric field can be obtained iteratively such that both the surface potential and electric field converge, as shown in Figure 3.31. The distance at which the potential and electric field are zero represents the edge of the depletion region (radius r_2). The charge on the TSV is equal to the total charge in the silicon and can be computed using Gauss's law as:

$$Q_{TSV} = \varepsilon_{Si} \oint_S \vec{E} \bullet dS = -2\pi r_1 L \varepsilon_{Si} \left. \frac{d\varphi}{dr} \right|_{r=r_1} \tag{3.59}$$

From (3.53), the voltage on the TSV (V$_{TSV}$) can be computed. Unlike the FDA, the electric field is non-uniform in the depletion region and hence (3.54) cannot be used to calculate the depletion capacitance based on the value for r_2 calculated from a numerical solution. Hence, the capacitance of the TSV is calculated using:

$$C_{TSV} = \left. \frac{dQ}{d\varphi} \right|_{r=r_1} = 2\pi r_1 \varepsilon_{Si} \frac{d}{d\varphi} \left(-\frac{d\varphi}{dr} \right) \Bigg|_{r=r_1} . \tag{3.60}$$

As in FDA, the threshold voltage is calculated by setting $\varphi_s = 2\varphi_B$. For the low frequency curve in Figure 3.28(b), (3.59) is used to calculate the TSV capacitance as a function of the bias voltage. For the high frequency curve, the capacitance beyond the threshold voltage remains constant at the threshold capacitance value. For the deep depletion curve, beyond the threshold voltage, the capacitance and bias voltage are calculated by neglecting the minority carrier density (n) in (3.58).

3.7.4 *Capacitance Vs. Bias Voltage*

As an example consider a copper TSV of radius 15μm, oxide thickness 0.1μm, length 100μm, oxide relative permittivity 3.9, silicon relative permittivity 11.9 and in a silicon substrate of conductivity 10S/m ($N_a = 0.137 \times 10^{22}/m^3$). The variation of capacitance (C_{tot}) with bias voltage (V_{TSV}) is shown in Figures 3.32(a), (b) and (c) using the FDA and numerical analysis for the low frequency, high frequency and deep depletion mode of operation, respectively. As a comparison, the oxide capacitance (C_{ox}) is also shown when the biasing effect is neglected. From the figures it is clear that only the oxide capacitance overestimates the TSV capacitance to the substrate, due to biasing.

In Figures 3.32 (a)–(c), even though FDA is inaccurate as compared to the numerical analysis, it provides a simple and quick way of estimating the capacitance for a given bias voltage. The accuracy of the numerical analysis is validated through measurements [Katti et al., 2010] in Figure 3.32(d) for a TSV with 5μm diameter, 20μm length, copper metallization, SiO_2 oxide liner of thickness 118.2nm and with a doping concentration of the p-type Si substrate of $2 \times 10^5 cm^{-3}$. The correlation is reasonably good given the limited information on the TSV parameters.

3.7.5 *Impact of Bias Voltage on Insertion Loss and Model Development*

If the bias voltage reduces the TSV capacitance to the substrate, then it should have an impact on the insertion loss of the signal.

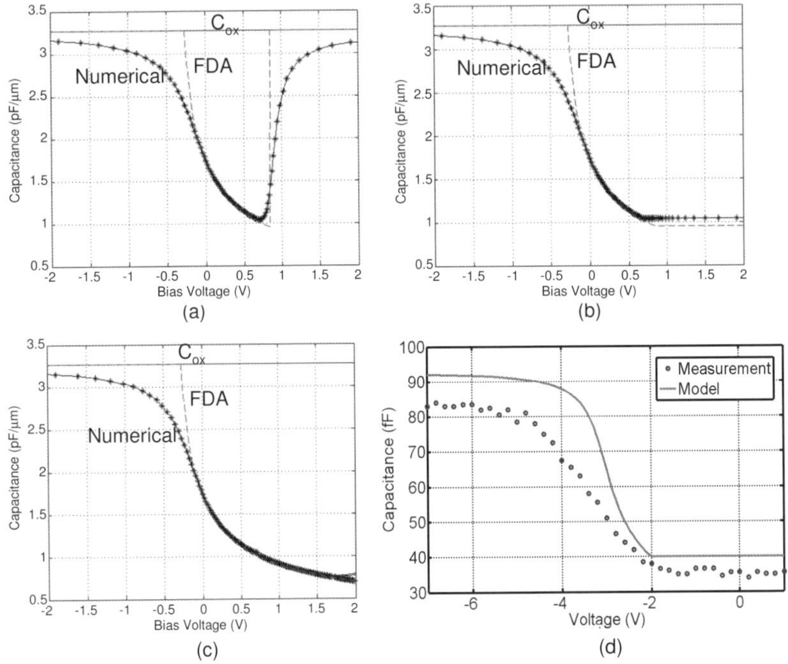

Figure 3.32: Capacitance Vs Bias Voltage (a) Low Frequency, (b) High Frequency, (c) Deep Depletion and (d) Model to Hardware correlation.

Consider a signal and ground return TSV shown in Figure 3.2(c) with $D = 100\mu m$, $R = 15\mu m$, $L = 100\mu m$, $d_{ox} = 0.1\mu m$, $\varepsilon_{ox} = 3.9$ and $\varepsilon_{Si} = 11.9$ and biased at 1.8059V. From Figure 3.32(c), the deep depletion curve results in a total capacitance (C_{tot}) of 0.7264pF. This is due to a depletion capacitance (C_{dep}) of 0.9343pF in series with the oxide capacitance (C_{ox}) of 3.2653pF. Using the model in Figure 3.8 and replacing C_1 with 0.7264pF, the insertion loss for a signal and ground return via referenced to 50Ω can be computed, which is shown in Figure 3.33 and compared to the case where the oxide capacitance of 3.2653pF is used for the capacitance C_1 (no bias). Clearly, with a lower bias capacitance, the insertion loss is lowered which results in a more gradual slope for the insertion loss during the slow wave to quasi-TEM transition phase.

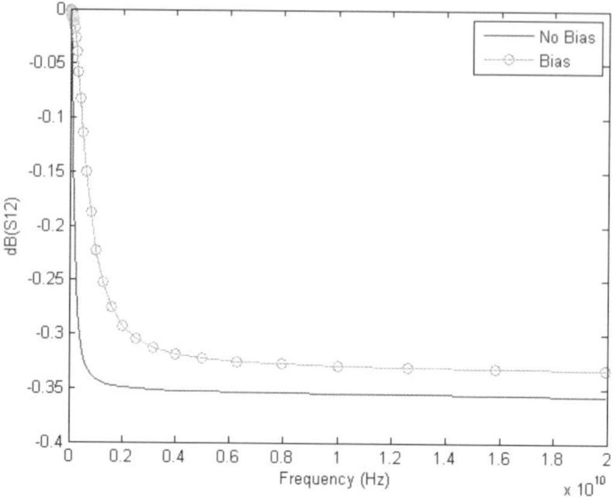

Figure 3.33: Insertion Loss with and without bias voltage.

3.8 Consideration of MOS Capacitance Effect in Electromagnetic Modeling

For general TSV configurations, the MOS capacitance effect discussed in the previous section can be included in the electromagnetic modeling method described earlier [Han et al., 2012]. The original modeling method is modified here by simply adding capacitance matrices related to the depletion region. Since the other elements except for the capacitances remain the same, we can focus primarily on the capacitance formulation when TSVs have depletion regions generated by DC bias voltages. Since the radial widths of depletion regions can be calculated by using formulas given earlier in this chapter, the procedure for calculating the depletion region width is not discussed in this section. Instead we assume that the depletion widths are already determined based on a specific DC bias condition. To simplify the problem, full-depletion approximation, which enables the direct calculation of the depletion width, is assumed here. In the modified formulation, TSVs having multiple depletion thicknesses can be considered as well.

Figure 3.34 shows the geometry of a TSV when depletion region is formed due to DC bias voltage. In addition to surfaces on copper conductor (S^C) and oxide liner (S^I), the surface on depletion region (S^D) boundary should be considered. Since S^C is a conductor surface, it should

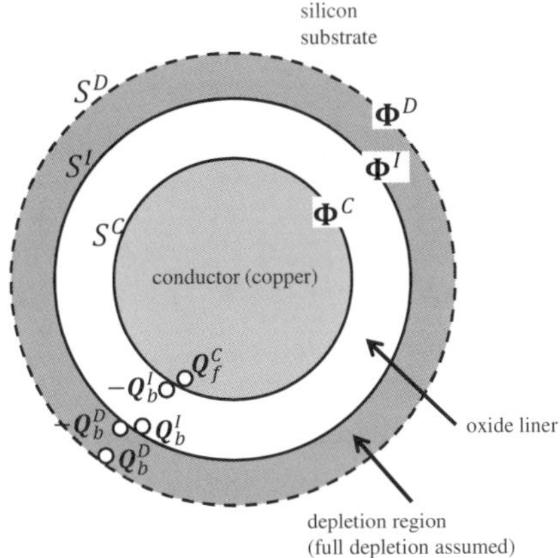

Figure 3.34: Description of a TSV with a depletion region. Potential and charge are defined in each surface.

satisfy Dirichlet boundary condition. For the other surfaces, Neumann boundary condition should be assigned. The three surfaces S^C, S^I, and S^D contain total charges $\mathbf{Q}^C$, $\mathbf{Q}^I$, and $\mathbf{Q}^D$, respectively. According to the boundary conditions assigned to the surfaces, each charge is composed of free or bound charge as follows:

$$\mathbf{Q}^C = \mathbf{Q}_f^C - \mathbf{Q}_b^I$$
$$\mathbf{Q}^I = \mathbf{Q}_b^I - \mathbf{Q}_b^D ,$$
$$\mathbf{Q}^D = \mathbf{Q}_b^D$$

(3.61)

where superscripts 'C', 'I', and 'D' represent conductor, insulator, and depletion region, respectively, and subscripts 'f' and 'b' represent free and bound charge, respectively. Actually, free volume charge, which originates from the governing Poisson's equation, also exists in the depletion region. However, the effect of free volume charges can be neglected when we focus on AC analysis since the free charges depend on the applied DC bias voltage. Therefore, all the charges in (3.61) are AC quantities. From (3.61), the free charge on the conductor surface S^C is

$$\mathbf{Q}_f^C = \mathbf{Q}^C + \mathbf{Q}_b^I = \mathbf{Q}^C + \mathbf{Q}^I + \mathbf{Q}^D .$$

(3.62)

The relation between total charges and potential on the surfaces can be described with the following matrix equation:

$$\begin{pmatrix} \mathbf{P}^C & \mathbf{P}^{CI} & \mathbf{P}^{CD} \\ \mathbf{P}^{IC} & \mathbf{P}^I & \mathbf{P}^{ID} \\ \mathbf{P}^{DC} & \mathbf{P}^{DI} & \mathbf{P}^D \end{pmatrix} \begin{pmatrix} \mathbf{Q}^C \\ \mathbf{Q}^I \\ \mathbf{Q}^D \end{pmatrix} = \begin{pmatrix} \mathbf{\Phi}^C \\ \mathbf{\Phi}^I \\ \mathbf{\Phi}^D \end{pmatrix}, \tag{3.63}$$

where $\mathbf{P}^{\{C,D,I\}}$'s are potential coefficient sub-matrices. Element in each sub-matrix can be calculated from the integral equation combined with cylindrical accumulation mode basis functions (AMBFs), as discussed in Chapter 2. Thus, each potential coefficient sub-matrix includes elements obtained from the interaction between the fundamental mode and higher order modes.

Polarization currents in the oxide liner region and the depletion region can be expressed as follows:

$$\begin{aligned} \mathbf{I}^{I,pol} &\approx j\omega \mathbf{C}^{I,ex}(\mathbf{\Phi}^C - \mathbf{\Phi}^I) \\ \mathbf{I}^{D,pol} &\approx j\omega \mathbf{C}^{D,ex}(\mathbf{\Phi}^I - \mathbf{\Phi}^D) \end{aligned}, \tag{3.64}$$

where $\mathbf{C}^{I,ex}$ and $\mathbf{C}^{D,ex}$ are excess capacitances in the oxide liner region and the depletion region, respectively. The values of capacitances can be obtained from the matrix relation (3.63). The polarization currents are related to bound charges by the following continuity equations:

$$\begin{aligned} \mathbf{I}^{I,pol} - j\omega \mathbf{Q}^I_b &= \mathbf{0} \\ \mathbf{I}^{D,pol} - j\omega \mathbf{Q}^D_b &= \mathbf{0} \end{aligned}. \tag{3.65}$$

By combining (3.64) and (3.65), we can obtain the relation between bound charges and potentials as:

$$\begin{aligned} \mathbf{Q}^I_b &= \mathbf{C}^{I,ex}(\mathbf{\Phi}^C - \mathbf{\Phi}^I) \\ \mathbf{Q}^D_b &= \mathbf{C}^{D,ex}(\mathbf{\Phi}^I - \mathbf{\Phi}^D) \end{aligned}. \tag{3.66}$$

With the above formulae, we can derive the following matrix equation that relates the free charge on the conductor surface and potentials on the conductor, the oxide liner, and the depletion region.

$$
\begin{pmatrix} \mathbf{Q}_f^C \\ \mathbf{0} \\ \mathbf{0} \end{pmatrix} = \begin{pmatrix} \mathbf{C}^C + \mathbf{C}^{I,ex} & \mathbf{C}^{CI} - \mathbf{C}^{I,ex} & \mathbf{C}^{CD} \\ \mathbf{C}^{IC} - \mathbf{C}^{I,ex} & \mathbf{C}^I + \mathbf{C}^{I,ex} + \mathbf{C}^{D,ex} & \mathbf{C}^{ID} - \mathbf{C}^{D,ex} \\ \mathbf{C}^{DC} & \mathbf{C}^{DI} - \mathbf{C}^{D,ex} & \mathbf{C}^D + \mathbf{C}^{D,ex} \end{pmatrix} \begin{pmatrix} \mathbf{\Phi}^C \\ \mathbf{\Phi}^I \\ \mathbf{\Phi}^D \end{pmatrix} \qquad (3.67)
$$

where
$$
\begin{pmatrix} \mathbf{C}^C & \mathbf{C}^{CI} & \mathbf{C}^{CD} \\ \mathbf{C}^{IC} & \mathbf{C}^I & \mathbf{C}^{ID} \\ \mathbf{C}^{DC} & \mathbf{C}^{DI} & \mathbf{C}^D \end{pmatrix} = \begin{pmatrix} \mathbf{P}^C & \mathbf{P}^{CI} & \mathbf{P}^{CD} \\ \mathbf{P}^{IC} & \mathbf{P}^I & \mathbf{P}^{ID} \\ \mathbf{P}^{DC} & \mathbf{P}^{DI} & \mathbf{P}^D \end{pmatrix}^{-1} .
$$

Equation (3.67) is the extension of the equivalent capacitance matrix equation shown in (3.41). Since the formulations for conductor impedance and the continuity relation on the conductor surface remain the same, we can modify the formulation for TSVs with depletion region by replacing the original capacitance matrix $\mathbf{C}^{eq}$ in (3.41) with the following:

$$
\mathbf{C}^{eq} = \begin{pmatrix} \mathbf{E} & \mathbf{0} \end{pmatrix} \begin{bmatrix} \mathbf{C}_C^{eq} - \mathbf{C}_{CX}^{eq} \mathbf{C}_X^{eq-1} \mathbf{C}_{XC}^{eq} \end{bmatrix} \begin{pmatrix} \mathbf{E} \\ \mathbf{0} \end{pmatrix}, \qquad (3.68)
$$

where $\mathbf{E}$ is an identity matrix and,

$$
\mathbf{C}_C^{eq} = \mathbf{C}^C + \mathbf{C}^{I,ex},
$$

$$
\mathbf{C}_{CX}^{eq} = \begin{pmatrix} \mathbf{C}^{CI} - \mathbf{C}^{I,ex} & \mathbf{C}^{CD} \end{pmatrix},
$$

$$
\mathbf{C}_X^{eq} = \begin{pmatrix} \mathbf{C}^I + \mathbf{C}^{I,ex} + \mathbf{C}^{D,ex} & \mathbf{C}^{ID} - \mathbf{C}^{D,ex} \\ \mathbf{C}^{DI} - \mathbf{C}^{D,ex} & \mathbf{C}^D + \mathbf{C}^{D,ex} \end{pmatrix}, \text{ and}
$$

$$
\mathbf{C}_{XC}^{eq} = \begin{pmatrix} \mathbf{C}^{IC} - \mathbf{C}^{I,ex} \\ \mathbf{C}^{DC} \end{pmatrix} .
$$

The equivalent capacitance (3.68) can be inserted to (3.41), which relates the terminal currents and potentials, by keeping the other matrices and vectors same. As discussed in Section 3.5, (3.41) leads to a Z-parameter matrix equation, which can be converted to the corresponding S-parameter matrix relation.

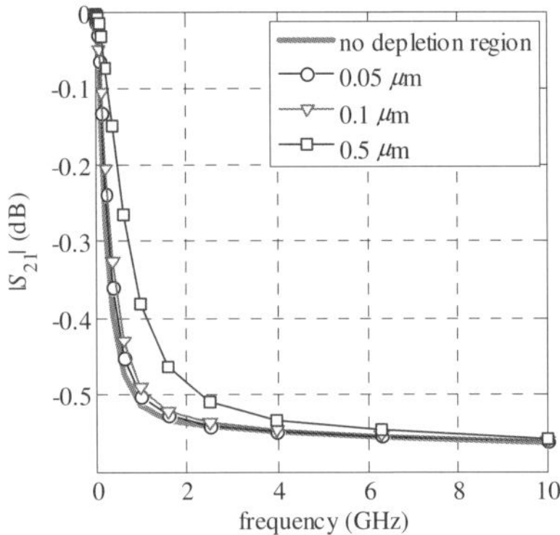

Figure 3.35: Insertion loss for two TSVs in Fig. 3.2(c) for different depletion depths.

The modified formulation for TSVs with depletion regions has been validated here through the simulation of a simple two-TSV configuration. We consider a TSV pair configuration similar to Figure 3.2(c), which is fabricated in a silicon substrate with conductivity 10S/m. The only difference from the structure in Figure 3.2(c) is the addition of the depletion region. The height and the diameter of TSVs are 100um and 30um, respectively, with a pitch between the two TSVs of 50um. The thickness of oxide (SiO$_2$) liner is 0.1um. With all the other geometry parameters fixed, four different depletion depths (0, 0.05, 0.1, and 0.5um) were analyzed. Figure 3.35 shows the insertion loss (S_{21}) of TSV pairs for different depletion depths, up to 10GHz. Compared to the case with no depletion region, the insertion losses of TSV pairs having depletion regions become smaller since the depletion region removes most of the mobile charge carriers and becomes closer to an insulator at AC frequencies. The increasing depletion depth results in an effect similar to increasing the oxide thickness, although the relative permittivity of depletion region is different from that of oxide liner. The insertion loss is significantly reduced in slow-wave mode, where the effects of oxide liner and depletion region become dominant.

Figures 3.36(a) and (b) show the equivalent line-to-line capacitance and conductance between two TSVs for different depletion depths. The effect of increase in depletion depth can be clearly seen in Figure 3.36(a), where the capacitances in slow-wave mode frequencies become smaller

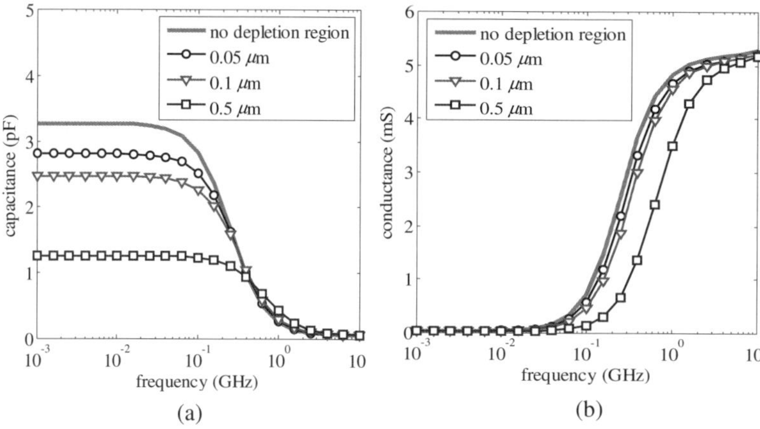

Figure 3.36: (a) Equivalent capacitance and (b) conductance of two TSVs for different depletion depths.

due to the series connection of oxide and depletion capacitance. However, the values of capacitances for quasi-TEM mode frequencies are almost the same. In contrast, the conductance in Figure 3.36(b) shows clear differences at high frequencies, where the larger depletion region results in smaller conductive substrate region.

3.9 Time Domain Response

So far, the frequency response of TSVs has been computed and represented in the form of S-parameters. Ultimately, a time domain response is necessary for digital applications. Often times, the response of TSVs in the presence of non-linear drivers is required. A SPICE netlist generation is therefore necessary to represent the TSV response in the form of a circuit in SPICE. This is possible through macro-modeling where a commercially available tool such as Idem [Idem, 2009] can be used to create a black box model. Idem accepts S-parameters and converts them into a SPICE circuit model through vector fitting. Issues such as passivity are satisfied after model construction to ensure a stable model when connected to non-linear drivers. Details on macro-modeling are available in [Swaminathan et al., 2007]. In this section, the time

    Design and Modeling for 3D ICs and Interposers

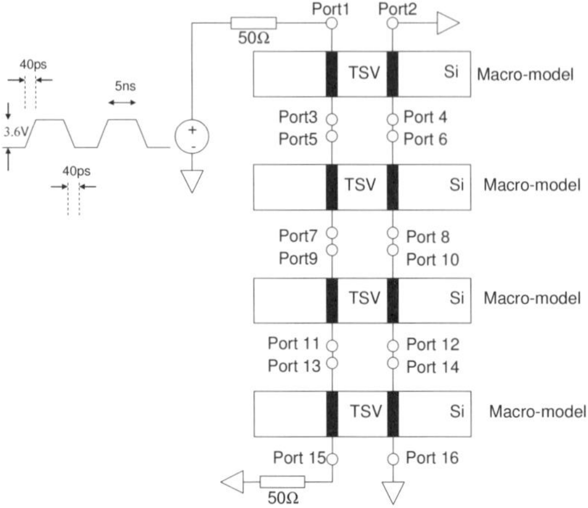

Figure 3.37: Spice schematic for time domain simulation of four silicon stack.

domain response of a four silicon stack is simulated using macro-modeling. The circuit schematic is shown in Figure 3.37 where a signal and ground return TSV pair ($D = 100\mu m$, $R = 15\mu m$, $L = 100\mu$, $d_{ox} = 0.1\mu m$, $\varepsilon_{ox} = 3.9$ and $\varepsilon_{Si} = 11.9$) has been modeled using the method described in Section 3.5 and then converted into a macro-model. The TSV pair for each silicon substrate has four ports as in Figure 3.16. A four silicon stack model is then constructed as shown in Figure 3.16 by concatenating the corresponding ports (a more exact model that accounts for the solder bumps and underfill material is provided in Chapter 4). A 3.6V voltage source with pulse width of 5ns, rise and fall time of 40ps and output impedance of 50Ω is connected to Port 1 with the output at Port 15 terminated in 50Ω. Ports 2 and 16 are grounded. The output waveform at Port 15 accounts for all the TSV effects as the signal propagates through the silicon stack.

The simulated time domain waveform for high resistivity (0.01S/m) and low resistivity (10S/m) silicon is shown in Figure 3.38. The 'RC' effect associated with the low resistivity silicon substrate can be seen due to the conductance of the substrate, since all other parameters are the same. A simple lumped element model has also been developed as in Figure 3.8 with $R = 30m\Omega$, $L = 75.8pH$, $C_1 = 13.04pF$, $C_2 = 0.0704pF$,

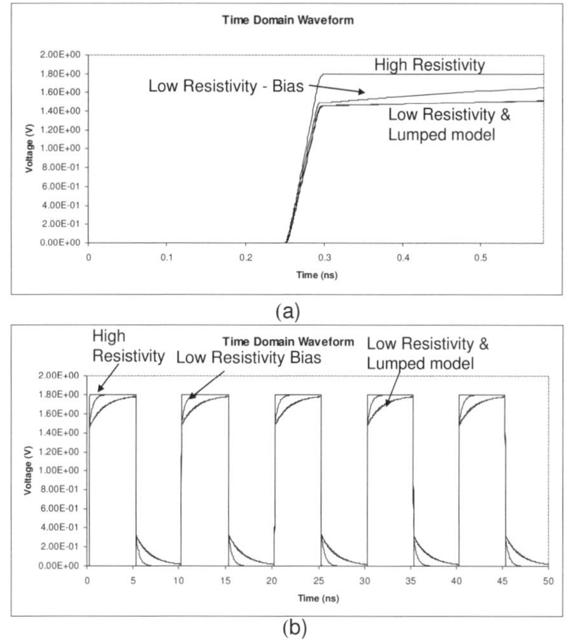

Figure 3.38: Time Domain Waveform (a) 0.58ns duration and (b) 50ns duration.

$R_2 = 105\Omega$ to mimic the four silicon stack through scaling using the parameters for a single silicon substrate (R and R_2 values were adjusted for fitting purposes) whose time domain response agrees well with the low resistivity silicon substrate macro-model. In the lumped element model, by increasing the value of R_2 the wave shape can be improved and made to coincide with the high resistivity silicon substrate waveform, indicating that the conductance of the silicon substrate is the cause for the 'RC' effect. The waveform when the silicon substrate is grounded ($V_{TSV} = 1.8V$) is also shown where the reduced capacitance ($C_{tot} = 0.7264pF$) for each stack improves the 'RC' effect.

3.10 Summary

This chapter described the modeling of through silicon vias (TSVs). Two approaches were described namely, 1) physics based and

2) rigorous electromagnetic modeling. In the physics based approach, analytical models were derived to show the effect of signal and ground return TSVs in the frequency domain. Though these models are simple to derive, they are limited to a TSV pair and can produce errors when the proximity-effect becomes important. In the electromagnetic modeling based approach, specialized basis functions have been used to represent the current and charge in the conductor and dielectric. Using a quasi-static approach, the frequency response for a TSV pair was extracted which agreed well with the physics based model. However, as the TSV density increases where proximity-effect becomes important and the number of TSVs is more than two, this method will provide more accurate results. A partial element equivalent circuit (PEEC) model has been derived which provides an easy method to define signal, power and ground assignments. This method has been extended to model tapered TSVs as well. When the silicon substrate is biased, the MOS capacitance due to bias voltage needs to be computed which has been discussed in this chapter. Two methods based on FDA and numerical analysis have been described for analyzing the TSVs in low frequency, high frequency and deep depletion modes of operation. A method for incorporating MOS capacitance effect into electromagnetic modeling has also been discussed in this chapter. Finally macro-modeling has been used to derive black box equivalent circuit models for the TSVs which have been used to compute time domain waveforms of a four silicon stack. The 'RC' effect seen in low resistivity silicon substrates have been highlighted.

References

1. [Kwon, 2011] Oh-Hyun Kwon, "Eco-Friendly Semiconductor Technologies for Healthy Living – Plenary Talk", International Solid State Circuits Conference (ISSCC), 2011.
2. [Kang et al., 2010] Uksong Kang, Hoe-Ju Chung, Seongmoo Heo, Duk-Ha Park, Hoon Lee, Jin Ho Kim, Soon-Hong Ahn, Soo-Ho Cha, Jaesung Ahn, DukMin Kwon, Jae-Wook Lee, Han-Sung Joo, Woo-Seop Kim, Dong Hyeon Jang, Nam Seog Kim, Jung-Hwan Choi, Tae-Gyeong Chung, Jei-Hwan Yoo, Joo Sun Choi, Changhyun Kim and Young-Hyun Jun, "8 Gb 3-D DDR3 DRAM Using Through-Silicon-Via

Technology", IEEE Journal of Solid-State Circuits, Vol. 45, No. 1, pp. 111–119, January 2010.

3. [Hasegawa et al., 1971] Hideki Hasegawa, Mieko Furukuwa and Hisayoshi Yanai, "Properties of Microstrip Line on Si – SiO_2 System", IEEE Transactions on Microwave Theory and Techniques, Vol. MTT-19, No. 11, pp. 869–881, November 1971.

4. [Ndip et al., 2011] Ivan Ndip, Brian Curran and Kai Lobbicke, "High-frequency modeling of TSVs for 3-D chip integration and silicon interposers considering skin-effect, dielectric and quasi-Tem modes", IEEE Transactions on CPMT, Vol. 1, No. 10, Oct. 2011, pp. 1627–1641.

5. [Barlea et al., 2008] N.M. Barlea, Sanzianaiulia Barlea and E. Culea, "Maxwell-Wagner Effect on the Human Skin", Romanian J. Biophys., Vol. 18, No. 1, pp. 87–98, Bucharest, 2008.

6. [Han et al., 2013] Ki Jin Han, Madhavan Swaminathan, "Modeling electrical interconnections in three-dimensional structures", United States Patent: 8,352,232, 1/08/2013.

7. [Han et al., 2010] Ki Jin Han, Madhavan Swaminathan, and Tapobrata Bandyopadhyay, "Electromagnetic Modeling of Through-Silicon Via (TSV) Interconnections Using Cylindrical Modal Basis Functions", IEEE Transactions on Advanced Packaging, Vol. 33, No. 4, pp. 804–817, November 2010.

8. [Kim et al., 2011] Joohee Kim, Jun So Pak, Jonghyun Cho, Eakhwan Song, Jeonghyeon Cho, Heegon Kim, Taigon Song, Junho Lee, Hyungdong Lee, Kunwoo Park, Seungtaek Yang, Min-Suk Suh, Kwang-Yoo Byun, and Joungho Kim, "High-Frequency Scalable Electrical Model and Analysis of a Through Silicon Via (TSV)", IEEE Transactions on Components, Packaging and Manufacturing Technology, Vol. 1, No. 2, pp. 181–195, February 2011.

9. [Ruehli et al., 1992] Albert E. Ruehli and Hansruedi Heeb, "Circuit Models for Three-Dimensional Geometrices Including Dielectrics", IEEE Transactions on Microwave Theory and Techniques, Vol. 40, No. 7, pp. 1507–1516, July 1992.

10. [Han, 2009] Ki Jin Han, "Electromagnetic Modeling of Interconnections in Three-Dimensional Integration", PhD Dissertation, Georgia Institute of Technology, 2009.

11. [Idem, 2009] IDEM R2009a [online], www.idemworks.com

12. [CST] CST Microwave Studio [Online]:
    http://www.cst.com/Content/Products/MWS/Overview.aspx

13. [Xie et al., 2012] Biancun Xie and Madhavan Swaminathan, "Electromagnetic Modeling of Complex Tapered Through-Silicon Via (TSV) Interconnections," Proc. 16th IEEE Workshop on Signal Propagation on Interconnects (SPI 2012), Sorrento, Italy, May 2012.

14. [Industrial Laser Application Note] Industrial Laser Application Note [Online]. Available: http://www.newport.com/laser-drilling-through-silicon-vias

15. [Band et al., 2009] T. Bandyopadhyay, R. Chatterjee, D. Chung, M. Swaminathan, & R. Tummala, "Electrical modeling of Through Silicon and Package Vias," in

2009 IEEE International Conference on 3D System Integration (3DIC), 28–30 Sept. 2009, San Francisco, CA, USA, 2009.

16. [Katti et al., 2010] G. Katti, M. Stucchi, K. De Meyer and W. Dehaene, "Electrical modeling and characterization of through silicon via for 3D ICs", IEEE Transactions on Electronc Devices, Vol. 57, No. 1, pp. 256–262, Jan. 2010.

17. [Band et al., 2011] T. Bandyopadhyay, K. J. Han, D. Chung, R. Chatterjee, M. Swaminathan, and R. Tummala, "Rigorous Electrical Modeling of Through Silicon Vias (TSV) with MOS Capacitance Effects," IEEE Transactions on Components, Packaging and Manufacturing Technology, Vol. 1, pp. 893–903, 2011.

18. [Band, 2011] T. Bandyopadhyay, "Modeling, Design and Characterization of Through Vias in Silicon and Glass Interposers", Ph.D. Dissertation, Georgia Institute of Technology, 2011.

19. [Han et al., 2012] K. J. Han, M. Swaminathan, "Consideration of MOS capacitance effect in TSV modeling based on cylindrical modal basis functions," Proc. IEEE Electrical Design of Advanced Packaging & Systems Symposium (EDAPS 2012), pp. 42–45, Taipei, Taiwan, Dec. 2012.

20. [Swaminathan et al., 2007] Madhavan Swaminathan and Ege Engin, "Power Integrity Modeling and Design for Semiconductors and Systems", Prentice Hall, 2007.

Chapter 4

Electrical Performance and Signal Integrity

In this chapter, the modeling methods developed are applied for the modeling of through silicon vias (TSV). The effect of process and material parameters on insertion loss is analyzed to develop guidelines for process optimization. The importance of cross talk between TSVs is illustrated through correlation with measurements and comparing to vias in low loss substrates. The electrical effects of TSVs are covered by modeling of via arrays. The impact of shielding, its effect on eye patterns, the importance of variability and effect of temperature are some of the topics addressed in this chapter for TSVs. With the industry considering silicon and glass as possible options for packaging of 3D ICs, a very important question remains as to which technology will prevail as the best interposer solution. Some of these questions have been answered here in the context of electrical performance. Packaging plays a very important role in 3D integration and therefore the challenges involved in TSV modeling and the interplay between silicon and packaging in the context of co-design has been used to conclude this chapter.

4.1 Process Optimization

Two measures have been used to evaluate the performance of TSVs as the process parameters are changed. In the first case, the insertion loss has been used as the measure for evaluating the performance of TSVs when the silicon substrate is unbiased, which typically occurs in interposers. Here a TSV pair has been analyzed (signal and ground via) as in Chapter 3. In the second case, the variation of via capacitance with bias is used as the measure for performance. Here, the total capacitance is considered which includes the oxide and depletion capacitance. Since, a lower capacitance leads to lower power requirement; this is a good measure for die stacks where the substrate is biased.

4.1.1 *Insertion Loss Variation for a TSV Pair*

Consider a pair of TSVs as shown in Figure 3.2(c) where the pitch D, length L, oxide thickness d_{ox}, radius R and silicon substrate conductivity are all process variables. The substrate is unbiased, as would be used in an interposer. In all of the computations in this sub-section, copper is used as via metallization with conductivity of 5.8×10^7 (S/m) and SiO_2 with a relative permittivity of 3.9 is used as the oxide liner material. Though tungsten is also being used as the via metallization due to its better coefficient of thermal expansion (CTE) match with silicon, it has not been considered here since its lower conductivity as compared to copper does not have a large effect on the insertion loss. The insertion loss is referenced to 50ohms.

The effect of varying the TSV parameters on the insertion loss is shown in Figure 4.1. In all of these results, the nominal is assumed to be $R = 10\mu m$, $D = 50\mu m$, $L = 100\mu m$ and $d_{ox} = 100nm$. So, all results are compared with the insertion loss corresponding to these dimensions. In Figure 4.1(a), the length is changed to $50\mu m$ and $150\mu m$ with all other parameters remaining the same. As can be seen, a shorter length decreases the insertion loss, which is a step in the right direction. Hence, thinner interposers are always desirable since a TSV length of $50\mu m$ can reduce the insertion loss by 0.2dB as compared to a $100\mu m$ length. In Figure 4.1(b), the oxide liner thickness has been changed to 50nm and 150nm. One would expect that a thicker oxide liner would reduce leakage and therefore significantly improve insertion loss, but this is true only in the transition region due to the Maxwell-Wagner effect described earlier in Chapter 3. In the low and high frequency range the effect of changing the oxide liner thickness is minimal and is not very advantageous in reducing the insertion loss. However, the oxide liner parameter is very important for managing cross talk between TSVs,

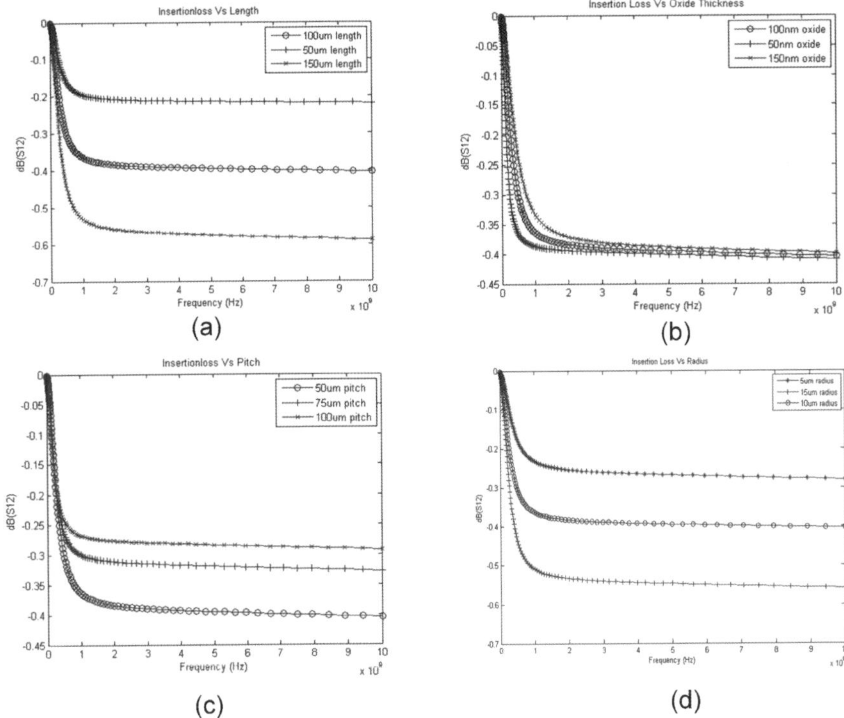

Figure 4.1: Insertion loss variation with process parameters (a) R = 10um, D = 50um, dox = 100nm, L = 50um, 100um, 150um; (b) R = 10um, D = 50um, L = 100um, dox = 50nm,100nm,150nm; (c) R = 50um, L = 100um, dox = 100nm, D = 50um, 75um, 100um; (d) L = 100um, D = 50um, dox = 100nm, R = 5um, 10um, 15um.

which is described later in this chapter. The pitch has been changed to 50μm and 75μm in Figure 4.1(c) with all other parameters remaining the same. A tighter pitch changes the impedance of the vias thereby changing its insertion loss due to poor matching. A tighter pitch for the TSVs is always desired since it enables easier routing and therefore the increased insertion loss should not be used as a factor for increasing pitch. Finally, the effect of varying radius is shown in Figure 4.1(d) for a radius of 5μm and 15μm with a constant pitch of 50μm. An increased radius results in an increased insertion loss due to its matching to 50ohms.

4.1.2 *Capacitance Variation of TSV*

The effect of different TSV physical and material parameters on its capacitance are discussed in this section based on the MOS capacitance analysis described in Chapter 3. The different TSV parameters addressed here are (i) TSV diameter, (ii) TSV oxide liner thickness, (iii) TSV oxide liner material, (iv) Silicon substrate resistivity, and (v) TSV metalization. The results shown are for the structure in Figure 3.29. The TSV capacitance-voltage curves for high frequency operation are plotted for these comparisons. In this section, 'Bias Voltage' refers to the difference in voltage between the TSV and the silicon substrate [Band, 2011]. The default parameters are copper metallization for the vias, silicon substrate of resistivity 10Ω-cm (conductivity of 10S/m), diameter of 15μm, SiO_2 oxide and oxide liner thickness of 0.1μm. In all of the examples, silicon is a p-type substrate biased to ground potential. Therefore the bias voltage is the voltage difference between the via and the silicon substrate.

The effect of varying the TSV diameter on its per unit length capacitance is shown in Figure 4.2(a) as a function of the bias voltage difference between the silicon substrate and TSV. This parameter corresponds to "$2r_0$" in Figure 3.29. From the figure, the capacitance decreases with decrease in the TSV diameter. Further, the difference between the TSV capacitance in the accumulation and inversion regions decrease with decrease in the TSV diameter. The TSV capacitance is a series combination of the oxide liner capacitance and the depletion capacitance, as described in Chapter 3. In the accumulation region, the TSV capacitance is very close to the oxide capacitance value. The capacitance in the depletion and inversion region is influenced by the depletion capacitance. As compared to the depletion capacitance, the liner capacitance reduces faster with reduction in the TSV diameter because the liner thickness is considered constant in this study. This results in smaller difference in TSV capacitance between accumulation and inversion regions with decrease in TSV diameter. In other words, a TSV with a smaller oxide liner thickness to diameter ratio leads to a larger difference between its capacitance in the accumulation and inversion regions.

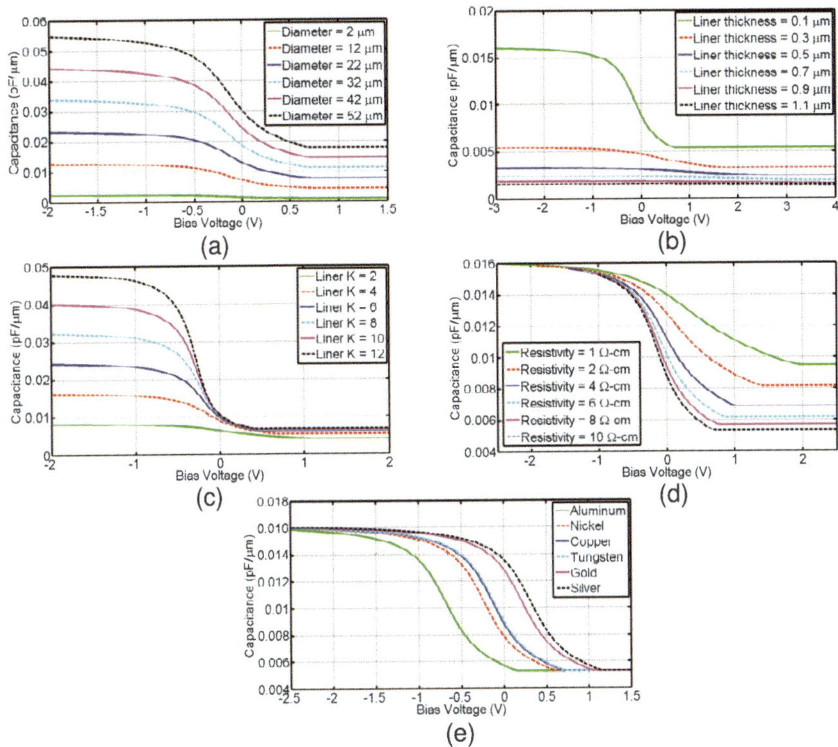

Figure 4.2: TSV capacitance variation with (a) Diameter, (b) Oxide liner thickness, (c) Liner dielectric constant, (d) Silicon resistivity and (e) Metal Courtesy: [Band, 2011].

The effect of varying the TSV oxide liner thickness ($r_1 - r_0$ in Figure 3.29) on its per unit length capacitance is shown in Figure 4.2(b). From the figure, a thicker oxide liner reduces the oxide capacitance. A thicker oxide liner also leads to a smaller electric field at the silicon - liner interface. This creates a smaller depletion region which leads to a higher depletion capacitance. Due to these effects, the difference between the TSV capacitance in the accumulation and inversion regions rapidly decrease with increase in the liner thickness. For TSVs with thick (>1μm) liner the MOS capacitance effect is negligible.

The effect of using different TSV oxide liner materials (shown as relative permittivity K) on its per unit length capacitance is shown in Figure 4.2(c). From the plot, the capacitance decreases with decrease in the dielectric constant of the liner material. Further, the difference between the capacitance in the accumulation and inversion regions decrease with decrease in the liner's dielectric constant. Decreasing the dielectric constant of the liner material decreases the liner capacitance but it has a negligible effect on the electric field in the silicon - liner interface. Thus, it has a negligible effect on the depletion capacitance. Due to these reasons the TSV capacitance in the inversion region is almost unaffected by the change in the dielectric constant of the liner although it has a large effect on the capacitance in the accumulation region.

The effect of silicon substrate resistivity on the per unit length capacitance is shown in Figure 4.2(d). From the plot, the TSV capacitance in the depletion and inversion regions decreases with increase in the silicon resistivity. The capacitance in the accumulation region remains unaffected by the change in the silicon resistivity. The liner capacitance depends only on the liner dielectric constant, the liner thickness, and the TSV diameter. It is unaffected by any change in the silicon resistivity. Hence, the capacitance in accumulation region is unaffected by change in the silicon resistivity. An increase in silicon resistivity implies a lower doping level in the silicon substrate. This leads to the formation of a wider depletion region around the TSV for the same voltage difference between the TSV and the silicon substrate. This in turn, reduces the depletion capacitance.

Finally, the effect of using different metals (for filling the TSV) on the per unit length capacitance is shown in Figure 4.2(e). The work function values of the different metals used in this parametric study is shown in Table 4.1. It is observed from the table that the TSV capacitance-voltage curve shape remains unchanged when the filler metal is changed. Changing the TSV filler metal offsets the Capacitance-Voltage (C-V) curve. Lower work function metals move the curve to the left whereas higher work function metals move the curve to the right. The oxide liner and depletion capacitances are independent of the work function of the TSV metal. However, the flatband and threshold voltages are directly related with the TSV metal work function. Hence changing the filler metal offsets the C-V curve without changing the capacitance values in accumulation or inversion regions.

Table 4.1: Work function of metals.

Metal	Work Function (eV)
Aluminum	4.1
Nickel	4.55
Copper	4.65
Tungsten	4.67
Gold	5
Silver	5.1

4.1.3 *Insertion Loss of a TSV Pair with Substrate Bias*

In certain cases, process optimization requires the computation of insertion loss of TSVs in the presence of a substrate bias voltage. Though this is possible by using analytical models as described in Chapter 3 with results in Figure 3.33, a more exact method is to model the insertion loss using an electromagnetic solver. The depletion region is mostly devoid of mobile charge carriers and therefore can be represented using silicon with zero loss, or as high resistivity silicon. Such a representation is possible in any electromagnetic solver. With the specialized basis functions described in Chapter 3, the depletion region can be modeled by modifying the capacitance as described in Chapter 3. Of course, to estimate the thickness of the depletion region (radius $r_2 - r_1$ in Figure 3.29), either the full depletion approximation or numerical analysis described in Chapter 3 can be used.

As an example consider a pair of TSVs with radius $R = 15\mu m$, pitch $D = 60\mu m$ and length $L = 100\mu m$. The oxide liner has thickness of $0.1\mu m$ with SiO_2 material surrounded by silicon with conductivity of $10S/m$. This structure was modeled in CST Microwave Studio [CST], a commercially available full wave solver, to compute the insertion loss for two cases, 1) with the silicon substrate floating and 2) with the silicon substrate biased creating a depletion region. The depletion region was modeled with silicon as a high resistivity substrate material. The results are shown in Figure 4.3. Similar to the analytical results in Figure 3.33,

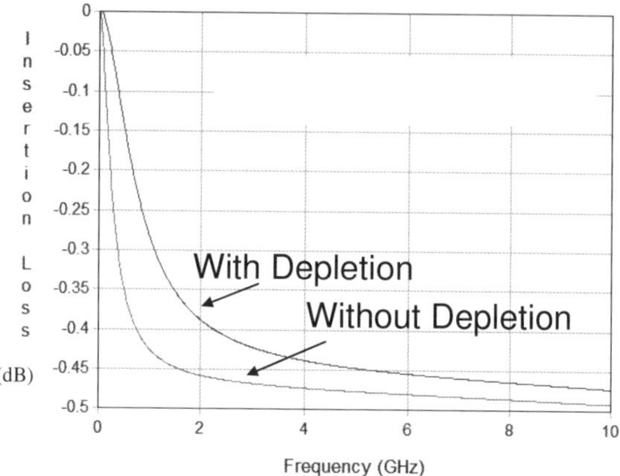

Figure 4.3: Insertion loss for a TSV pair with and without depletion region [Band, 2011].

the insertion loss with depletion improves due to the presence of the high resistivity silicon ring between the oxide liner and silicon substrate, thereby reducing the leakage into the silicon substrate.

4.2 Cross Talk in Interposers

So far, we focused on a TSV pair for extracting its insertion loss. When the silicon substrate is not grounded, the silicon substrate can increase the coupling between TSVs as compared to vias in an insulator, leading to cross talk being a major problem. The signature of the cross talk waveform in TSVs for standard CMOS grade silicon (conductivity of 10S/m) is quite different as compared to high resistivity silicon substrate with conductivity of 0.01S/m or less. This topic is addressed in detail here starting with a coupled TSV pair and then extending it to TSV arrays.

4.2.1 *Cross Talk Measurements and Correlation to Model*

In this section, the coupling between TSV pairs is compared to measurements in the time domain showing the importance of the TSV effect as compared to vias in the inter layer dielectric (ILD) layers. ILD layers are oxide layers that separate the silicon substrate from pads and

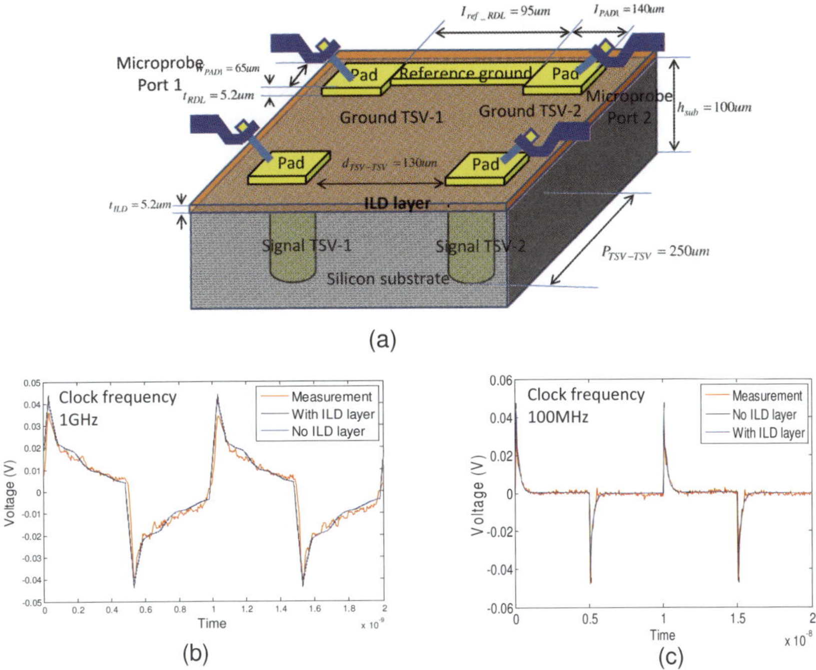

Figure 4.4: Cross talk measurements (a) Probe set-up and dimensions, (b) Cross talk waveform at 1GHz clock frequency and (c) Cross talk waveform at 100MHz clock frequency.

interconnections used for routing. The structure of the TSV pair is shown in Figure 4.4(a) consisting of two TSVs (TSV-1 and TSV-2) with their adjacent ground vias (ground TSV-1 and TSV-2) [Cho et al., 2011]. The two ground vias are tied together using a ground strap. The cross section consists of the silicon substrate containing the TSV with the ILD on top. The dimensions of the structure are shown in Figure 4.4(a). The TSV has a diameter of 33µm and oxide thickness of 0.52µm. Since this is a two layer structure (silicon substrate and ILD), each layer was modeled separately. The four TSVs were modeled using the cylindrical basis functions described in Chapter 3 while the vias, pads and ground straps in the ILD layer were modeled using the methods described in Chapter 2. All of the coupling between structures in a layer was included in the modeling. For each layer the S-parameters were computed and converted to a spice netlist using Idem [Idem, 2009]. Idem provides a method whereby S-parameters in a touchstone file can be converted into a spice netlist by satisfying passivity and causality [Swaminathan et al., 2007]. The corresponding ports were tied together in Spice to enable continuity of voltages and currents. As described in [Cho et al., 2011], a TDR source was used to excite the structure at Port 1 and the coupled waveform was measured at Port 2. A 50ps TDR pulse source with 2V amplitude was used with 50Ω source and load termination, which included the loss due to the cables. The resulting coupled waveforms (measured and modeled) for pulse periods of 1ns (1GHz) and 10ns (100MHz) are shown in Figures 4.4(b) and (c), respectively. The model agrees well with measurements. The modeling results with and without the ILD layer is shown in the figure showing little difference between the two, indicating that the TSV coupling is the dominant mechanism. Hence, most of the coupling occurs between the TSVs with a small coupling effect being generated by the vias in the ILD layers. In Figures 4.4(b) and (c), the cross talk waveform consists of an initial peak followed by a slowly decaying waveform. The slowly decaying waveform is due to the RC effect coming from the silicon substrate, which would be absent in vias surrounded by good insulating dielectric material. In addition, the finite silicon conductivity increases the peak cross talk voltage as compared to a good insulating material.

To better understand cross talk behavior, the coupling between TSVs has been analyzed in detail in the following section both in frequency and time domain.

4.2.2 *Coupling between TSVs*

A pair of TSVs is shown in Figure 4.5(a) along with its partial element equivalent circuit (PEEC) model in Figure 4.5(b). A PEEC model can be viewed as a circuit representation of the structure without assigning any current return paths. PEEC models can be derived directly from Maxwell's equations as has been done here using the rigorous electromagnetic modeling method described in Chapter 3 using specialized basis functions and implemented in 3DPF [3DPF, 2012]. The equivalent circuit in Figure 4.5(b) provides a direct correlation with the structure in Figure 4.5(a) where frequency dependent partial resistances

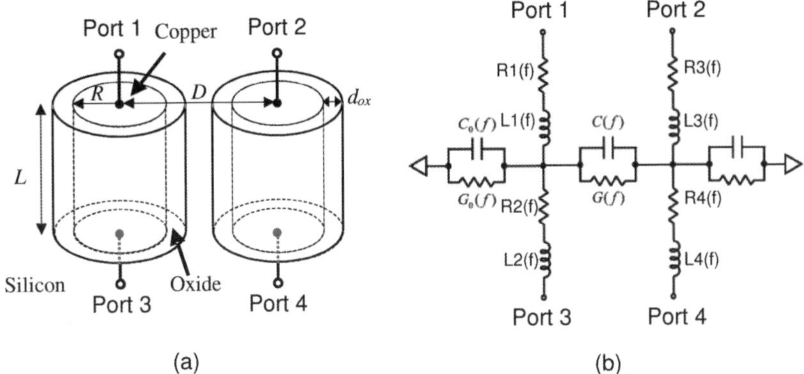

Figure 4.5: TSV Pair (a) Structure with port definitions and (b) PEEC model.

R1, R2, R3, R4 and partial self inductances L1, L2, L3, L4 are assigned to each TSV as shown. In addition, partial mutual resistances and inductances are also assigned (not shown). The capacitance and conductance shown in parallel are of two types namely, the capacitance ($C_0(f)$) and conductance ($G_0(f)$) to ground (at infinity) and coupling capacitance ($C(f)$) and conductance ($G(f)$). Though PEEC models have little validity without specifying the current return path, it is a useful model to use in this section to better understand the coupling between TSVs. The PEEC model can be converted to an S-parameter representation, as has been done in this section.

Consider a TSV pair with radius $R = 10\mu m$, pitch $D = 50\mu m$ and length $L = 100\mu m$ in a silicon substrate with conductivity of 10S/m. The oxide liner d_{ox} is of thickness 100nm, 500nm and 1000nm. In addition consider the same TSV pair with oxide liner thickness of 100nm in a high resistivity silicon substrate of conductivity 0.01S/m. The TSVs use copper as the conductor with oxide liner made of SiO_2. To generate the PEEC model, the ports are as defined in Figure 4.5(a) where the coupling between Port 1 and 2 is the near end cross talk (NEXT) and between Port 1 and Port 4 is the far end cross talk (FEXT).

The computed S-parameters (referenced to 50ohms) for NEXT and FEXT coupling are shown in Figure 4.6. The results are quite interesting. With an increase in the oxide thickness for a silicon conductivity of 10S/m, both NEXT and FEXT decrease, but only in the transition region.

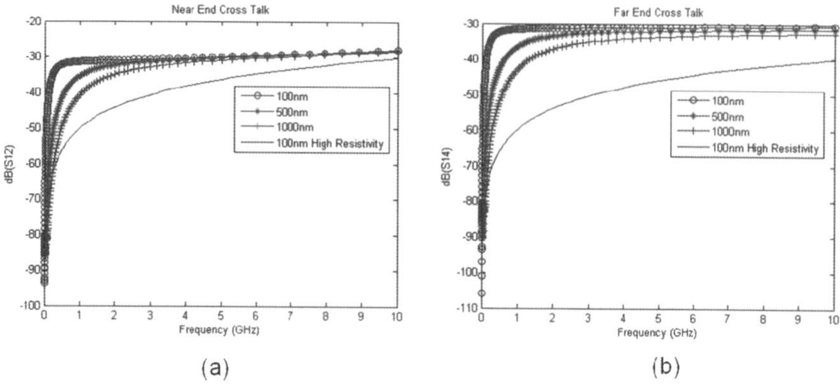

Figure 4.6: Coupling between TSVs (a) NEXT and (b) FEXT.

At low and high frequencies the dB levels are quite similar indicating that the benefit in using a thicker oxide liner occurs over a narrow band of frequencies. With the use of 0.01S/m high resistivity silicon substrate, even with an oxide thickness of 100nm, a significant improvement in NEXT and FEXT can be seen over a broad frequency band. This is because the high resistivity silicon substrate (0.01S/m) acts as a low loss dielectric medium, with the primary mode of coupling being capacitive, as opposed to the low resistivity silicon substrate (10S/m) where both conductance and capacitance contribute to the coupling.

Let's now look at the cause for the effect in Figure 4.6 more carefully by analyzing the behavior of the circuit components in Figure 4.5(b) as a function of frequency [3DPF, 2012]. The self resistance, self inductance, conductance $G_0(f)$ and capacitance $C_0(f)$ are shown in Figure 4.7. Since resistances R1, R2, R3, R4 are the same and not affected by the oxide thickness or silicon conductivity, only a single curve is shown. Same is true with the inductances L1, L2, L3 and L4. The frequency dependence for resistance and inductance show the typical skin effect behavior where the resistance increases with frequency while the inductance decreases with frequency. The resistance increases from 2.5mohms at 1MHz to 22.5mohms at 10GHz which is substantial. In contrast, the inductance decreases from 17.25pH at 1MHz to 15.5pH, which is not as drastic. The important behavior to note is the self conductance and self capacitance to ground. As the oxide thickness is increased, the

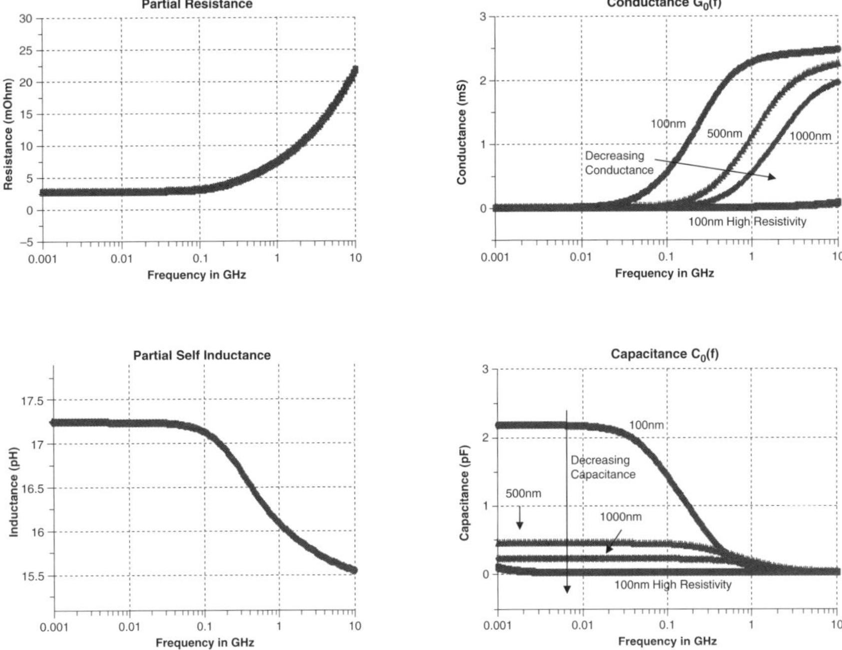

Figure 4.7: Self resistance, inductance, conductance and capacitance variation with frequency from 0.001 to 10GHz.

conductance decreases where between 100nm and 1000nm thick oxide (silicon conductivity of 10S/m) the difference is around 2mS at 500MHz, and 0.5mS at 10GHz. Therefore the difference in leakage is larger in the transition region. The conductance for high resistivity silicon substrate is almost non-existent. In contrast, the difference in self capacitance between 100nm and 1000nm thick oxide is largest in the low frequency region (around 2pF), gradually decreases in the transition region and has no difference in the high frequency region for the low resistivity silicon substrate. The high resistivity silicon substrate has the least capacitance in the low frequency region and has a value similar to the low resistivity silicon at high frequencies.

Let's next look at the coupling parameters, namely mutual resistance, inductance, conductance G(f) and capacitance C(f). The mutual resistance is the voltage measured on one conductor when the other conductor has current flowing through it. It often times occurs due to

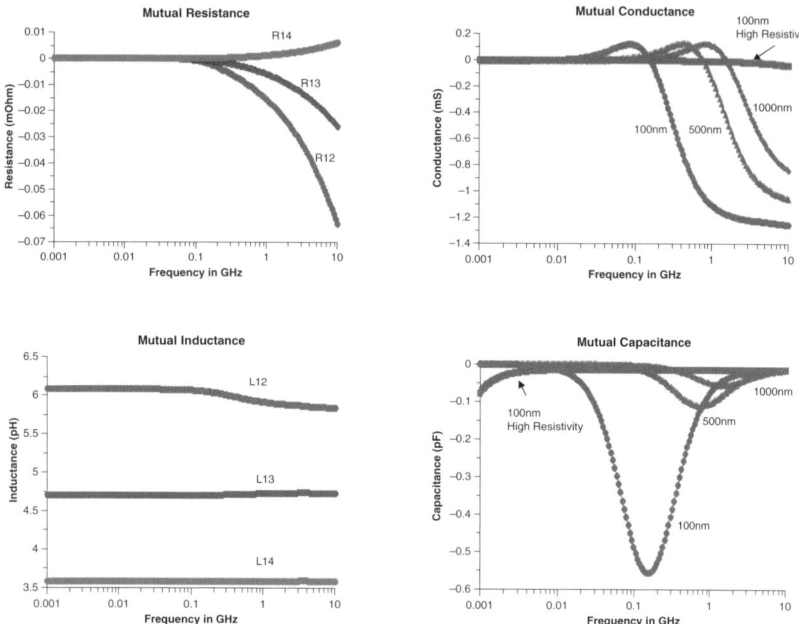

Figure 4.8: Mutual resistance, inductance, conductance and capacitance.

current crowding when the current in one conductor crowds near an adjacent conductor. In Figure 4.8, the mutual resistance is negligible and can be ignored. The mutual partial inductance occurs between all the self inductances in the circuit shown in Figure 4.5(b). Due to symmetry, only three of the mutual inductances are shown in Figure 4.8. The frequency dependence is minimal and varies between 3.5pH to 6pH. In Figure 4.8, the mutual conductance is negligible at low frequencies irrespective of the oxide thickness but increases significantly as the oxide thickness reduces in the transition region. From the figure, the mutual conductance is negligible for high resistivity silicon substrate. The mutual capacitance in Figure 4.8 shows a very interesting behavior for low resistivity silicon substrate (10S/m). The mutual capacitance is small at low frequencies (tends towards zero at DC due to slow wave effect), increases to a maximum at roughly the center frequency of the transition region and decreases to a small value at high frequencies. With an increase in oxide thickness, the peak value of capacitance decreases and the transition frequency range shifts to a higher frequency. The mutual capacitance for the high resistivity substrate does not exhibit this 'V' shaped behavior and has a negligible value. Comparing the mutual conductance and capacitance curves in Figure 4.8, it is apparent that the mode of coupling is capacitive at lower frequencies (in the transition region) and resistive at higher frequencies, an effect that is important for understanding the temperature effect discussed later.

The frequency dependent circuit parameters lead to the S-parameters shown in Figure 4.6. As indicated earlier, an increase in oxide thickness has a larger impact on reducing coupling than insertion loss. This effect however, only is valid in the transition region.

4.2.3 *Effect of Temperature*

One of the questions to ask is the following: "Does temperature affect the electrical response of TSVs?". The reason this is so important is because with stacking of ICs, heat will not be able to readily escape causing hot spots in the system. We will provide the theory and results for a simple structure as in Figure 4.4(a) in this section with more detailed results in a later section.

As described earlier, through silicon vias consist of copper or tungsten metallization with an oxide layer surrounded by a silicon substrate. The dependence of the copper or tungsten resistivity with temperature is given by [Xie et al., 2011], [Zhao et al., 2011]:

$$\rho_m(T) = \rho_0\left[1 + \alpha(T - T_0)\right] \quad (\Omega - m) \tag{4.1}$$

where, ρ_0 is the electrical resistivity at temperature T_0, and α is the temperature coefficient of electrical resistance. On the other hand the conductivity of silicon is affected by the doping density and temperature. The temperature-dependent silicon conductivity can be described by [Yin et al., 2007]:

$$\sigma_{Si}(T) = 1.602 \times 10^{-17} N_a \mu_p(T) \quad (S/m) \tag{4.2}$$

where the parameter N_a represents the concentration of substrate dopant impurity, and $\mu_p(T)$ represents the temperature-dependent carrier mobility. In addition, the variation of the permittivity of silicon with temperature can be written as:

$$\varepsilon_{Si}(T) = \varepsilon_0 \varepsilon_{Si} (1 - j \tan \delta - j \frac{\sigma_{Si}(T)}{\omega \varepsilon_0 \varepsilon_{Si}}) \quad (F/m) \quad\quad (4.3)$$

where ε_{Si} is the real part of relative permittivity at room temperature, $\tan\delta$ is the loss tangent and ω is the angular frequency in radians/sec.

A plot of the variation of copper (inverse of (4.1)) and silicon conductivity with temperature is shown from 20 to 120 degrees centigrade. From the figure, both conductivities decrease with temperature causing an increase in the resistance of both copper and silicon. As shown in this section for TSVs, the effect of resistance increase for silicon far out weighs the effect of increased copper resistance at high frequencies, which causes the overall electrical performance to improve with increase in temperature. This is because silicon becomes a better insulator with increase in temperature.

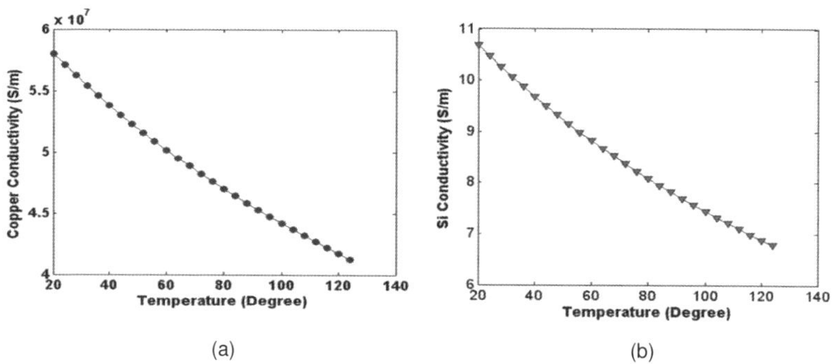

Figure 4.9: Variation of conductivity with temperature (a) copper and (b) silicon.

4.2.3.1 *Modeling Temperature Effects*

The electric field integral equation (EFIE) and scalar potential integral equation (SPIE) used in Chapter 3 to model TSVs can be suitably modified to include temperature effects. These equations are provided here with temperature dependence included. The substitution of the conduction mode basis functions (CMBF), accumulation mode basis functions (AMBF) and polarization mode basis functions (PMBF) for the current and charge are identical as in Chapter 3. The temperature dependent equations can be written as:

$$\frac{\overrightarrow{J^C}(r,w)}{\sigma(T)} + j\frac{\omega u}{4\pi}\int_{V'} G(\vec{r},\vec{r}')\overrightarrow{J^C}(\vec{r}',\omega)dV' = -\nabla\Phi(\vec{r},\omega) \qquad (4.4)$$

$$\frac{1}{4\pi\varepsilon_{Si}(T)}\int_{V'} G(\vec{r},\vec{r}')q(\vec{r}',\omega)dV' = \Phi(\vec{r},\omega) \qquad (4.5)$$

$$\frac{\overrightarrow{J^P}(r,\omega)}{j\omega\varepsilon_0(\varepsilon_{SiO_2} - \varepsilon_{si}(T))} + j\omega\frac{u}{4\pi}\int_{V'} G(\vec{r},\vec{r}')\overrightarrow{J^P}(\vec{r}',\omega)dV' = -\nabla\Phi(\vec{r},\omega) \qquad (4.6)$$

where the definition of the parameters are identical to the definitions in Chapter 3. The system of equations to be solved reduces to an equation similar in form to (3.41) which can be written as [Xie et al., 2012]:

$$\left[Y_C(T) + j\omega C^{eq}(T)\right]\Phi^C = \begin{bmatrix} I_t \\ 0 \end{bmatrix} \qquad (4.7)$$

where the definition of the parameters are the same as in (3.41) except for the temperature dependence.

As described in Chapter 3 and this chapter, the coupled system of equations can be represented using equivalent resistance (R), inductance (L), conductance (G) and capacitance (C) parameters. So, which parameters does the temperature variation affect? This is illustrated in Figure 4.10 by constructing the network for a 3×3 TSV array showing R, G and C parameters (L not shown). As shown in the figure the temperature dependence affects the resistance R and conductance G in the network where the temperature at each point in the network is shown in parenthesis.

4.2.3.2 *Effect of Temperature on Cross Talk*

Here we consider a simple example by computing the cross talk waveforms for the structure in Figure 4.4(a) as a function of temperature. The S-parameter coupling between the two pads in Figure 4.4(a) is

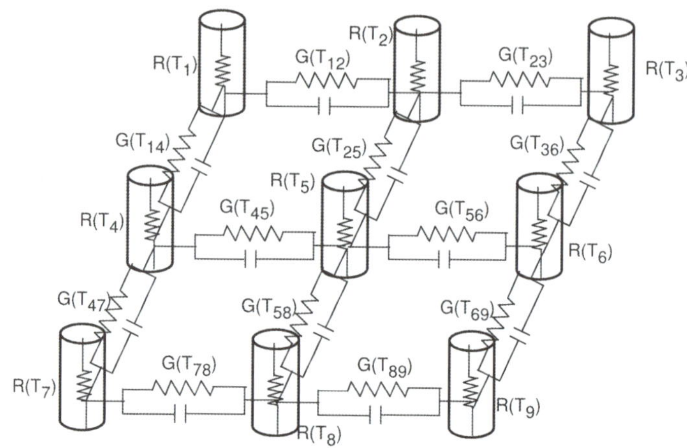

Figure 4.10: TSV array showing temperature dependence for resistance and conductance [Xie et al., 2012].

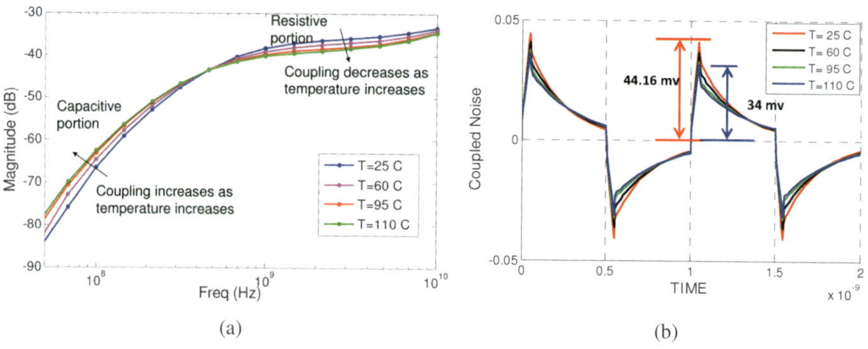

(a)   (b)

Figure 4.11: Cross talk variation with temperature (a) S-parameter and (b) time domain waveform.

shown in Figure 4.11(a) as a function of temperature. As can be seen the coupling can be divided into two parts, a frequency range below 0.5GHz where capacitance dominates and a frequency range beyond 0.5GHz where the resistance dominates. The coupling increases with temperature in the capacitive region while it decreases with temperature in the resistive region. Let's now propagate a 1GHz clock as in Figure 4.4(a) into one of the pads and compute the cross talk on the other pad, which is shown in Figure 4.11(b) as a function of temperature. As can be seen, the peak cross talk waveform decreases with an increase in temperature with a value of 44.16mV at 25 degrees C and 34mV at 110 deg C, resulting in a 23% decrease in cross talk. Of course this value would vary based on the frequency content of the pulse. However, as the frequency increases, one can expect the peak cross talk waveform to decrease. This is a very important effect for TSVs since the silicon becomes a better insulator at high frequencies as the temperature increases. The temperature effects described in this section have also been confirmed through measurements in [Lee et al., 2011], though on a different structure.

4.2.4 *Revisiting Crosstalk*

Based on the analysis described most of the coupling occurs between TSVs and not through the vias in the ILD layer. In addition the oxide thickness provides better isolation in the transition region. Let's next look at the cross talk between TSVs in the time domain based on the structure in Figure 4.12, which is similar to Figure 4.4 but with the ILD layer removed. The TSVs have a length $L = 100\mu m$, pitch $D = 50\mu m$ and radius $R = 10\mu m$. Two of the TSVs are signal (numbered 1 and 2) while the other two are ground (labeled G), providing the necessary reference. Four cases are considered namely, oxide thickness $d_{ox} = 100nm$, 500nm, 1000nm in a silicon substrate of conductivity 10S/m (low resistivity) and $d_{ox} = 100nm$ in a silicon substrate of conductivity 0.01S/m (high resistivity). The signal TSV 1 is excited using a pulse train of amplitude 1V, frequency of 1GHz and rise/fall time of 50ps through a 50ohm source resistance. In Figure 4.12(a), signal TSV 2 is terminated in 50ohms with the far end for both TSV 1 and 2 unterminated (Figure 4.12(a)) and terminated in 50 ohms (Figure 4.12(b)). The resulting cross talk waveforms coupled into the near end of TSV 2 is shown in Figure 4.12(c) (unterminated) and Figure 4.12(d) (terminated).

As expected, the cross talk waveform in Figure 4.12(d) is smaller as compared to Figure 4.12(c) due to termination, which eliminates reflections from the far end of the signal TSVs. In Figures 4.12(c) and (d), case 1, case 2, case 3 represent the waveforms for oxide thickness $d_{ox} = 100nm$, 500nm, 1000nm in a silicon substrate with conductivity

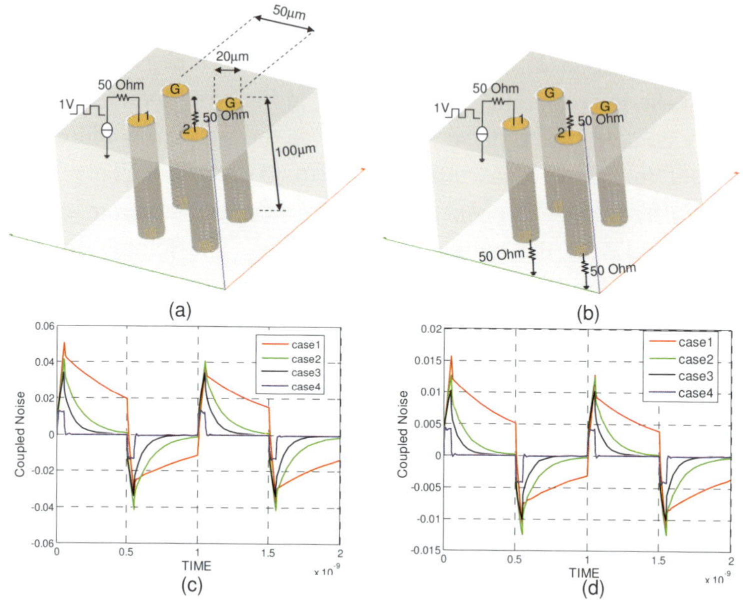

Figure 4.12: TSV structure showing Signal and Ground TSVs (a) unterminated, (b) terminated and resulting cross talk waveforms (c) unterminated, (d) terminated.

10S/m, respectively, while case 4 is for $d_{ox} = 100$nm with silicon conductivity of 0.01S/m. There are two very important effects that become apparent from the cross talk waveforms in Figure 4.12 namely, 1) the peak value of cross talk which decreases as the oxide thickness increases for silicon substrate of conductivity 10S/m, which is expected due to lower leakage as the oxide thickness increases. However, this reduction is only around 30% as seen from Figure 4.12 when the oxide thickness is increased from 100nm to 1000nm. This small decrease can be attributed to the smaller difference in the coupled S-parameters at high frequencies and 2) the slow decaying waveform after the peak cross talk which decays at a slower rate for $d_{ox} = 100$nm as compared to $d_{ox} = 1000$nm for silicon substrate conductivity of 10S/m. The slow decaying waveform can be attributed to the 'RC' effect arising from the conductance and capacitance between signal TSVs in the silicon substrate.

The reason for the difference in the rate of decay as a function of oxide thickness is due to the larger difference in the coupled S-parameters in the transition region. From Figures 4.12(c) and (d), in both cases 2 and 3 (500nm and 1000nm oxide thickness), the cross talk waveforms have reached the 'zero' level as compared to case 1 (100nm) where significant level of cross talk still remains on the signal TSV when the next transition occurs. This "long tail effect" can be detrimental to the performance of the system since the cross talk tail from the previous pulse can interfere with the present pulse creating inter symbol interference (ISI). Increasing the oxide thickness can have a large impact on the "long tail", leading to a faster decay in the 'RC' waveform, which is desired. In contrast to the 10S/m silicon conductivity, the 0.01S/m silicon conductivity even with a thin oxide of 100nm results in a large reduction in the peak cross talk and completely eliminates the "long tail effect", leading to a sharp cross talk pulse, as shown in Figures 4.12(c) and (d) which is easier to manage for signal integrity. Since the silicon substrate with conductivity of 0.01S/m behaves like a low loss dielectric, it leads to much lower coupled S-parameters both at low and high frequencies, leading to manageable cross talk levels.

Since, packaging engineers are more used to working with low loss dielectrics, they need to be aware that cross talk can be a major problem when silicon interposers are used. To illustrate the importance of cross talk in silicon interposers, consider two TSVs as in Figure 4.13(a) with $R = 10\mu m$, $L = 100\mu m$ and $d_{ox} = 100nm$ with varying pitches of $D = 50\mu m$, $100\mu m$, $150\mu m$ and $200\mu m$. The S-parameter coupling between ports 1 and 2 (S12) based on Figure 4.13(a) is plotted in Figure 4.13(b) as a function of the pitch for two frequencies namely, 1GHz and 10GHz.

As a comparison, the coupling S12 between vias of comparable dimensions used in organic packages (called through organic vias TOV) is shown in Figure 4.13(b). The organic vias are surrounded by a dielectric medium with relative permittivity of 3.9. From Figure 4.13(b), at 1GHz, TSVs have around 25dB higher coupling for a given pitch while at 10GHz, the coupling between TSVs is around 7dB higher, as compared to TOVs. This difference in coupling levels can be attributed to the TSVs being in the transition region at 1GHz as compared to the quasi-TEM region at 10GHz, for the structure shown.

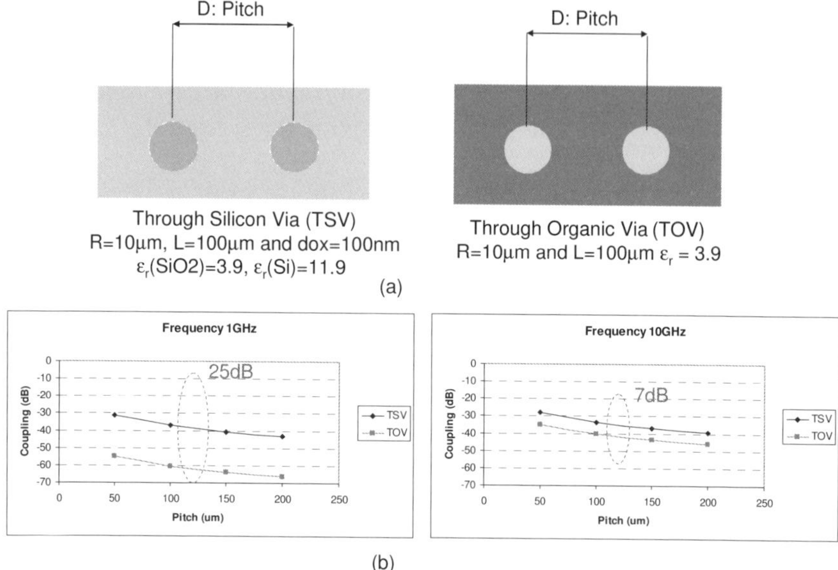

Figure 4.13: Coupling Vs Pitch for TSV and TOV (a) Structure and (b) Coupling at 1GHz and 10GHz [3DPF, 2012].

4.3 Via Arrays

From the previous section, cross talk can be large between adjacent TSVs, especially when the silicon substrate is unbiased as in an interposer. One would then expect coupling to be significant even for TSVs farther away and hence it becomes important to model TSV arrays as opposed to just a few TSVs. This coupling can induce variability of the signals propagating through the TSVs. This is discussed in this section by starting with via arrays, elaborating further on the "long tail problem" and then determining its effect on Pseudo-random bit streams. The coupling effect is then extended to a larger array and the importance of keep out zones (KOZ) and their effect on TSV performance is addressed. Finally the temperature effect on TSV arrays is discussed as well.

4.3.1 *5 × 5 TSV Array*

Let's start with a 5×5 array as shown in Figure 4.14(a) consisting of $R = 10\mu m$, $L = 200\mu$, $dox = 0.1\mu m$ and $D = 100\mu m$ in a slab of silicon with conductivity 10S/m and 0.01S/m. The oxide is SiO_2 with relative permittivity of $\varepsilon_r = 3.9$ and silicon has a relative permittivity of $\varepsilon_r = 11.9$. The array consists of ground vias interspersed between the signal vias (not shown). Three vias are shown numbered 1, 2 and 13. The near end S-parameter coupling between TSV1 and TSV2 is shown in Figure 4.14(b) over a 10GHz bandwidth, showing the difference in coupling between high conductivity (10S/m) and low conductivity (0.01S/m) silicon substrate, similar to what was covered in the previous section.

Consider a pulse with rise time of 100ps and amplitude 2V propagating through TSV1. The far end of TSV1 and both sides of TSV2 are terminated in 50Ω. The near end cross talk waveform on TSV2 is plotted in Figure 4.15(b). As can be seen, the 10S/m conductivity silicon substrate leads to 5X larger peak voltage as compared to the 0.01S/m conductivity silicon substrate. Moreover, the low resistivity substrate results in a waveform that has 8X longer coupled noise duration, which can be a significant problem. This difference can be explained by looking at the coupled S-parameters as shown in Figure 4.14(b) where the oxide thickness plays a large role in increasing coupling in the transition region of the TSV. A larger oxide thickness would therefore help in reducing cross talk for low resistivity substrates especially when the silicon substrate is not grounded. Let's next consider what happens when the rise time is increased to 1000ps, as shown in Figure 4.15(c). Here, the peak coupled noise increases by 12X for the 10S/m substrate as opposed to the 0.01S/m substrate. This increase in coupling as the rise time increases can be explained from Figure 4.14(b) where the difference in coupling between the two substrates is larger at lower frequencies. However, the duration of the pulse is still roughly 1ns, similar to the 100ps rise time example. Let's next consider the cross talk when a pulse of 10ps rise time is propagated through TSV1, as shown in Figure 4.15(a). Now the peak value of cross talk is only 1.4X larger for 10S/m substrate as opposed to 0.01S/m substrate though the duration of the waveform is still ~1ns. The reduced peak noise difference can be

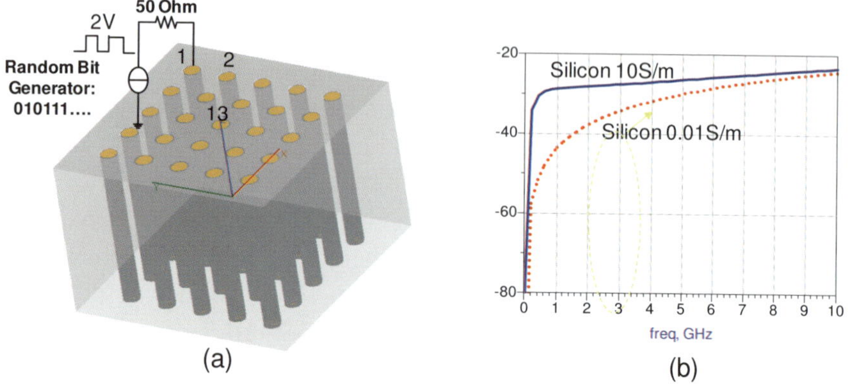

(a)   (b)

Figure 4.14: 5 × 5 TSV Array (a) Structure (b) Near end coupling between vias 1 and 2.

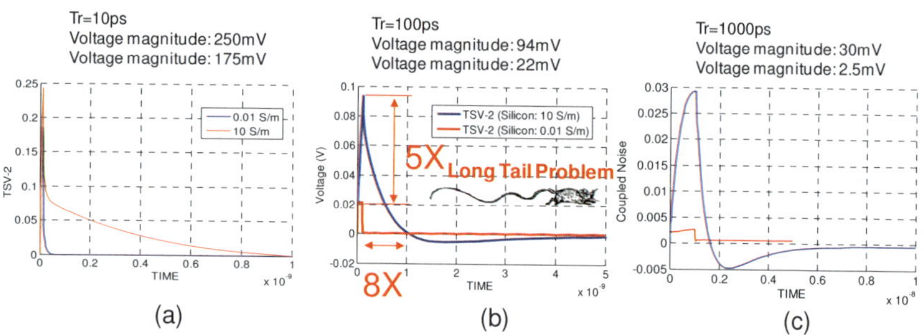

(a)   (b)   (c)

Figure 4.15: Cross talk waveforms (a) Rise time = 10ps, (b) Rise time = 100ps and (c) Rise time = 1000ps.

attributed to less coupling between the two substrates at high frequencies as illustrated in Figure 4.14(b). A very important result from this example is the duration of the cross talk pulse which is comparable irrespective of the rise time of the pulse due to the resistive behavior of the substrate when the silicon conductivity is 10S/m. *This is called as the 'long tail problem' (Figure 4.15(b)) in this book where a mouse with a long tail can easily get into trouble, similar to pulse deterioration due to inter symbol interference (ISI).*

Due to excessive cross talk, shielding techniques are therefore desired. A common method pursued in a package for shielding is to surround a noise prone via with ground vias, thereby creating a Faraday cage. This may not work well for silicon interposers as shown in Figure 4.16(a) where a quiet TSV (Q) at the center is surrounded by ground TSVs (G).

The dimensions of the array are identical to Figure 4.14(a). Consider a switching scenario as in Figure 4.16(a) where the remaining signal TSVs (S), are switching with a rise time of 100ps. All of the TSVs are terminated in 50Ω on both sides. The resulting near end coupled noise waveform on the quiet TSV (Q) is shown in Figure 4.16(b). A peak coupling of 120mV results once again with a long tail due to the 'RC' effect of the silicon substrate, in spite of the surrounding ground vias. This is an example which shows that coupling occurs through the silicon substrate which cannot be mitigated easily using a Faraday cage, and therefore cross talk needs to be evaluated carefully in a silicon interposer design.

As is well known, a digital system can contain two kinds of bit streams namely, 1) periodic stream such as a clock signal and 2) a random stream which is more representative of the signals propagating in a system. When an interconnect channel is tested for performance, often times a pseudo-random bit stream (PRBS) data is propagated through the interconnections, the eye diagram at the far end is recorded from which the eye-mask and jitter are evaluated.

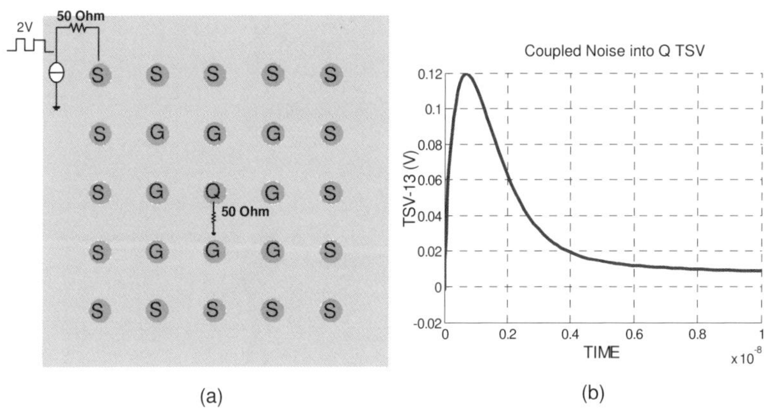

Figure 4.16: 5 × 5 TSV array showing (a) Shielding structure and (b) Cross talk on Q TSV.

A major issue for any system is variability, where the performance of the interconnections cannot be determined apriori since each interconnection behaves differently. With enhanced coupling between TSVs in a silicon interposer, variability can be an issue as illustrated through the following example. Consider the 5×5 TSV array in Figure 4.14(a) where each signal TSV is supporting a PRBS at 5Gbps. Due to the short length of the TSVs, the eye diagram obtained at the far end of each TSV should be identical.

The far end eye diagrams for TSV1 and TSV13 are shown in Figure 4.17. From the figure TSV1 has an eye height of 2.01V and jitter of 7.71ps while TSV13 has an eye height of 1.8V and jitter of 9.78ps, showing variability over a short diagonal separation pitch of 282μm. This provides another reason for the need for managing cross talk in silicon interposer designs [Xie et al., 2011a].

A silicon substrate can contain hundreds and thousands of TSVs depending on the size of the interposer and the pitch of the TSVs. With the substrate not grounded, a designer needs to understand the coupling between large arrays of TSVs to understand variability. Depending on

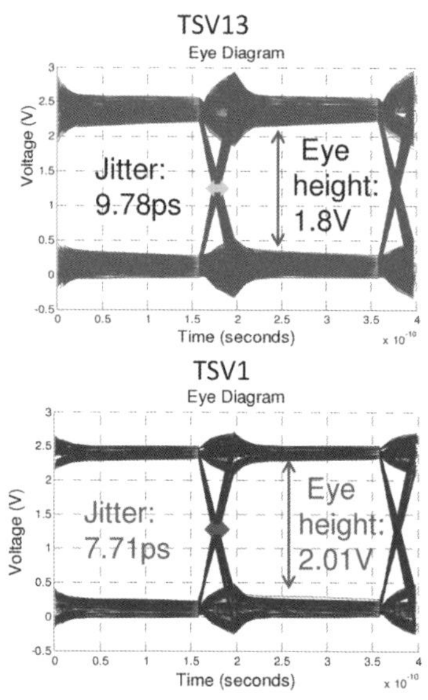

Figure 4.17: Eye height and Jitter at far end of TSV13 and TSV1.

the presence of keep out zones (KOZ), the TSV array may not be uniform, which can complicate the computation of the coupling between TSVs since a larger area may need to be meshed. In such scenarios, the methods described in Chapter 3 on the use of modal basis functions can be very useful and efficient since the silicon substrate is never meshed and few basis functions are required to represent the current and charge.

4.3.2 *Suppressing Cross Talk*

As described in this chapter, cross talk between TSVs can be a major problem. One way to suppress cross talk is by increasing the oxide thickness or by separating the vias. In this section, we provide two other ways to reduce cross talk. In Chapter 3, the effect of biasing on the insertion loss of TSVs was discussed. By creating a bias voltage between the copper vias and silicon substrate, a depletion layer is created that better shields the vias from the substrate. This can help reduce cross talk which has been discussed here (not considered in earlier cross talk simulations).

Consider the structure in Figure 4.18 consisting of three TSVs with a ground via (G) between two signal vias (S). The vias are of radius $r_0 = 10\mu m$ with oxide radius $r_1 = 10.1\mu m$, pitch 50μm and length 100μm. A 1.8V voltage swing is applied to one of the signal TSVs and the cross talk is measured on the quiet TSV. The silicon substrate has conductivity 10S/m, has a permittivity of 11.9, carrier concentration of $1.2 \times 10^{15}/cm^3$ with the oxide having a relative permittivity of 3.9. As shown in Figure 4.18, the p-type silicon substrate is grounded and therefore the voltage between the via and substrate causes a depletion region as shown, which can be modeled as an annulus with zero conductivity (devoid of charge carriers). In this example, the depletion region for each of the TSVs is different based on the voltage on the via, shown as Depletion regions 1, 2 and 3 in Figure 4.18. Assuming that the active signal TSV has a voltage transition of 0 to 1.8V, with 0.9V as the switching level, this corresponds to a value much larger than the threshold voltage calculated from the equations in Chapter 3 and hence the TSV is in the deep depletion region. Using the full depletion approximation the Depletion 1 layer thickness can be calculated as 0.69μm with $r_2 = 10.69\mu m$ for this TSV.

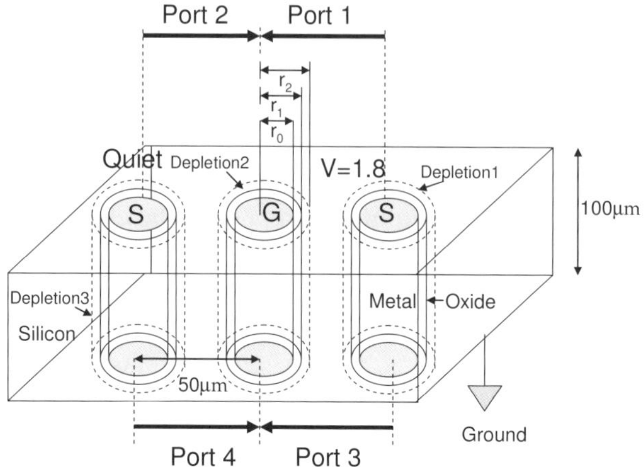

Figure 4.18: Signal-Ground-Signal TSV with silicon substrate grounded

However, for both the ground and quiet signal TSV in Figure 4.18, the via is close to zero volts and hence the thicknesses for Depletion 2 and 3 need to be recalculated based on the threshold voltage, which results in a value of 0.27μm with $r_2 = 10.27$μm for these two vias. Ports on the signal lines are referenced to the ground via and numbered as shown in Figure 4.18. The near end coupling between the two signal TSVs (S12 expressed in dB) is shown in Figure 4.19(a) referenced to 50 ohms. From the figure, biasing (GTSV_Ox_MOS) improves the near end coupling by ~8dB @ 100MHz but as the frequency increases beyond 1GHz, the coupling with and without (GTSV_Ox) biasing remains the same. This is because the impedance of the oxide and depletion capacitance decreases with increasing frequency and their isolation levels decrease as the frequency increases. As described in Chapter 3, biasing also improves the insertion loss by around 0.2dB at 300MHz (S13 in dB) in this example, as shown in Figure 4.19(b). Comparing Figures 4.19(a) and (b), the benefit in reducing cross talk is far greater than improving insertion loss, when the substrate is grounded, especially at lower frequencies.

The near end cross talk for a 1.8V step with 1ns and 0.1ns rise time are shown in Figures 4.20(a) and (b) respectively. For the 0.1ns rise time,

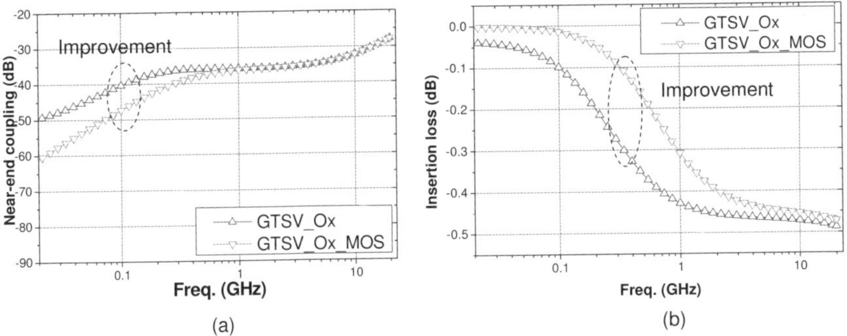

Figure 4.19: (a) Near End Coupling and (b) Insertion Loss.

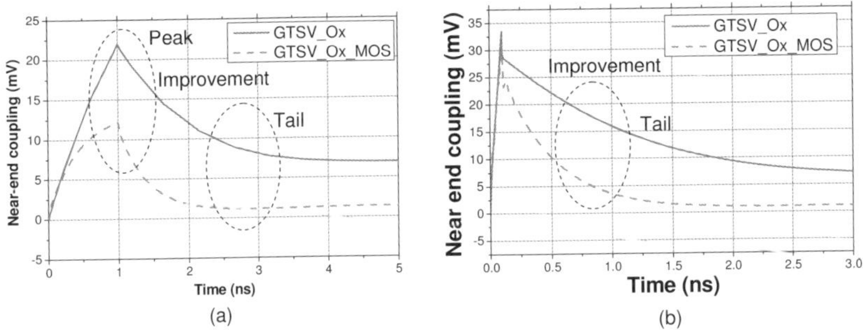

Figure 4.20: Near End Cross Talk (a) 1ns rise time and (b) 0.1ns rise time signal.

the maximum amplitude of cross talk with and without biasing are similar while with the 1ns rise time, the maximum amplitude reduces from 22mV to 12mV, a 45% reduction in cross talk. In both cases, the RC time constant is greatly reduced with biasing and therefore the duration of the long tail waveform described earlier is reduced as well. This can be a benefit to better manage interference between bits in a channel.

With ICs, ohmic contacts can be created by connection to high diffusion areas, thereby enabling the biasing of the silicon substrate. However, interposers are passive structures created through BEOL processes and therefore it is not easy to create these ohmic contacts. Hence the silicon interposer is floating.

In this section we consider a case where the side walls of the vias are heavily doped with the majority carriers in the substrate. For example, a p-substrate or n-substrate will have p^+ or n^+ surface layer along the via walls, allowing for an ohmic contact to be created at the metal-semiconductor junction. This is possible using diffusion [Streetman et al., 2000] or by other means [Bartley et al., 2013]. Assuming this is possible at a reasonable cost, then the ground vias can be selectively treated to create an ohmic contact to the silicon substrate, thereby grounding the silicon interposer. The signal vias will continue to have the oxide layer to minimize leakage. This structure is shown in Figure 4.21 and though the process steps increase, has benefits not only in reducing cross talk but also can help to minimize Electromagnetic Interference (EMI) since the interposer is grounded rather than floating. It is important to note that the metal-semiconductor junction is important for the ground vias since if this turns into a Schottky barrier junction, then the substrate is no longer grounded.

The results in this section is based on the rigorous analysis in [Yang et al., 2013] though the concept can also be found in [Engin et al., 2012]. Similar to the previous example, 1.8V is applied to the signal TSV creating a Depletion 1 and Depletion 3 thickness of 0.69µm and 0.27µm, respectively. Since the depletion region is very narrow in an ohmic contact, Depletion 2 thickness can be assumed to be 0µm. The resulting near end coupling (S12 in dB) and insertion loss (S13 in dB) are shown in Figure 4.22 for two cases namely 1) with (GTSV_noOx_MOS) and 2) without (GTSV_noOx) the depletion region respectively for the active and quiet TSVs, to demonstrate the importance of depletion region modeling. Compared to Figure 4.19(a), the near end coupling increases by 20dB @ 100MHz when the interposer is floating (GTSV_Ox) as compared to the interposer grounded (GTSV_noOx_MOS). However, the insertion loss worsens by 0.05dB @ 300MHz (GTSV_noOx_MOS) when compared to Figure 4.19(b) (GTSV_Ox). Beyond 1GHz, the floating and grounded interposer provides similar results for near end coupling. The resulting near end cross talk on the quiet signal via is shown in Figures 4.23(a) and (b) for rise time transition of 1ns and 0.1ns respectively on the active via, both when the Depletion regions 1 and 3

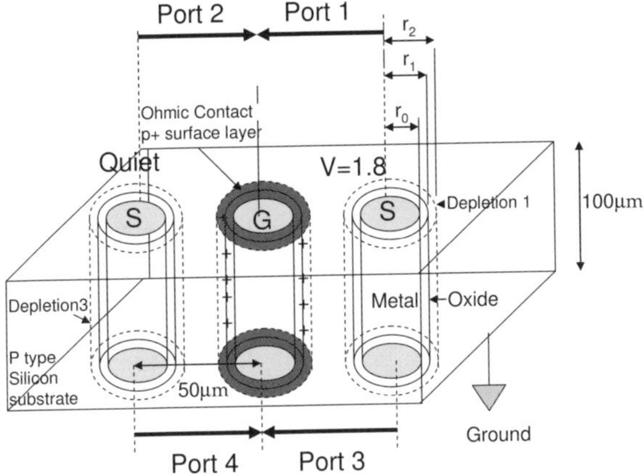

Figure 4.21: Signal-Ground-Signal TSV with substrate grounded using ohmic contact along the ground via walls.

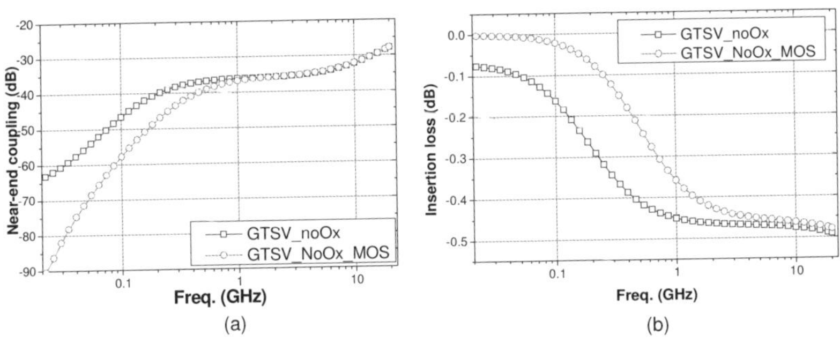

Figure 4.22: (a) Near end coupling and (b) Insertion Loss with ohmic contact for ground via.

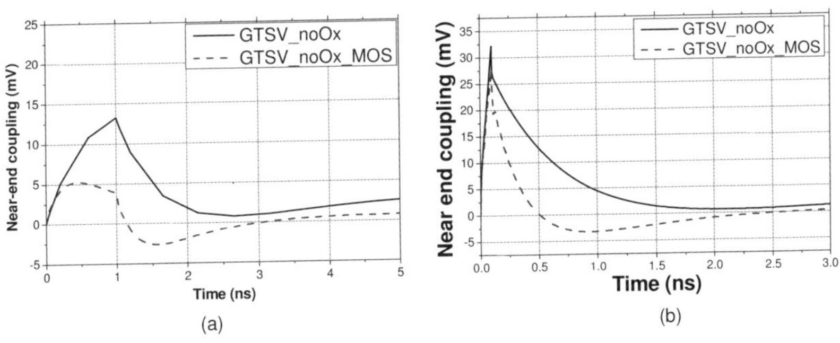

Figure 4.23: Near end coupling for (a) 1ns rise time and (b) 0.1ns rise time signal with ohmic contact on the ground TSV.

are present (GTSV_noOx_MOS) and absent (GTSV_noOx). Compared to the floating interposer (GTSV_Ox), the near end peak coupling in Figure 4.23(a) reduces from 22mV to 5mV, corresponding to a 77% reduction. For a 0.1ns rise time signal, the advantage of using the grounded versus the floating interposer is minimal for peak near end coupling. However, for both the 1ns and 0.1ns rise time signals, the RC time constant effect is greatly reduced, which can be very advantageous.

From this section two important items emerge for cross talk estimation and mitigation, namely 1) if the silicon substrate is grounded, the importance of depletion region modeling becomes important since without it cross talk will be over estimated and 2) shielding the interposer provides advantages for cross talk management.

4.3.3 *20×20 TSV Array*

The electromagnetic method using modal basis functions as described in Chapter 3 has been used in this section to compute the S-parameters for a 20×20 uniform TSV array with dimensions of $R = 10\mu m$, $L = 100\mu m$, $D = 50\mu m$ and $d_{ox} = 0.1\mu m$. The conductor used is copper with conductivity $\sigma_{Cu} = 5.8 \times 10^7 S/m$ and silicon conductivity of 10S/m. The oxide used is SiO_2 with relative permittivity of 3.9 and silicon substrate of relative permittivity 11.9. The insertion loss of the 400 TSVs

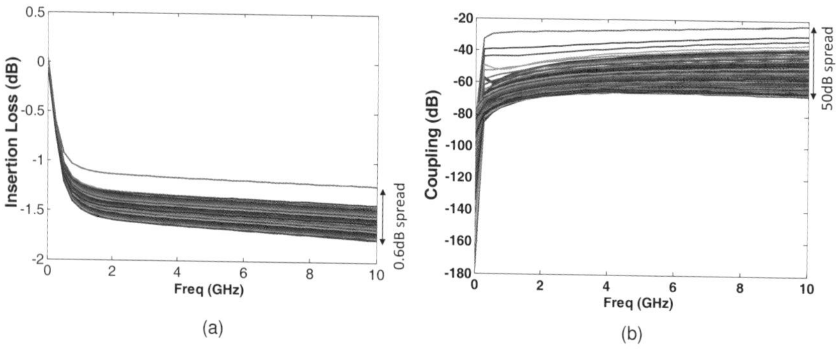

Figure 4.24: 20×20 TSV array (a) Insertion loss Vs frequency over 10GHz and (b) Near end coupling between TSV1 and other TSVs Vs frequency over 10GHz.

referenced to 50Ω and ground reference at infinity is shown in Figure 4.24(a). From the figure the insertion loss has a 0.6dB spread among the 400 TSVs with a range in insertion loss between 1.2dB to 1.8dB at 10GHz, which represents 30%–50% variability across the silicon substrate. The near end coupling between TSV1 (similar in position as in Figure 4.14(a)) and the remaining TSVs is shown in Figure 4.24(b) with a coupling range of 20dB–70dB. To capture the variability requires that arrays of TSVs be modeled as a group.

4.3.4 *TSV Array and Keep Out Zone*

One of the major problems in designing silicon substrates with through silicon vias are the mechanical stresses involved. A qualitative explanation providing the reason for the mechanical stresses and their affect on the electrical response of the TSVs is briefly described in this section [Jayantha et al., 2008].

Consider a TSV surrounded by silicon substrate, as shown in Figure 4.25(a). The TSV is made of copper conductor which has a coefficient of thermal expansion (CTE) of 17ppm/C at a temperature of 20C. On the other hand, silicon has a CTE which is much lower at 3ppm/C at the same temperature. When a chip is powered up using TSVs through a silicon substrate, due to the flow of heat, temperature in the system increases and correspondingly decreases when the chip is powered down, causing varying amounts of expansion and contraction between the

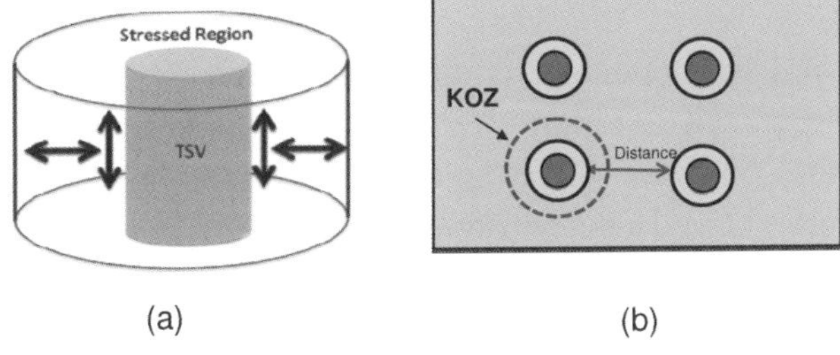

(a)   (b)

Figure 4.25: (a) Mechanical Stresses for a TSV (Courtesy: G. Subbarayan, Purdue Univ.) and (b) Defining keep out zones (KOZ).

copper via and silicon substrate. This results in excessive mechanical stresses around the TSVs. This can create cracks near the TSV whose probability of occurrence is higher if the metal loading around the TSV increases. This is one reason why some TSV processes prefer the use of tungsten with CTE of 4.5ppm/C, which is better matched to silicon, though the resistivity of tungsten is larger than copper. Due to the CTE mismatch, the mechanical stresses are the largest at the TSV-silicon interface and decreases as the distance from the TSV increases. The presence of the oxide reduced the stresses. Hence, a physical design rule that is used to manage mechanical stresses is to ensure that no metal such as from wiring or vias exist in the vicinity of the rogue TSV by defining keep out zones (KOZ), as shown in Figure 4.25(b). The use of KOZs can make the array non-uniform, resulting in a change in the electrical parasitics of the TSVs.

As an example, consider a silicon substrate of size $200\mu m \times 200\mu m$ containing eight TSVs numbered from 1 to 8, with non-uniform pitch, as shown in Figure 4.26. The TSVs have physical dimensions $R = 10\mu m$, $L = 100\mu m$ and $d_{ox} = 0.1\mu m$ with relative permittivity of 3.9 and 11.9 for the oxide and substrate. In the figure the TSVs have a varying pitch of $50\mu m$, $100\mu m$ and higher. These TSVs were analyzed using the electromagnetic method described in Chapter 3 to first obtain the current density in the conductors to illustrate the importance of proximity effect, where the current distribution in the conductor is affected by the current flowing in neighboring conductors. A plot of the normalized current density in each conductor is shown in Figure 4.26 for a 1V excitation on all the conductors. From the figure, it can be seen that the normalized current density is not the same on all the conductors and current distribution in each conductor is different as well. For example, TSV6 shows the prominence of skin effect where the current is maximum on the skin of the conductor while the current distributions in TSV1, TSV3 and TSV4 are skewed and non-uniform due to a combination of skin and proximity effect. From the figure, skin and proximity effect are prominent on TSVs that are closer together ($50\mu m$ pitch) while skin effect is more prominent on TSVs that are farther apart ($100\mu m$). This variation will cause the parasitics of the TSVs to be non-uniform as well.

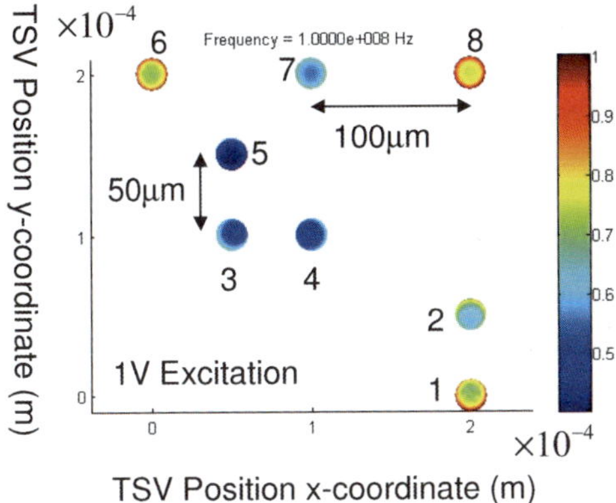

Figure 4.26: Current density in non-uniform TSV array @ 100MHz.

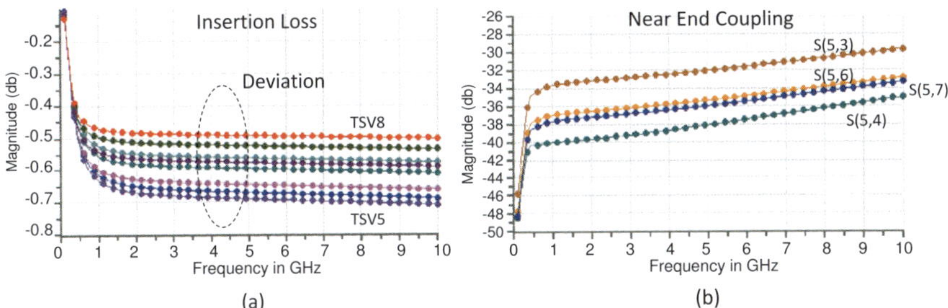

Figure 4.27: Non-uniform TSV array (a) Insertion loss and (b) Near end coupling.

The S-parameters of the TSVs (assuming ground at infinity) referenced to 50Ω is shown in Figure 4.27. As can be seen from Figure 4.27(a), the insertion loss between the input and output terminals of the TSVs has a 0.2dB spread with TSV8 having the lowest insertion loss of 0.5dB and TSV5 having the highest insertion loss of 0.7dB. This can be discerned from the placement of the TSVs in Figure 4.26 where TSV5 has more neighbors than TSV8, causing increased variation due to coupling and proximity effect. The near end coupling between the input terminals of the TSVs is shown in Figure 4.27(b) between TSV5 and its nearest neighbors. As expected, the coupling to TSV3 is the largest due to its close proximity. However, coupling to TSV6 and TSV7 is higher as compared to TSV4 even though they are equidistant, due to the presence of TSV3.

 Design and Modeling for 3D ICs and Interposers

4.3.5 *High Frequency Electrical – Thermal Modeling*

As mentioned earlier, temperature profiles affect the electrical response of TSVs. Since the temperature gradients in a system are a function of the materials, their interfaces and the boundary conditions, system level modeling methods are required to compute the gradients. Similarly, since the electrical response of TSVs is affected by the presence or absence of neighboring TSVs, modeling of TSV arrays is also required.

Since the electrical effects depend on thermal profiles, a methodology that supports electrical-thermal modeling over a broad range of frequencies is therefore required. This is illustrated through an example for a silicon interposer in this section with details on the computational method used for modeling the thermal effects provided in Chapter 5. Here the focus is primarily on the electrical effects. More details on the computational part of thermal modeling for this example are available in [Xie et al., 2012].

A 3D system consisting of two stacked dies, thermal interface material (TIM), 4-layer package, micro-bumps, silicon interposer and under-fill are shown in Figure 4.28. The dies in the stack are of size 8mm × 8mm. The silicon interposer shown in the figure is of size 30mm × 30mm and the package size is 60mm × 60mm. The 120 × 120 TSV array is distributed in the center of the silicon interposer, as shown in Figure 4.28. The TSV diameter is 20 micron with a pitch of 66.7 microns. The detailed geometrical parameters and material thermal conductivity are listed in Table 4.2.

Two design cases are studied here namely, 1) Case-1 where the power consumption of Die 1 and Die 2 are 8 W and 2 W, respectively and 2) Case-2, where the power consumption of Die 1 and Die 2 are

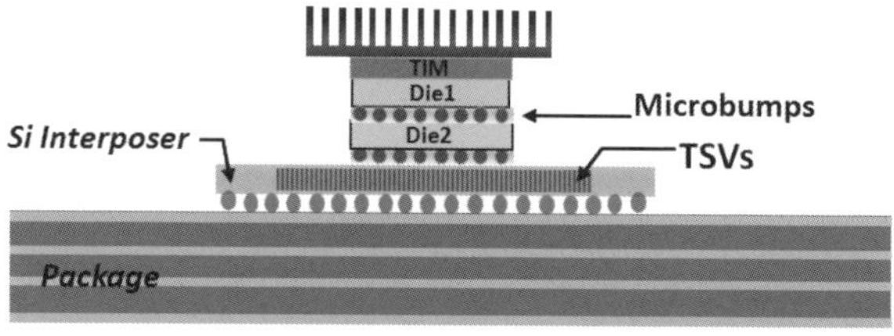

Figure 4.28: 3D system with two stacked dies.

Table 4.2: Geometrical and material parameters.

	Material Thickness (mm)	Thermal Conductivity (W/mK)
Glass-ceramic	0.35	5
Copper Plane	0.03	400
Die	0.2	110
Underfill	0.2	0.4
Micro-bump	0.2	60
TIM	0.2	2.0
TSV	0.2	400

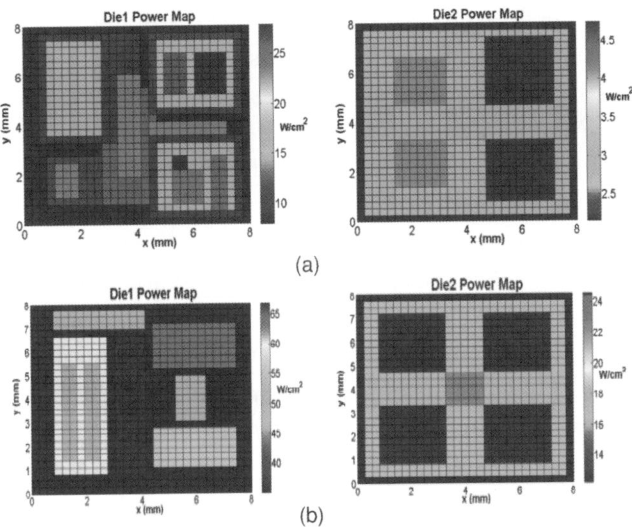

Figure 4.29: Die power maps for (a) Case-1 and (b) Case-2 [Xie et al., 2012].

30W and 12.5W, respectively. Their non-uniform die power maps are shown in Figures 4.29(a) and (b), respectively. From the 120×120 TSV array, a 5×5 TSV array which is located at the center of the silicon interposer is used, as shown in Figure 4.30. The interposer thickness and oxide thickness are $200 \mu m$ and $0.1 \mu m$, respectively. The TSV filling material is copper. The conductivity of silicon interposer is $10.4 S/m$ at room temperature with a doping density of $1.32 \times 10^{15} cm^{-3}$. Air convection with convection coefficient of $10 W/(m^2 K)$ is applied to the top surface of silicon interposer and both sides of the package.

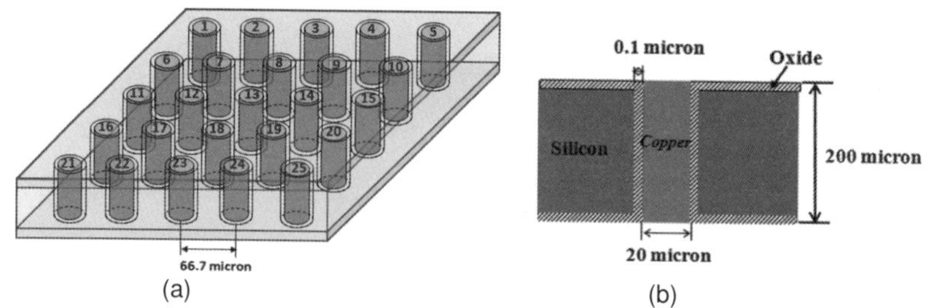

Figure 4.30: 5 × 5 TSV array showing numbering for the TSVs (a) perspective view and (b) cross section.

The simulated steady state temperature distributions of the silicon interposer for the two design cases are shown in Figure 4.31. From the figure, the interposer temperature varies from 36.5 to 40.5 Celsius for Case-1 and 76 to 92 Celsius for Case-2 due to the non-uniform die power map used in Figure 4.29. As expected, the temperature is lower for Case-1 as compared to Case-2 due to lower power dissipation with the hot areas concentrated beneath the die stack. Clearly, these temperature variations can introduce electrical performance variations. There are two questions to be answered here: 1) How dominant is the temperature effect? and 2) What temperature granularity is required in the interposer to compute its effect on the electrical response of TSVs? In a 3D system, the pitch between adjacent TSVs is usually in the range of 50–100 microns, depending on the process used. For a 5×5 or 10×10 TSV array, it covers an area of less than 1 mm^2. Due to the high thermal conductivity of silicon interposer, the temperature variation in the 5×5 TSV array region is usually very small (less than one degree based on simulations). As a result, large granularity can be used to compute average temperatures in regions of the interposer using which the corresponding electrical effects can be computed. Based on simulations, our experience is that average temperature in an area that measures 1mm^2 can be used while computing the TSV response. An average temperature of 40 deg C for Case-1 and 92 deg C for Case-2 in the TSV array will therefore be used to compute the electrical response which will be compared to the results at 25 deg C (room temperature).

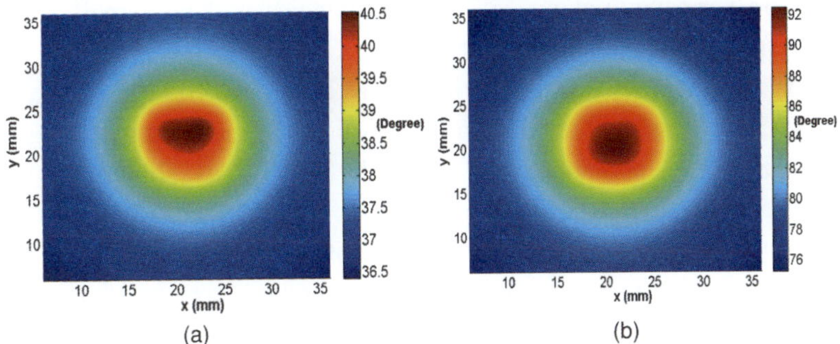

Figure 4.31: Temperature Distribution in the silicon interposer (a) Case-1 and (b) Case-2.

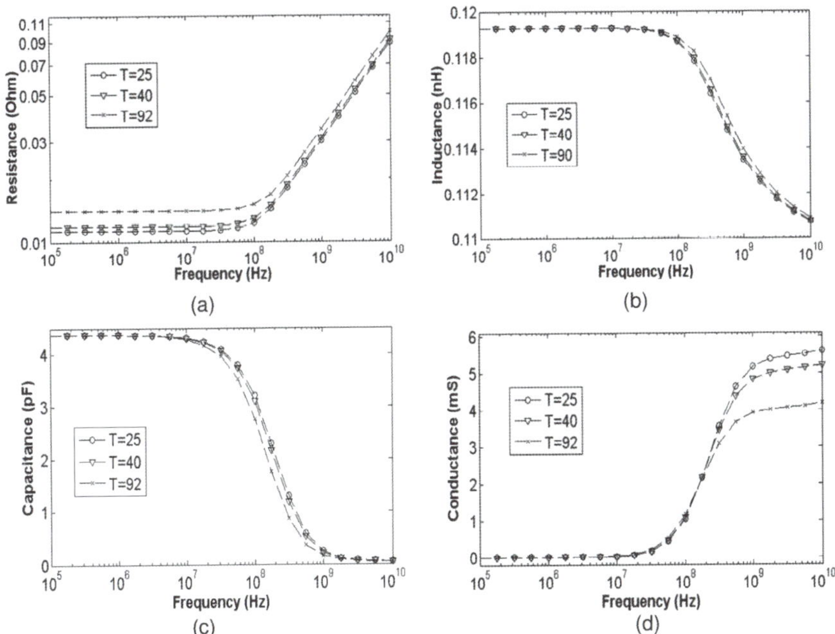

Figure 4.32: RLGC self parameters for TSV1 (a) Resistance, (b) Inductance, (c) Capacitance and (d) Conductance.

As shown in Figure 4.32(a), the series resistance of TSV1 from Figure 4.30(a) increases with temperature based on (4.1). In this example, temperature coefficient of electrical resistance of $0.0039K^{-1}$ for copper has been used. From Figure 4.32(b), the temperature increase has little effect on series inductance at low frequencies and has small effect at higher frequencies due to skin effect.

For TSV capacitance, the temperature effect can be seen in the range 0.05–1GHz, as shown in Figure 4.32(c). With increasing temperature, the equivalent capacitance is reduced. Finally, as shown in Figure 4.32(d), in the low frequency range, the conductance does not change with temperature. However, in the frequency range 0.2–10GHz, the conductance decreases with increase in temperature, which is due to decrease in silicon conductivity, as depicted by (4.2).

The near end cross talk on TSV7 is shown in Figure 4.33 when TSV1, TSV3, TSV11 and TSV13 are each excited by a synchronous clock signal of amplitude 2V and frequency 1GHz at the top side in Figure 4.30(a). The bottom side of these TSVs, are each terminated to 50ohms. In this example, TSV7 is surrounded by ground vias by connecting TSV2, TSV6, TSV8 and TSV12 to ground. Hence, TSV7 is completely shielded in this example. The near end cross talk on TSV7 is shown in Figure 4.33 for three temperatures namely, 25 deg C (room temperature), 40 deg C (Case-1) and 92 deg C (Case-2). The peak value of cross talk is 59.6mV at 25 deg C, 58.3mV at 40 deg C and 51.8mV at 92 deg C resulting in 2.2% and 13% reduction respectively as compared to room temperature. It is interesting to note that unlike the waveforms in Figures 4.4 and 4.11, the cross talk waveform in Figure 4.33 appears

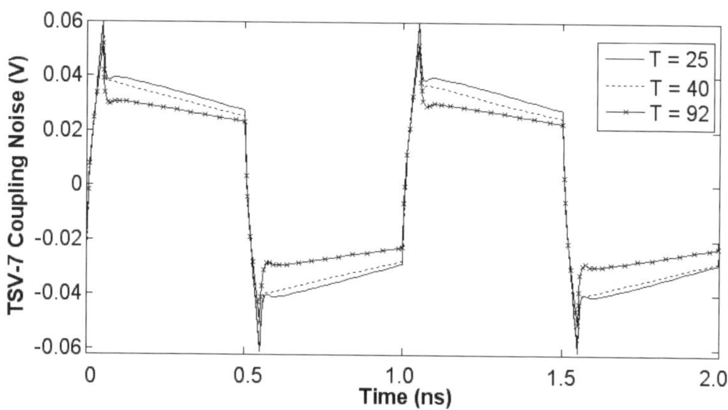

Figure 4.33: Near End Cross Talk on TSV7 as a function of temperature.

discontinuous after the initial peak. This can be attributed to the two different macro-modeling methods used. In Figures 4.4 and 4.11, Idem [Idem, 2009] was used to fit the S-parameters while in Figure 4.33, ADS [ADS, 2009] was used. From Figure 4.33, the effect of temperature on electrical performance can be seen only for large temperature increases of the order of 67 deg C in this example which is related to the amount of power being dissipated.

Based on simulations, the temperature dependence therefore becomes important when stacked dies dissipate large amounts of power (high flux density) creating hot spots where the temperature increase is substantial.

4.4 Interposers

So far, the through silicon via (TSV) characteristics have been discussed in detail where the vias are made of copper metallization which pass through a silicon substrate with an oxide liner used to isolate the via from the semi-conducting substrate. The interposer solution for two stacked ICs is shown in Figure 4.34(a) where the silicon interposer is mounted on the printed circuit board (PCB). Hence, both the signal and power are supplied to the stacked ICs through the interposer. It was shown earlier that TSVs can be lossy and can produce excessive coupling in a low resistivity (high conductivity) silicon substrate. These issues can be managed by using a high resistivity (low conductivity) silicon substrate or by using a thick oxide liner (to minimize coupling). Another possibility is to use a glass interposer where through vias are created in a glass substrate, as shown in Figure 4.34(b). Since glass is a low loss

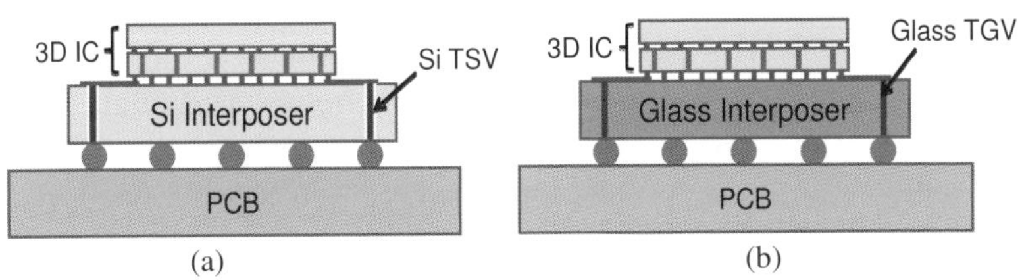

Figure 4.34: Interposer solutions for stacked ICs (a) Silicon and (b) Glass (Courtesy [Band, 2011]).

dielectric medium, the vias do not require an insulating layer and hence copper vias can directly be created in the glass substrate. Recently, vias in glass substrates have been demonstrated with 30μm pitch in an ultra thin substrate with thicknesses less than 200μm [Sukumaran et al., 2011]. Glass as an interposer solution has several advantages such as the possibility of matched CTE to silicon (if suitably tailored), excellent surface flatness, dimensional stability, high resistivity and availability in large thin panels. Glass interposers also can provide low cost solutions. If glass interposers are indeed low loss, low cost and provide similar fine pitch via capability as silicon interposers, then why not just use glass as the interposer solution? The glass and silicon interposer solutions are compared here by addressing the electrical performance issues and temperature related effects in this section. This analysis forms the basis for some of the conclusions in Table 1.4 of Chapter 1.

4.4.1 *Through Glass Vias*

Let's start by looking at the electrical performance of through glass vias (TGV) and comparing its response with TSVs. Consider the glass interposer structure shown in Figure 4.35(a) consisting of a 200μm glass substrate with 20μm thick organic polymer (RXP with properties described in the next section) on either side. The borosilicate glass substrate has a relative permittivity of 4.7–4.8 with a dielectric loss tangent of 0.002–0.003 over a 20GHz bandwidth [Sukumaran et al., 2011]. The laser drilled vias have a diameter of 35μm (R = 17.5μm) and pitch of 250μm (D). The vias are plated with Copper of conductivity 5.8×10^7S/m. Similar dimensions are also used for the silicon interposer with RXP polymer on either side to provide the necessary insulation, in this study. Silicon with varying oxide liner thickness t_{liner} (d_{ox}) and conductivity was used to compare the results between through silicon and through glass vias using a Ground-Signal-Ground (G-S-G) configuration, as shown in Figure 4.35(b).

The insertion loss, near end coupling and far end coupling between the signal vias shown in Figure 4.35(b) are shown in Figure 4.36 based on electromagnetic simulations [CST]. Three examples of silicon

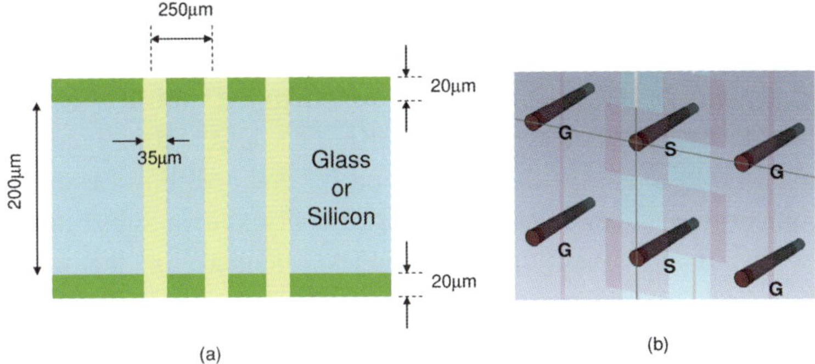

Figure 4.35: Interposer solutions for stacked ICs (a) Silicon or Glass and (b) Glass (Courtesy [Band, 2011]).

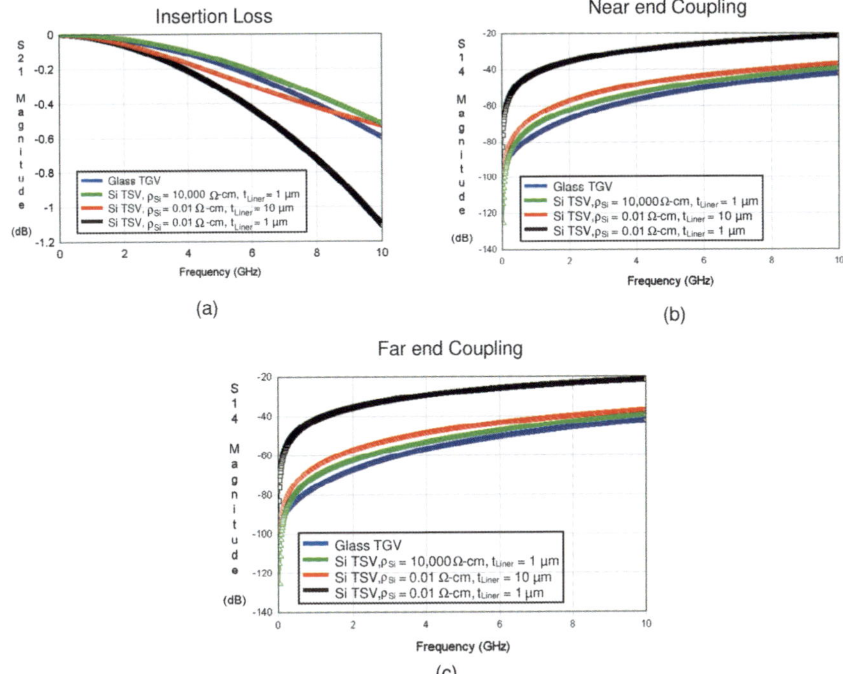

Figure 4.36: Comparison between TGV and TSV (a) Insertion Loss, (b) Near end coupling and (c) Far end coupling [Band, 2011].

substrates are compared namely with resistivity of 10,000Ω-cm (0.01S/m conductivity; high resistivity) and oxide liner thickness $d_{ox} = 1\mu$m and with resistivity 0.01Ω-cm (10000S/m conductivity; ultra low resistivity) with oxide liner thicknesses $d_{ox} = 1\mu$m and 10μm. From Figure 4.36, the insertion loss, near end coupling and far end coupling for the TGVs are comparable to the TSVs in high resistivity silicon substrate. These results are much lower than the values obtained for the ultra low resistivity silicon substrate (with 1μm thick oxide liner) as expected, due to lower substrate coupling. Therefore, with CMOS grade silicon having a conductivity of 10S/m for the silicon interposer, one would expect that the electrical performance of Through Glass Vias (as measured by its insertion loss) would be superior to through silicon vias. Then why even consider using the silicon substrate as the interposer solution? To answer this question, we will look at the signal and power integrity aspects of the problem together by introducing interconnects for fan out and power and ground planes for supplying power in the interposer.

4.4.2 *Response of Power and Ground Planes*

Power and ground planes are often times used in the package and PCB for supplying power to the ICs. The interposer is therefore expected to support power and ground planes for high performance applications. Though providing power to the ICs is covered in more detail in Chapter 5, in this section we will compare the response of the power and ground planes in glass and silicon interposers. Along with this comparison, we will also provide some modeling details on methods used to compute this response, which is an extension of the methods compiled in [Swaminathan et al., 2007].

Power and ground planes are guiding structures which can be modeled by solving Helmholtz equation with open circuit or magnetic wall boundary condition at the plane periphery [Swaminathan et al., 2007]. The Helmholtz equation can be written as:

$$(\nabla_T^2 + k^2)u = -j\omega\mu J_z \tag{4.8}$$

where ∇_T^2 is the transverse Laplace operator parallel to the planar structure, k is the wavenumber, u is the voltage, ω is the angular frequency, μ is the permeability, d is the distance between the planes, and J_Z is the current density excitation injected normal to the planes. Now consider the cross section shown in Figure 4.35(a) and assume that either side of the structure contains metal planes, with the top plane assigned to power and the bottom plane assigned to ground. Let's start with assuming that the interposer is made of silicon with conductivity σ_{Si}.

To model the electromagnetic wave propagation between the power and ground planes, the structure can be approximated by dividing it into square unit cells where the dimension of each cell is less than $\lambda/10$, where λ is the wavelength corresponding to the highest frequency signal. The model for a unit cell of the power and ground plane of size $h \times h$, is shown in Figure 4.37 where the model consists of impedance (Z) and admittance (Y) elements. The impedance and admittance elements in Figure 4.38 can be computed as [Sridharan et al., 2011]:

$$Z = \frac{R_s (R + j\omega L)}{R_s + R + j\omega L} \quad ; Y = \frac{(G_1 + j\omega C_1)(G_{Si} + G'_{Si} + j\omega C_{Si})}{2(G_{Si} + G'_{Si} + j\omega C_{Si}) + G_1 + j\omega C_1} \tag{4.9}$$

where the per unit cell (pul) parameters for the silicon interposer can be expressed as:

$$R = \frac{2}{\sigma t} + 2\sqrt{\frac{j\omega\mu}{\sigma}} \tag{4.10}$$

$$R_s = 2\sqrt{\frac{j\omega\mu}{\sigma_{Si}}} \tag{4.11}$$

$$L = \mu d \tag{4.12}$$

$$C_1 = \frac{\varepsilon_1 h^2}{d_1} \tag{4.13}$$

$$G_1 = \omega C_1 \tan \delta_1 \tag{4.14}$$

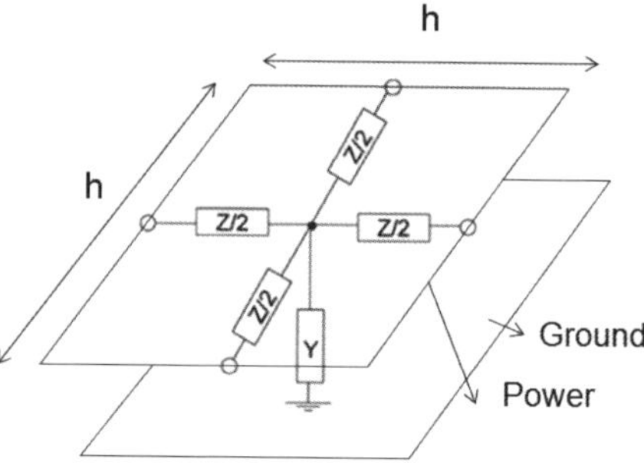

Figure 4.37: Power and ground plane model for a unit cell [Sridharan et al., 2011].

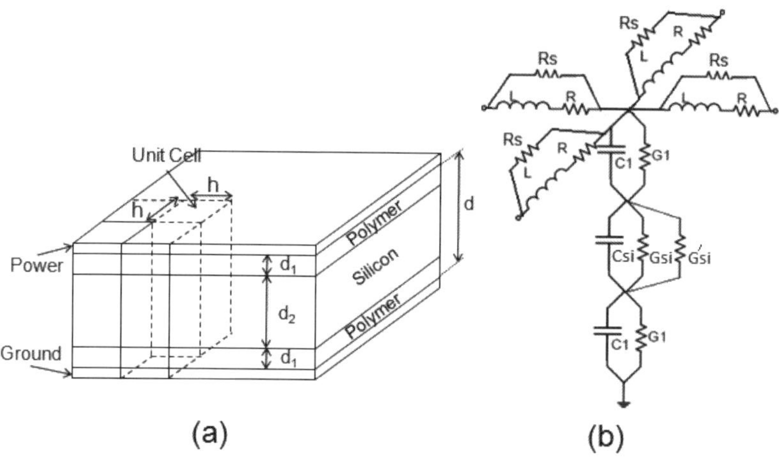

(a)

(b)

Figure 4.38: (a) Discretization of the power and ground planes into square unit cells of size h and (b) Model of unit cell with detailed elements [Sridharan et al., 2011].

$$C_{Si} = \frac{\varepsilon_{Si} h^2}{d_2} \tag{4.15}$$

$$G_{Si} = \omega C_{Si} \tan \delta_{Si} \tag{4.16}$$

$$G'_{Si} = \frac{C_{Si} \sigma_{Si}}{\varepsilon_{Si}} \tag{4.17}$$

for unit cell size h, permeability μ, polymer permittivity ε_1, polymer loss tangent $\tan\delta_1$, silicon permittivity ε_{Si}, and silicon dielectric loss tangent $\tan\delta_{Si}$. In the above equations σ_{Si} and σ denote the conductivity of silicon and the metal layers respectively. The thicknesses of the polymer and silicon are d_1 and d_2 respectively. G_1 and G_{Si} are polymer and silicon conductance terms respectively. G'_{Si} is an additional term which is expressed from the complex permittivity to account for the losses due to the finite conductivity of silicon. In (4.10) R includes the DC and skin effect resistance of the conductor. In (4.11) R_s has been included to account for the losses due to longitudinal current in conductive silicon [Verma et al., 2004]. The effect of R_s is negligible in the quasi-TEM and slow-wave modes of silicon. However, as the conductivity increases, in the transition region between slow-wave and skin effect mode, R_s becomes dominant and has to be accounted for. The approximation of the planes into unit cells for the silicon interposer is shown in Figure 4.38(a) with a detailed unit cell model in Figure 4.38(b) showing the component elements described in the equations above. The model of a unit cell can now be cascaded to model the entire power/ground plane pair. The unit cell model in Figure 4.38(b) can be simplified to model the glass interposer by eliminating resistance R_s, capacitance C_{Si} and conductances G_{Si}, G'_{Si}. Details on the model for the glass interposer is similar to the modeling approach described in [Swaminathan et al., 2007] for low loss dielectrics based on the Multi-layered Finite Different Method (MFDM). Sphinx [ESD, 2011], a commercially available software package based on MFDM is also available for modeling power and ground planes. Assuming that the power and ground plane pair is approximated with M unit cells in x-direction and N unit cells in Y direction, after computation of the per-unit-cell parameters, a (MxN)x(MxN) admittance matrix Y can be constructed as described in [Swaminathan et al., 2007] to compute the impedances of the power and ground planes.

Consider a 470μm thick 35mm × 32mm silicon interposer (relative permittivity $\varepsilon_r = 11.9$ and conductivity $\sigma = 10S/m$ conductivity) with 20μm thick RXP polymer (relative permittivity $\varepsilon_r = 2.51$ and loss tangent $\tan\delta = 0.004$) on both sides, as shown in Figure 4.35(a). Using MFDM, the structure was simulated to compute the transfer impedance Z_21 between ports 1 and 2, with port locations as shown in Figure 4.39(a).

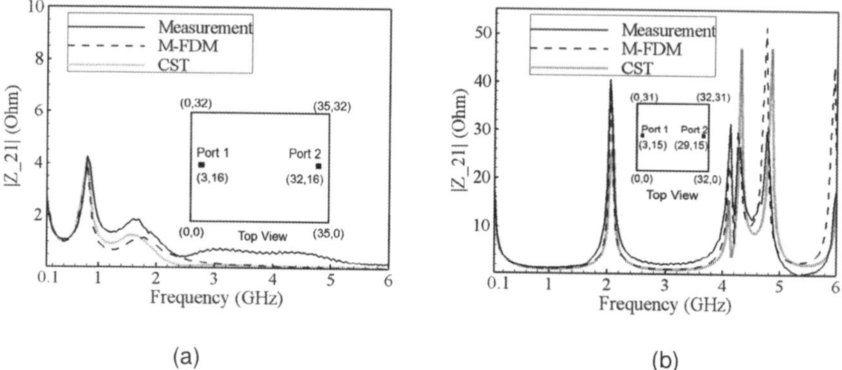

Figure 4.39: Transfer impedance between power and ground planes (a) Silicon and (b) Glass Interposer [Sridharan et al., 2011].

The results have been compared with both measurements and full wave simulation [CST] in Figure 4.39(a), indicating the accuracy of the models derived for the silicon interposer and their use in the multi-layered finite difference method to compute the port impedances. Similarly, MFDM was also applied to compute the power and ground plane transfer impedance Z_21 for a 175μm thick, 32mm × 31mm glass interposer (relative permittivity $\varepsilon_r = 6.7$ and $\tan\delta = 0.006$) with 20μm thick RXP polymer on both sides. The location of the ports is similar to the silicon interposer, as shown in Figure 4.39(b). The results have once again been correlated with measurements and full wave simulation [CST], showing the accuracy of MFDM. So, what is the purpose of MFDM if it provides the same accuracy as full wave simulation? Since MFDM meshes only the power and ground planes and not the dielectric, it provides significant advantages in terms of memory consumption and computation time. Hence, when interposers and PCBs become very complex containing several large power and ground plane layers, MFDM is a very efficient way of solving the problem.

Now, let's get back to the transfer impedance responses shown in Figure 4.39. Since $V = ZI$ based on Ohm's law, for a 1A current excitation, the voltage V measured at any point between the power and ground plane is the impedance at that point. Therefore, if the 1A current source is excited at port 1 in Figure 4.39 along with a frequency sweep that extends from 0.1GHz–6GHz, the resulting voltage measured at port 2 is plotted in Figure 4.39 as a function of frequency. For any high frequency design, the voltage fluctuations between the power and ground planes should be kept to a minimum. Therefore in Figure 4.39, high impedance at any frequency represents a problem since excessive voltage fluctuations can occur at that frequency. Comparing Figures 4.39(a) and (b), the silicon interposer has much lower impedance across the frequency spectrum (suppresses anti-resonances) as compared to the glass interposer, indicating that the silicon interposer is better in managing power supply fluctuations between the power and ground planes. This can be attributed to the additional substrate loss of silicon as compared to glass interposer, which dampens the propagation of the guided electromagnetic wave between the planes [Swaminathan et al., 2007].

This is a clear advantage for the use of silicon as the interposer material as compared to glass. One may argue that the thickness of glass is much lower than the silicon interposer in this example, and hence the conclusions reached are not fair. It is therefore important to note that as the thickness of the glass interposer between the power and ground planes increase, the impedance will only increase and not decrease, making the impedances higher than what is shown in Figure 4.39(b).

4.4.3 *Signal Integrity*

Signal integrity represents the ability to propagate signals through interconnections with high fidelity to ensure first incidence switching of the receiver circuits. Managing signal integrity involves minimizing reflections, cross talk and power supply noise along the signal path so that a clean waveform can be received at the receiver circuit. In the interposer and printed circuit boards containing power and ground planes, the signal lines are often times routed in their presence causing return path discontinuities (RPD) [Swaminathan et al., 2010].

During the charging and discharging of the signal lines (which behave as transmission lines at high frequencies), return currents are induced on the reference planes, which follow the path of least impedance. Any interruption of the return current due to change in reference planes cause return-path-discontinuities (RPDs), which in turn induces excessive jitter and noise on the signal waveform. In such scenarios, the power distribution impedance at the RPD location is an important parameter to control for managing signal integrity [Swaminathan et al., 2010]. A microstrip–to–microstrip transition is an example of a RPD which occurs in most packages and printed circuit boards. In this section, we will evaluate the effect of silicon and glass interposer on a microstrip–to–microstrip transition in the presence of a power and ground plane. This transition is representative of the routing structures typically used in interposers. Other types of transitions are also possible which are covered in detail in [Swaminathan et al., 2007].

Consider Figure 4.40(a), where signal lines on top of the power plane connect to signal lines below the ground plane through two vias. The signal lines have air on one side and therefore can be categorized as microstrip transmission lines. The dimensions of the structure are as shown in Figure 4.40(a). The electrical properties of silicon, glass, and polymer are the same as described in the previous section. A 1 µm thick sidewall liner made of RXP polymer was used for the through silicon via (TSV), to insulate the copper via from the semi-conducting silicon substrate. From the figure, the microstrip–to–microstrip transitions cause a change in the reference plane for the signal line, creating RPDs at the via locations. When the power and ground planes resonate, their impedances increase as in Figure 4.39, leading to large noise disturbances between the planes, as the signal lines switch states. With more signal lines switching, more noise is generated between the power and ground planes, and therefore Simultaneous Switching Noise (SSN) is an important effect to consider while designing interposers.

The effect of RPDs can be ascertained by either measuring or computing the insertion loss of the signal line. In this section we will rely on full wave simulation [CST] for computing the insertion loss of the signal line between the input (port 1) and output (port 2) of the microstrip–to–microstrip transition, as shown in Figure 4.40(a). Figure 4.40(b) shows the insertion loss (S21) for silicon and glass, assuming silicon is replaced with glass, without changing other layers in the stack-up shown in Figure 4.40(a). The results are quite interesting. The

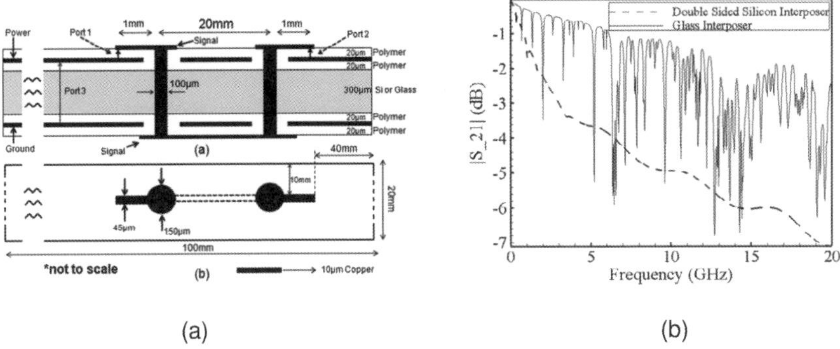

(a)   (b)

Figure 4.40: Signal lines in silicon and glass interposer (a) Microstrip–to–Microstrip transition and (b) Insertion loss [Sridharan et al., 2011].

insertion loss of the signal line in glass interposer is lower than silicon interposer, if computed as an average across the frequency spectrum of 20GHz. This is expected since glass material has lower loss than silicon. However, the glass interposer produces sharp notches in the insertion loss of the signal line at specific frequencies, which can be attributed to the power and ground plane resonances described in Figure 4.39(b). In fact, the power and ground plane impedances at the two via discontinuities in Figure 4.40(a) will exhibit large values at frequencies which are coincident with the sharp notches in the insertion loss in Figure 4.40(b) [Swaminathan et al., 2010]. These sharp notches in the insertion loss of the signal line are absent in the silicon interposer since the power plane resonances are suppressed due to higher losses, resulting in a smooth insertion loss, as shown in Figure 4.40(b). As a designer, the question to ask ourselves is the following: Is it preferable to propagate signals on a signal line with low loss and high variability (glass) as compared to one with high loss but low variability (silicon)? This can be answered by propagating a pseudo-random bit stream through the signal line input and computing the eye diagrams at the output, to assess the quality of the transmitted signal waveform.

The frequency spectrum of an input bit stream is important, while analyzing the effects of frequency dependent losses. The frequency spectrum of a pseudo-random bit stream (PRBS) consists of harmonics distributed across multiple frequencies based on the data pattern, as

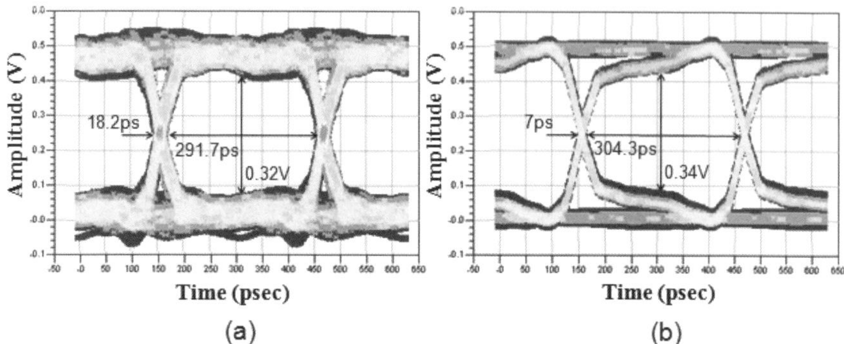

Figure 4.41: Eye diagrams for (a) Glass Interposer and (b) Silicon Interposer [Sridharan et al., 2011].

opposed to a simple clock signal (see Chapter 5). Since the harmonics across multiple frequencies can experience varying insertion losses for glass as shown in Figure 4.40(b), it can result in uncertainties in signal amplitudes and also uncertainties in rise/fall times, leading to excessive jitter and reduced eye opening. Figure 4.41 shows the eye diagram comparison obtained using a circuit simulator [ADS], between glass and silicon interposer for a 3.2 Gbps 2^{10}-1 PRBS stream transmitted between port 1 and port 2, as shown in Figure 4.40(a). It can be seen that jitter (7ps Vs 18.2ps) and eye opening (0.34V Vs 0.32V) are improved in the silicon interposer, as compared to glass interposer. Hence, often times a predictable insertion loss albeit with higher loss is preferable to a lower insertion loss with large unpredictability. In Figure 4.41(b), the RC effect of the silicon interposer can be clearly seen due to the TSVs, which is absent in Figure 4.41(a) for the glass interposer. Of course, the results for the glass interposer can be greatly improved by adding decoupling capacitors at via locations between the power and ground plane, enabling continuity in the return current paths [Swaminathan et al., 2010], which is not covered in this section. So, in a nut shell, for the example provided, the silicon interposer though lossy as compared to the glass interposer, provides better signal integrity with lower decoupling requirements!

4.4.4 *A More Rigorous Look at Simultaneous Switching Noise and Signal Integrity*

The previous section looked at a single interconnect with via transitions. What happens when multiple signal lines transition through power and ground planes? In an earlier section we had mentioned that coupling between vias in silicon interposer is much larger than for the glass interposer. Wouldn't this coupling then create problems for signal integrity?

To answer this question we consider a Simultaneous Switching Noise (SSN) experiment which consists of an 8bit microstrip-via-microstrip transition, with dimensions and port definition as shown in Figure 4.42. The cross section of the structure is similar to Figure 4.40(a) which consists of four metal layers with microstrip lines (length = 10mm, width = 40um, thickness = 10um, spacing = 80um) on the top and bottom layer and power and ground planes on second and third layer, respectively. Silicon (with 10S/m conductivity and $\varepsilon_r = 11.9$) and glass (with $\varepsilon_r = 6.7$ and $\tan\delta = 0.006$) interposer are used in the structure to investigate the noise effects of silicon and glass interposer.

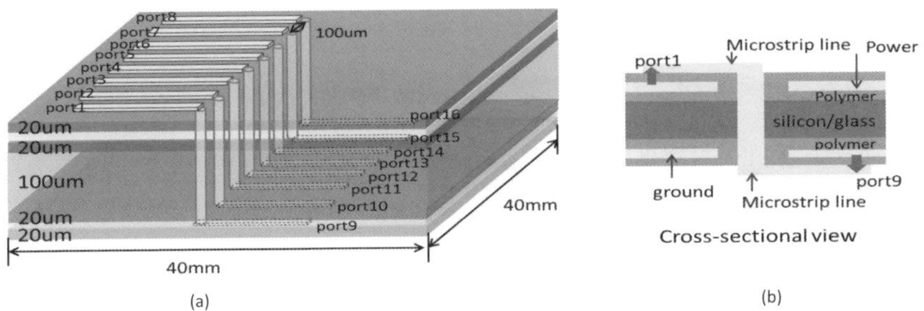

Figure 4.42: Eight bit microstrip-via-microstrip transition between power and ground planes.

Let's start by looking at just the through vias in silicon and glass as shown in Figure 4.43 without the power/ground planes and microstrip lines. The geometry of the 8×1 TSV/TGV array model with 16 ports defined is shown in Figure 4.43. The coupling between all vias was extracted using the methods described in Chapters 2 and 3 using the specialized basis functions.

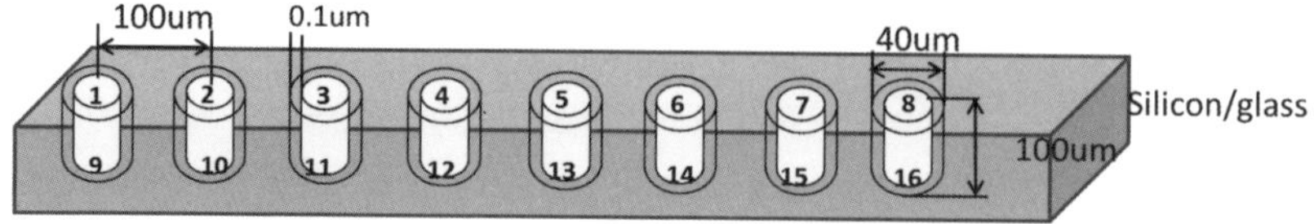

Figure 4.43: 8 × 1 Via Array for Silicon and Glass.

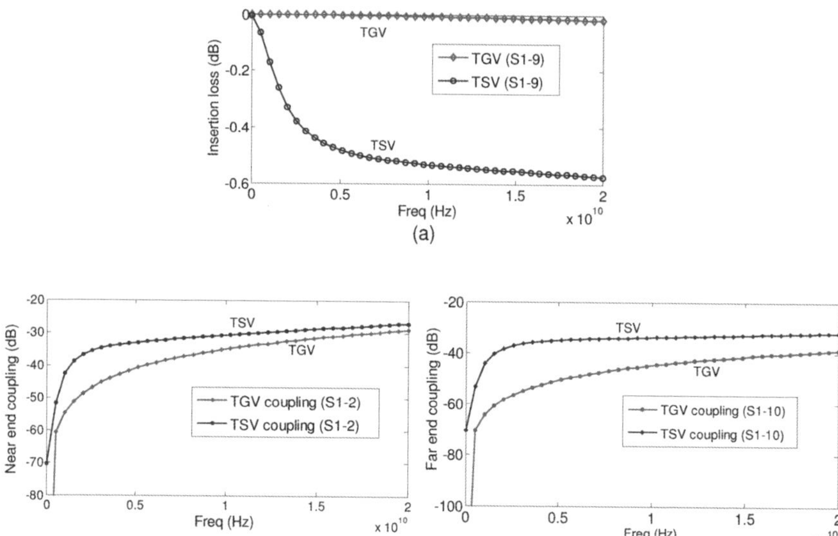

Figure 4.44: Insertion loss and coupling for silicon and glass vias (a) insertion loss, (b) near end coupling and (c) far end coupling.

The near end and far end coupling between the first two vias is shown in Figure 4.44 for both silicon and glass interposer. The insertion loss between ports 1 and 9 for the first via is shown in Figure 4.44(a) indicating that the glass interposer has a much lower insertion loss which is expected due to the low loss of the medium. As shown in Figure 4.44(b), the near end coupling between ports 1 and 2 (between the first and second via) indicates that at the transition frequency, the glass vias have much lower coupling as compared to the via in silicon. A similar effect can be seen in Figure 4.44(c) where the far end coupling between ports 1 and 10 are shown where the coupling in the glass interposer is substantially lower than the silicon interposer. Similar responses have been described in earlier sections in this chapter due to the lower loss in glass, and therefore these results are not surprising.

Let's now go back to the original structure in Figure 4.42 where power/ground planes are added and signal lines are connected to the vias. The insertion loss between ports 1 and 9 (line at edge) and ports 4 and 12 (line at center) over a 20GHz bandwidth is shown both for the silicon and glass interposer in Figures 4.45(a) and (b), respectively. Similar to the response shown earlier, the insertion loss for the glass interposer in Figure 4.45(b) shows sudden change in insertion loss due to the return path discontinuity (RPD) effect described earlier. Though the insertion loss for the silicon interposer is higher overall in Figure 4.45(a) as compared to the glass interposer, the insertion loss is smooth. The near end and far end coupling between ports 1 to 2 and ports 1 to 10 is shown in Figures 4.45(c) and (d) respectively, for both the silicon and glass interposer. Surprisingly, the coupling levels are now similar as opposed

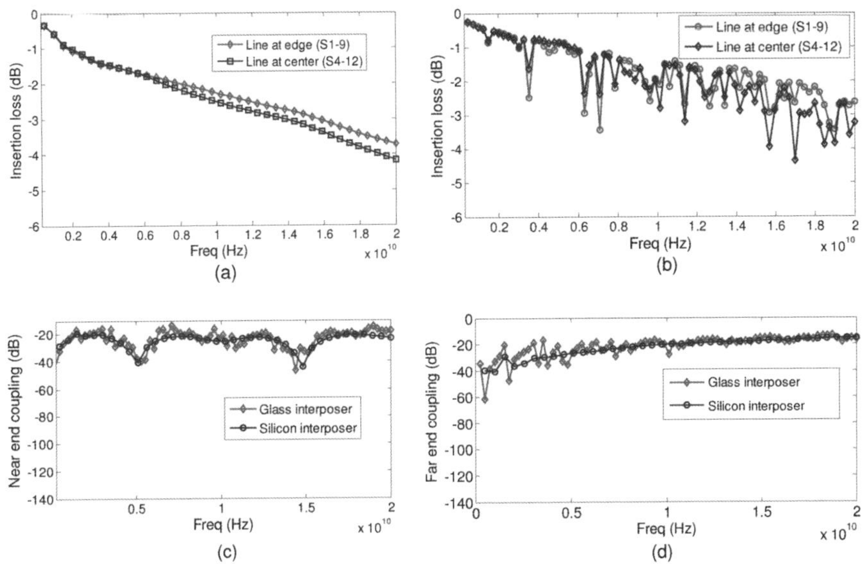

Figure 4.45: Insertion loss and coupling with power/ground planes and signal lines (a) insertion loss for the silicon interposer, (b) insertion loss for the glass interposer, (c) near end coupling and (d) far end coupling.

to the results in Figures 4.44(b) and (c). This is because the RPD excites the cavity modes between the power and ground planes which in turn increases via to via coupling for the glass interposer. Since, losses are higher for silicon, this effect is smaller. Hence, with power and ground planes, the coupling between lines due to the presence of vias is similar for both the silicon and glass interposer (silicon is actually better), as can be seen from Figures 4.45(c) and (d).

Figure 4.46 shows the eye diagram comparison obtained using ADS [ADS, 2009], between glass and silicon interposer for a 5Gbps random bit signal transmitted from port1~port8 to port9~port16, with 8 bits switching simultaneously. Since the via coupling effects are included, the eye diagram at the far end of the lines includes the effect of via coupling. The via coupling includes coupling through the cavity modes and direct electromagnetic coupling through the near field. Since the harmonics across multiple frequencies can experience varying insertion losses for glass as shown in Figure 4.45(b), it can cause uncertainties in signal rise/fall times and also uncertainties in signal amplitudes, resulting in a larger jitter and reduced eye height as shown in Figure 4.46(b). The eye

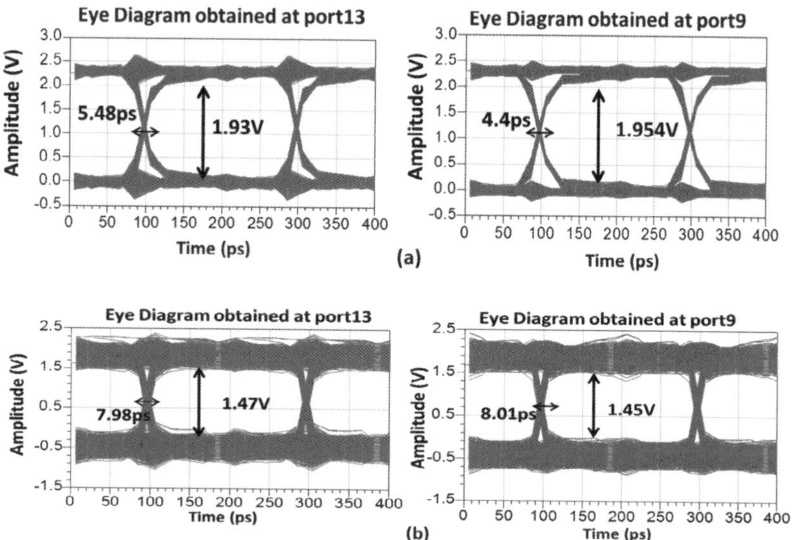

Figure 4.46: Eye diagrams for 8 bits switching (a) Silicon and (b) Glass interposer.

diagrams obtained at ports 9 and 13 in Figure 4.42(a) for silicon and glass interposers are compared in Figure 4.46. Even with all via couplings included, silicon provides a larger eye height and smaller jitter as compared to glass.

As discussed earlier, with any temperature increase, the cross talk in silicon interposers decreases. From a thermal standpoint, the thermal conductivity of silicon (149W/m/K) is much higher than glass (1W/m/K). Therefore silicon interposers are much better at spreading the heat, thereby minimizing the occurrence of very large temperature increase in the substrate (simulation studies have shown that the temperature difference in silicon and glass interposer with the same boundary conditions is ~4–5 deg C with details in Chapter 5). Hence, at high frequencies when large power needs to be dissipated, silicon is a better material than glass as the interposer material for managing signal integrity, though the insertion loss may require compensation.

4.5 Modeling and Design Challenges

It appears that the semiconductor and system companies are relying more on the silicon interposer solution due to infrastructure and supply chain related issues. However, the packaging industry seems to be leaning more towards the glass interposer due to better properties and potential cost benefits from the use of larger size panels for processing. Clearly, both solutions are viable depending on the application. Irrespective of the solution used, there is a need for the electronic design automation (EDA) industry to step up and provide solutions that enable the design of 3D systems. Here, we briefly discuss five major issues that need to be addressed from an electrical standpoint for 3D stacking to be viable namely [Swaminathan, 2012],

(a) *Parasitic Extraction*: Though simple analytical solutions can be used, a question still remains as to whether parasitics are being computed correctly especially when the TSVs come closer together. This requires either an extension to the simple analytical models by incorporating other electromagnetic effects or numerical methods that solve Maxwell's equations directly on the TSV geometry. These numerical methods need to handle multi-scale dimensions since the geometries of the structures used for the TSVs can have a scale ratio of 1:1000 (smallest:largest dimension that requires meshing) or more.

(b) *Silicon Interposer Type*: Scalability is a very important issue that needs to be addressed. As an example, a floating silicon interposer can lead to coupling between vias. With the presence of KOZ, extracting the TSV parasitics can become quite challenging, especially when an electromagnetic solver is used. Either solvers that capture the coupling between vias in large arrays with reasonable memory and CPU time are required, or methods for grounding the silicon substrate and the corresponding modeling methods need to be developed. Both are challenging problems that need to be addressed.

(c) *Interposer and Package Interactions*: In a 3D stack, chip-package co-design is an area of utmost importance. This is obvious since more current needs to be passed through the package as dies are stacked, requiring a better understanding of the power integrity related issues in the interposer, as discussed in Chapter 5. With the interposer serving as the first level package, the signals from the stacked die communicate with the outside world through the package. So, a question to consider is if the TSV parasitics on the die can propagate into the package? To further explain this problem, consider Figure 4.47 with two dies of thickness 100μm (10S/m silicon conductivity) containing TSVs (0.1μm oxide thickness) bonded to each other through micro bumps using copper with height of 20μm. The TSVs have a pitch of 50μm and form a 3×3 array, as shown in the figure. The stacked dies are assembled onto a low loss organic package (forming the interposer) using solder bumps with height 50μm and pitch 100μm with an underfill material of relative permittivity 3.9. The organic package is a flip chip, multi-layered package with the top layer routing as shown in Figure 4.47. Since this is an example for illustrative purposes only and not a fabricated package, all the details are not provided. However, the complexity of the problem is captured quite well in the figure. In Figure 4.47, the input pulse is propagated from the top on TSV a1 as shown in the

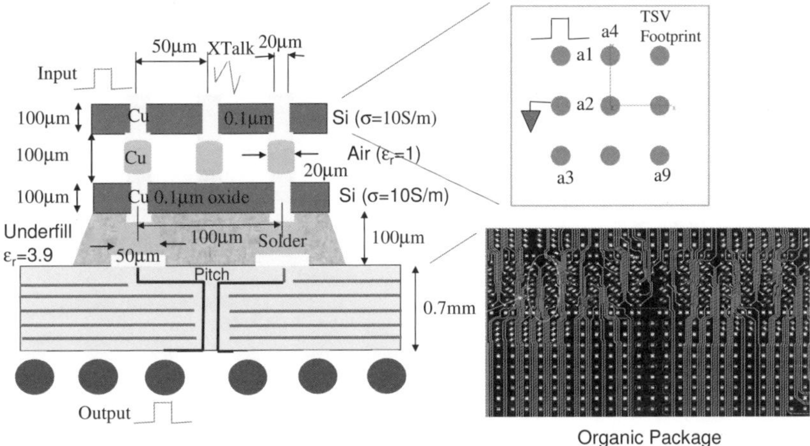

Figure 4.47: Two stacked dies on an organic package showing TSV footprint and first layer of flip chip organic package [Swaminathan, 2012].

footprint with TSV a2 grounded. Due to the difference between the micro bump and solder pitch, not all TSVs exit into the package. In the figure, TSVs a1 and a3 connect to interconnections that exit from the bottom of the package. TSV a4 terminates on the bottom of first die while TSV a9 terminates at the bottom of second die. The goal is to simulate the signal integrity of the waveforms at the bottom of the package and on TSVs that terminate within the die. To enable simulation, models were developed and concatenated together to create a model in Spice. For modeling the stacked die and package 3DPF and Sphinx were used [ESD, 2011] and for concatenating the models, macro-models of the individual blocks were developed using Idem [Idem, 2009]. These were then simulated in Spice using 1V signal swing with a rise time of 40ps at TSV a1. All of the ports were terminated in 50ohms. The simulated waveforms are shown in Figure 4.48. The waveform at a1, on top of the die stack, experiences reflections due to the mismatch between the TSV, package and 50ohm termination impedance. The waveform exiting the package clearly has the RC effect of the TSV causing rounding of the pulse edge. This can be a problem for package designers since it can affect delay based on the switching levels required. The cross talk to TSV

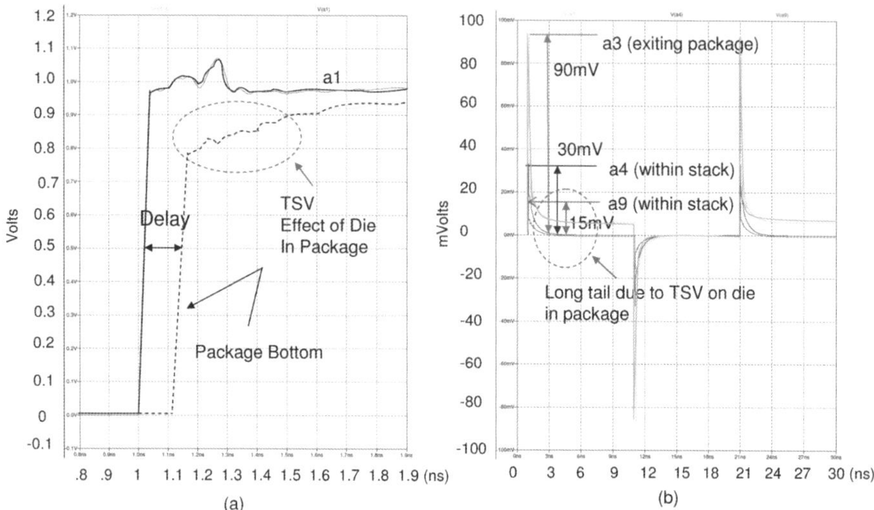

Figure 4.48: (a) Signal waveforms on TSV a1 at the top and bottom and (b) Cross talk waveforms.

a3 exiting the package is 90mV which includes the coupling between TSVs and the interconnections in the package. The important behavior to note is that the cross talk waveform has the long tail effect that is typically seen in TSVs, as described earlier in this chapter. Hence, the RC behavior propagates into the package, even though the package is low loss. As described earlier, this can cause issues with jitter due to interference between bits in a PRBS. Though the cross talk to TSV a4 and a9 are smaller at values of 30mV and 15mV respectively, they are still large considering the distance between the aggressor (a1) and victim TSVs (a4, a9) and exhibit the long tail effect as well. Since, the waveforms in the package are affected by the TSVs in the die and vice versa (when signal and power are supplied to the die), chip-package co-design approaches become necessary.

(d) *Temperature Dependence*: Temperature represents a very important parameter that can change the electrical performance results especially for a silicon interposer based solution. This is more so in a 3D system where heat has trouble escaping easily, creating localized hot spots. Hence, exchange of information between the electrical and thermal modeling tools is required.

(e) *Path Finding or Verification*: Finally, EDA tools that support both path finding and verification need to be developed. For 3D technology to mature, tools that provide constant feedback is required that allows designers to modify layouts during the design process based on assessment of either signal or power integrity. This will minimize re-spins of the design. In addition, assessment of the technology and its effect on electrical performance is required as well, as part of the sizing process. Similarly, after design completion, verification of the entire design to ensure that it meets signal and power integrity specifications are required. Though numerous full wave and fast solvers are already commercially available, the authors believe that there is still a need for innovation on the EDA side for extracting and analyzing the electrical performance of 3D systems. At the time of writing this book, the authors believe that the need for a path finding tool is a lot greater than a verification tool, since 3D technology is still in its infancy.

An example of a structure for path finding is shown in Figure 4.49(a) which is composed of multiple tiers interconnected through bumps,

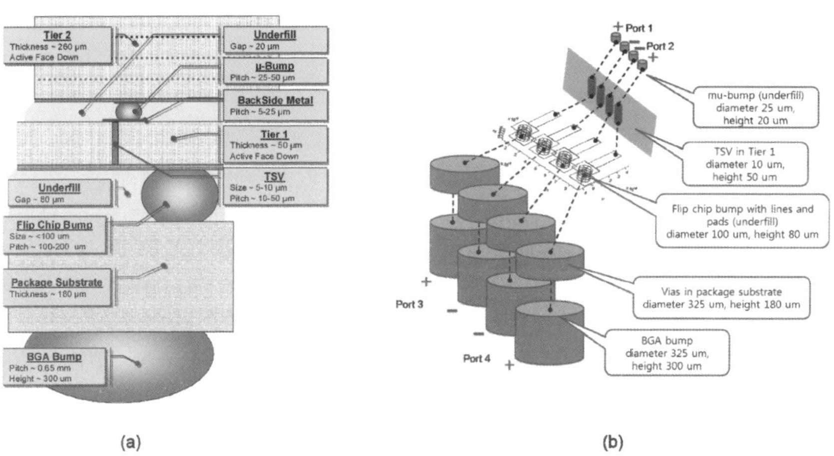

Figure 4.49: (a) Cross-section of 3D Structure (Courtesy: Sematech) and (b) Modeling of structure for path finding.

TSVs, through vias and solder balls. To finalize the geometric detail of the structure, numerous design options and stack up need to be tested for estimating the electrical performance. As an example, one option could use wirebonds to connect the IC to the interposer containing TSVs. In this section, 3DPF [3DPF, 2012] has been used to analyze the structure in Figure 4.49(a) where the original structure is approximated as shown in Figure 4.49(b). In Figure 4.49(b), the entire model is subdivided into bump arrays (in the underfill layer), package modules, and silicon interposer modules. The electrical performance of each sub module is then obtained using the methods described in Chapters 2 and 3. Since most of the performance determining structures in Figure 4.49(b) has cylindrical cross section, the modeling can be done very efficiently using the cylindrical basis functions described in previous chapters.

For modeling large bumps such as solder balls, the spheres can be replaced by a collection of cylinders. Since, the solder balls are not perfect spheres due to the presence of the pads on either side, the approximation can be done quite well with a few cylinders. As an example consider a pair of solder balls with height 300μm, diameter 325μm and pitch 650μm, as shown in Figure 4.50(a). The partial spheres can be approximated using a 3 cylinder, 5 cylinder or 7 cylinder model, as shown in Figure 4.50(a) where the insertion loss and return loss are shown assuming that one of the partial spheres as the return conductor. As can be seen, with an increase in the cylinder segments, the insertion and return loss converge with a 15GHz bandwidth obtained using a 3 cylinder model when compared to a full wave solver [CST].

By connecting the simulated sub-modules, the insertion loss for the entire stack can be obtained. Using this procedure the effect of modifying a sub-module and the resulting electrical performance can be gauged. The return loss, near end cross talk (NEXT) and far end cross talk (FEXT) of the entire stacked structure are shown in Figure 4.51, where the cross talk comes from the electrical coupling between two differential channels defined in Figure 4.49(b). In Figure 4.51(a), the presence of the package via and BGA bump improves the insertion loss due to better matching.

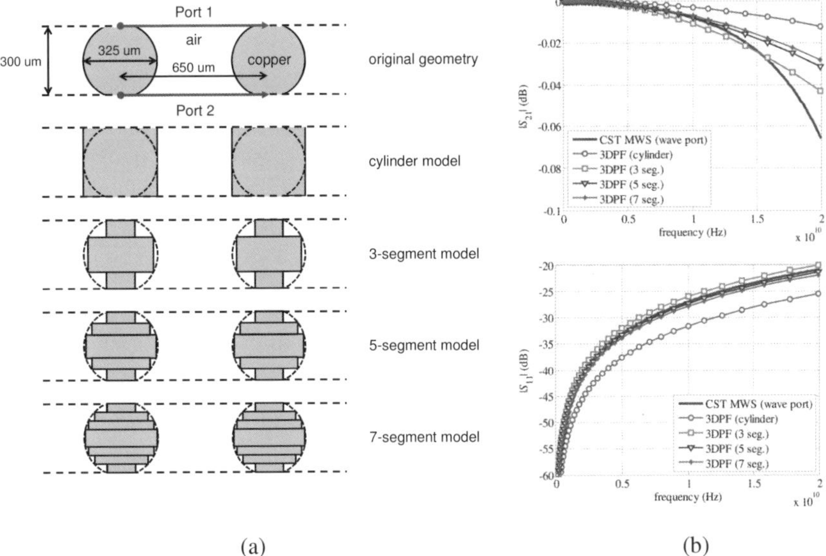

Figure 4.50: (a) Representation of partial spheres using cylinders and (b) S-parameters of approximate ball models and correlation with full-wave EM model.

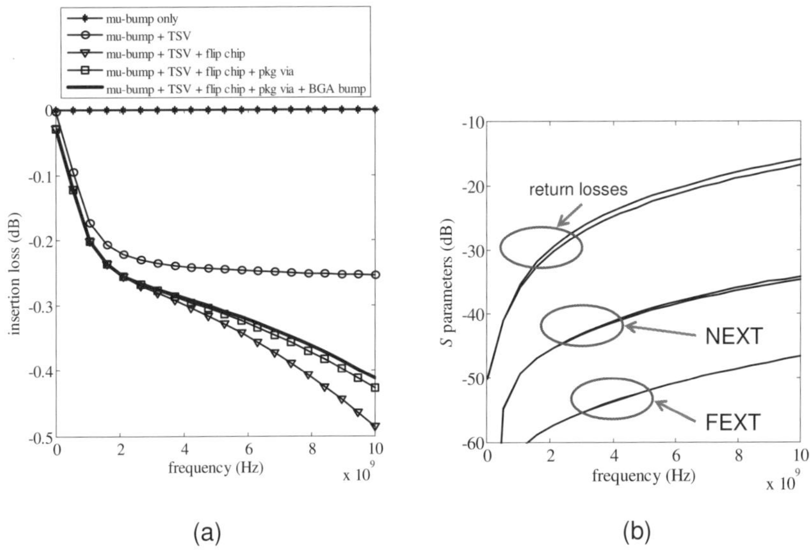

Figure 4.51: (a) Insertion loss as the stack-up is being created and (b) return loss, NEXT and FEXT of entire structure.

4.6 Summary

This chapter started as an extension to Chapter 3 where the methods developed were used to analyze a TSV pair to compute its insertion loss. The effect of physical variations on the insertion loss of the TSV pair was analyzed to determine the important parameters for technology tuning. This was followed by the analysis of the TSV capacitance due to substrate bias as a function of physical and material properties. The TSV pair was then extended to four TSVs to analyze cross talk which was further extended to TSV arrays. The importance of coupling between TSVs was illustrated through several examples including the effect of shielding (using a Faraday cage or by grounding the substrate) and RC effect on jitter and eye height. Temperature as an important parameter for the analysis of TSVs was discussed using a simple structure followed by system level analysis where the need for exchanging thermal and electrical data was discussed. Silicon and glass interposers were then compared to determine which had better electrical performance. The importance of low impedance between the power and ground planes in the interposer was illustrated by first looking at power distribution followed by signal integrity. The role of return path discontinuities due to via transitions (covered in detail in Chapter 5) was addressed and used as a test structure to compare the performance of silicon and glass interposer both for one bit and eight bits switching. The effect of loss in dampening the power distribution anti-resonances thereby improving signal integrity for the silicon interposer was covered in detail. Finally, a few challenges arising in the modeling and design of interposers were discussed briefly followed by the need for a path finding tool for estimating technology tradeoffs and electrical performance.

References

1. [Band, 2011] Tapobrata Bandhyopadhyay, "Modeling, Design, and Characterization of Through Vias in Silicon and Glass Interposers", PhD Dissertation, Georgia Institute of Technology, 2011.
2. [Cho et al., 2011] Jonghyun Cho, Eakhwan Song, Kihyun Yoon, Jun So Pak, Joohee Kim, Woojin Lee, Taigon Song, Kiyeong Kim, Junho Lee, Hyungdong Lee,

Kunwoo Park, Seungtaek Yang, Minsuk Suh, Kwangyoo Byun and Joungho Kim, "Modeling and Analysis of Through-Silicon Via (TSV) Noise Coupling and Suppression Using a Guard Ring", IEEE Transactions on Components, Packaging and Manufacturing Technology, Volume: 1, Issue: 2, pp. 220–233, Feb. 2011.

3.　[3DPF, 2012] 3DPF, E-System Design, www.e-systemdesign.com

4.　[Jayantha et al., 2008] S. M. Sri-Jayantha, et al., "Thermomechanical modeling of 3D electronic packages," IBM Journal of Research and Development, Vol. 52, No. 6, Nov. 2008.

5.　[Sukumaran et al., 2011] V. Sukumaran, T. Bandyopadhyay, Q. Chen, N. Kumbhat, F. Liu, P. Pucha, Y. Sato, M. Watanabe, K. Kitaoka, M. Ono, Y. Suzuki, C. Karoui, C. Nopper, M. Swaminathan, V. Sundaram, and R. Tummala, "Design, Fabrication and Charactererization of low-cost glass interposers with fine-pitch through package vias", 61st IEEE Electronic Components and Technology Conference (ECTC), pp. 583–588, Orlando, FL, June 2011.

6.　[Bartley et al., 2013] Gerald K. Bartley, Philip R. German, David p. Paulsen and John Sheets II, "Through Silicon Via Direct FET Signal Gating", United States Patent Application 20130001676, Date Filed: Jan. 3, 2013.

7.　[Streetman et al., 2000] Ben G. Streetman and Sanjay Banerjee, "Solid State Electronic Devices", Fifth Edition, Prentice Hall, 2000.

8.　[Yang et al., 2013] Decao Yang, Jianyong Xie, Madhavan Swaminathan, Xing-Cheng Wei and Er-Ping Li, "A Rigorous Model for Through Silicon Vias (TSV) with Ohmic Contact in Silicon Interposer". To be published in IEEE Microwave and Wireless Component Letters, 2013.

9.　[Engin et al., 2012] A. E. Engin and N. S. Raghavan, "Metal Semiconductor (MES) TSVs in 3D ICs: Electrical Modeling and Design", IEEE 3D Systems Integration Conference (3DIC), pp. 1–4, 2012.

10.　[CST] CST Microwave Studio [Online], http://www.cst.com/Content/Products/MWS/Overview.aspx

11.　[Sridharan et al., 2011] Vivek Sridharan, Madhavan Swaminathan and Tapobrata Bandyopadhyay, "Enhancing Signal and Power Integrity using Silicon Interposer", IEEE Microwave and Wireless Component Letters, Vol. 21, Issue: 11, pp. 598–600, 2011.

12.　[Verma et al., 2004] Verma, A.K, Nasimuddin, and Sharma, E.K., "Analysis and circuit model of a multilayer semiconductor slow-wave microstrip line," IEE Proc. Microwaves Antenn. Propag., Vol. 151, No. 5, pp. 441–449, 20 Oct. 2004.

13.　[Swaminathan et al., 2010] Swaminathan, M, Daehyun Chung, Grivet-Talocia, S, Bharath, K, Laddha, V, and Jianyong Xie; "Designing and Modeling for Power Integrity," IEEE Transactions on Electromagnetic Compatibility, Vol. 52, No. 2, pp. 288–310, May 2010.

14.　[Swaminathan et al., 2007] Madhavan Swaminathan and Ege Engin, "Power Integrity Modeling and Design for Semconductors and Systems", Prentice Hall, 2007.

15. [ESD, 2011] Sphinx, www.e-systemdesign.com

16. [Idem, 2009] IDEM R2009a [online], www.idemworks.com

17. [ADS, 2009] Advanced Design System 2009U1, Agilent Technologies.

18. [Xie et al., 2011a] Biancun Xie, Madhavan Swaminathan, KiJin Han and Jianyong Xie, "Coupling Analysis of Through-Silicon Via (TSV) Arrays in Silicon Interposers for 3D Systems", IEEE International Symposium on Electromagnetic Compatibility, pp. 16–21, Long Beach, CA, Aug. 2011.

19. [Swaminathan, 2012] Madhavan Swaminathan, "Electrical Design and Modeling Challenges for 3D System Integration", Half Day Tutorial, Designcon, Santa Clara, CA, Jan. 2012.

20. [Xie et al., 2011] Xie J, Swaminathan M. Electrical-thermal co-simulation of 3D integrated systems with micro-fluidic cooling and Joule heating effects. IEEE Transactions on Components, Packaging and Manufacturing Technology 2011; 1(2): 234–246.

21. [Zhao et al., 2011] Zhao W-S, Wang X-P, Yin W-Y. Electrothermal effects in high density through silicon via (TSV) array. Progress In Electromagnetics Research 2011; 115: 223–242.

22. [Yin et al., 2007] Yin W-Y, Kang K, Mao J-F. Electromagnetic-thermal characterization of on-chip coupled (a) symmetrical interconnects. IEEE Trans. Adv. Packag. 2007; 30(4): 851–863.

23. [Xie et al., 2012] Jianyong Xie, Biancun Xie and M. Swaminathan, "Electrical-thermal modeling of through-silicon-via (TSV) arrays in interposer," accepted for publication in International Journal of Numerical Modeling: Electronic Networks, Devices and Fields, 2012.

24. [Lee et al., 2011] Lee M, Cho J, Kim J, Pak J. S, Kim J, Lee H, Lee J, Park K. Temperature-dependent through-silicon via (TSV) model and noise coupling. IEEE 20th Conference on Electrical Performance of Electronic Packaging and Systems (EPEPS), 247–250, 2011.

Chapter 5

**Power Distribution, Return Path Discontinuities
and Thermal Management**

In the last chapter, signal and power integrity for interposers were briefly reviewed and the effect of return path discontinuities (RPD) was introduced. In addition the effect of temperature on the high frequency TSV response was discussed.

A very important element for 3D ICs is power distribution since the current density in the package will increase due to the stacking of ICs. With the large communication bandwidth between the logic and memory IC, it is also expected that the power consumed by the input/output (I/O) terminals of the IC will increase, thereby increasing the importance of I/O power distribution.

In the interposer and package/PCB, power and ground planes are invariably used to supply power to the integrated circuits. It is therefore important to understand the effect of power and ground planes on signal integrity. The interaction between signal and power distribution in the interposer, package and printed circuit board (PCB) occurs through RPDs, as covered briefly in Chapter 4. If RPDs result in increased jitter and reduced eye height as described in Chapter 4, then clearly the RPD effects need to be investigated and mitigated. This is possible through two methods namely, 1) using bypass capacitors and 2) using power transmission lines (discussed in Chapter 6).

With the stacking of ICs, a major problem is power dissipation since large amounts of heat can get trapped in the system, generating hot spot areas. In the power distribution network (PDN), large current densities and temperature gradients can cause increased voltage drops due to Joule heating, an effect that is very important for 3D systems especially when the power levels for the ICs increase. Since heat dissipation and power delivery are related to each other, we have combined them in this chapter along with an analysis of water based cooling for managing the temperature increase.

This chapter covers details on power distribution with a focus on the interposer and package/PCB. Therefore it is not the intention of this chapter to analyze on-chip power grids in detail – instead the power integrity part of the problem is discussed in the context of I/O signaling. Since TSVs have lower inductance as compared to wirebond or package-on-package (see Chapter 1), it is expected that TSV technology will help reduce power supply noise. In this chapter our focus is therefore on addressing the power distribution and thermal management related challenges for 3D ICs and interposers based on TSV technology.

5.1 Power Distribution – An Overview

As is well known, a chip has two kinds of transistors to be powered namely a) transistors within a single chip that communicate with each other also called the core and b) transistors on different chips that communicate with each other also called the I/O. Since the I/O power distribution is often times more noisy (since the signal lines exit the chip), it is decoupled from the core power distribution to maintain low noise levels. In the mid-1990s, a concept called the Target Impedance was developed that helped streamline the design of power distribution networks [Swaminathan et al., 2007]. Today, the target impedance as a design parameter has been adopted quite extensively by both industry and academia around the world and several methodologies have been developed for designing power distribution networks based on the target impedance concept. A brief description of target impedance and power distribution impedance is provided in this chapter for completeness, with a more extensive coverage available in [Swaminathan et al., 2007].

Consider a packaged chip mounted on a printed circuit board (PCB), as shown in Figure 5.1. The chip is powered by a voltage regulator module (VRM) mounted on the PCB. The VRM contains the DC supply voltage as well as the necessary regulatory circuit to manage voltage fluctuations when the chip draws current. The VRM supplies current through the power and ground planes in the PCB and package to the power and ground terminals of the chip. Decoupling capacitors serve as reservoirs and storage of charge in the power distribution network when there is a sudden surge of current due to transistors switching. At any point in the system, impedances can be calculated or measured between the power and ground terminals, also called as the power distribution impedance. Since, the functionality and reliability of transistors depend on the voltage fluctuations between the power and ground terminals, these fluctuations should not be allowed to exceed a maximum limit. This limit is determined by calculating the target impedance, a frequency domain parameter that sets an upper limit for the maximum impedance allowed in the power distribution network. The value of target impedance is calculated based on Ohm's law, as described in Figure 5.1. From Ohm's law, the voltage fluctuation (ΔV) due to a current source (I) can be calculated as:

$$\Delta V = ZI \tag{5.1}$$

where Z is the impedance. For a given chip, the current drawn by the chip is either known or can be computed. The transistors have a specification in terms of the maximum voltage fluctuation they can tolerate, as a tolerance of the DC supply voltage. Hence, the target impedance for the power distribution can be calculated as:

$$Z = \frac{\Delta V}{I} \tag{5.2}$$

In the above equation, the target impedance Z is expressed in the frequency domain. Hence, as the current "I" varies with frequency, the target impedance Z also varies with frequency. For a moment, let's assume that the current "I" is constant with frequency, which makes the target impedance Z constant with frequency as well. The constant target impedance curve is shown in Figure 5.1 where the vertical axis represents power distribution impedance and the horizontal axis is frequency. Over the last two decades, due to scaling of the transistor, the voltage has been decreasing while the current has been steadily increasing leading to target impedances steadily decreasing from one computer generation to the next. Hence, low target impedance is always required and preferred to minimize power supply fluctuations in any digital system. Here, power supply fluctuation refers to the variation of voltage with time across the power supply terminals of the IC.

In Figure 5.1, Target Impedance is what is desired, while the actual power distribution impedance seen by the chip (shown as a circle in Figure 5.1) may vary. The goal is to ensure that the power distribution impedance is always less than the target impedance.

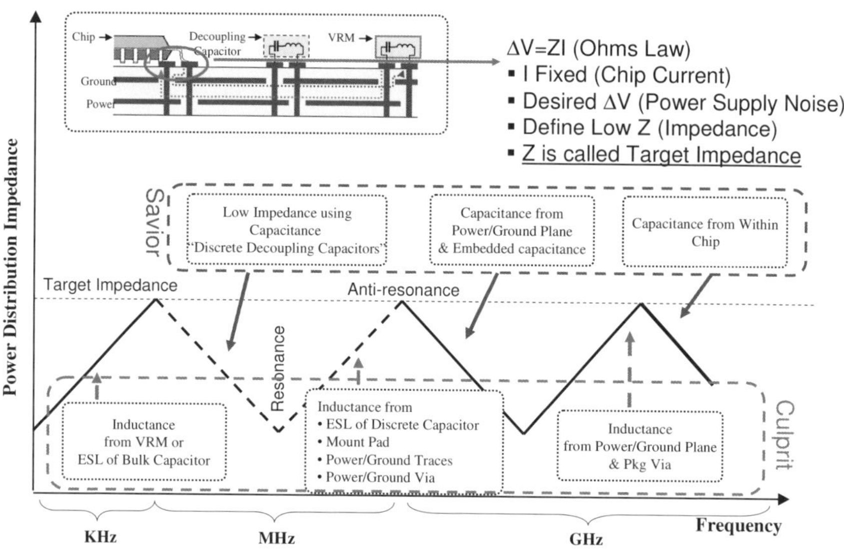

Figure 5.1: Power Distribution Impedance and Target Impedance.

In a system, parasitics in the power distribution manifest themselves as resistances, inductances and capacitances. The resistances lead to a flat impedance curve with "zero" slope while the inductances and capacitances result in a positive and negative slope respectively for the impedance, when plotted on a log scale with frequency. Since the goal is to maintain power distribution impedance less than the target impedance, inductances are always the "culprit" while capacitances are the "saviors" while designing the power distribution network, as depicted in Figure 5.1.

Various parts of the system affect the power distribution impedance over a different frequency spectrum. For example, the VRM inductance has an effect in the KHz frequency range while the plane capacitance affects the impedance in the upper MHz frequency range. Inductances and capacitances in the network lead to resonances (impedance null) and anti-resonances (impedance peak), where the latter needs to be managed using decoupling capacitors. The decoupling capacitors themselves have inductance (equivalent series inductance ESL) which needs to be accounted for, while designing the power distribution network. Clearly, designing the power distribution network is complex, and thankfully over the last decade, several papers and books have been published that has made the design process much easier. We will use this overview as a starting point to delve deeper into the power distribution details associated with silicon and glass interposers and revisit the issue of maintaining low target impedance for the I/O power distribution. For more details on power distribution design, the readers are referred to [Swaminathan et al., 2007] and [Novak et al., 2007].

5.2 Power Distribution for 3D Integration

Various embodiments for 3D integration are being proposed today with almost an infinite number of possibilities in terms of how the stacked ICs will communicate with each other. These include 3D stacked dies (which include both face to face and face to back connections), 3D stacked dies side by side on a logic die, stacked dies side by side on an interposer (also called 2.5D), dies on either side of an interposer, stacked dies on an interposer side by side on a PCB to name a few. Three embodiments of stacked dies in a system are shown in Figure 5.2 that capture the main issues associated with power distribution, no matter how the dies are re-arranged.

In all of the embodiments in Figure 5.2, a VRM supplies power through the voltage (Vdd) and ground (Gnd) planes in the PCB and interposer. The term interposer has been used loosely to also define the package since additional packaging may be required in between the interposer and PCB for many applications. The voltage and ground

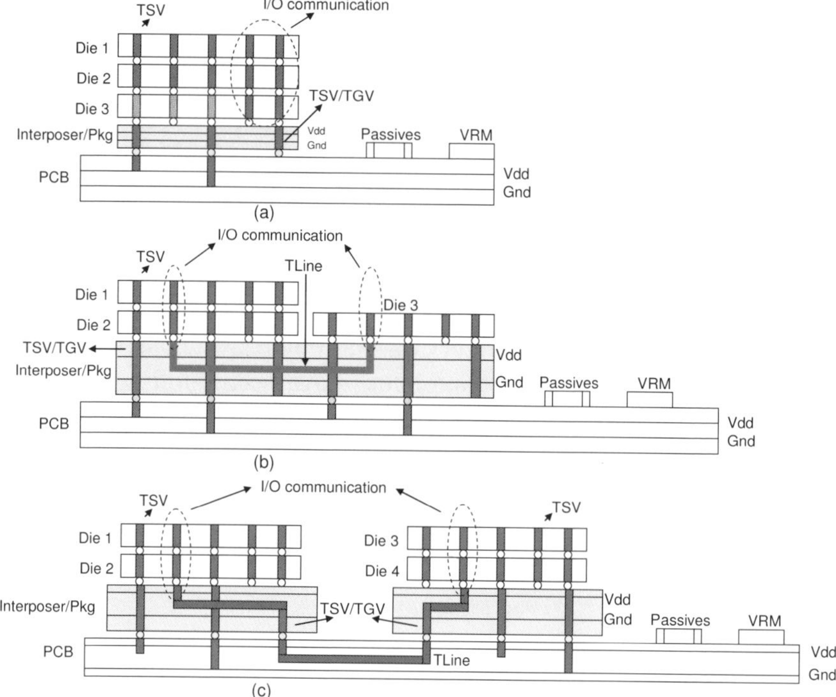

Figure 5.2: 3D Integration (a) Stacked dies on an interposer, (b) Stacked dies side by side on an interposer and (c) Stacked dies side by side on a PCB.

planes are connected through vias to the on-chip power grid (not shown) which in turn supplies power to the transistors in the die. The vias consist of through silicon vias (TSV) in the die, TSV or through glass via (TGV) in the interposer and PCB vias. The passives mainly constitute decoupling capacitors attached to the power distribution network (PDN) which are present both on the PCB (surface mount capacitors) and within the die (on chip capacitors). Clearly, when circuits switch they draw current from the PDN. But where does this current go? It obviously has to go towards charging the interconnections used to communicate between the circuits. Since current flows as a loop, the current drawn from the PDN has to complete its loop through the interconnections being charged or discharged and therefore the PDN response will have a direct effect on Signal Integrity (SI). The current loop when interrupted causes a return path discontinuity (RPD) which will be explained in detail later. Let's now consider the I/O communication as shown in Figure 5.2. In Figure 5.2(a), the communication is between dies 1, 2 and 3 through the vias and hence the current loop formed by the voltage, ground and signal vias with the rest of the PDN will affect Signal Integrity. With on-chip capacitors providing local capacitance, the current loop needs to be completed between the three dies through the on-chip capacitance for obtaining the best signal integrity. However, this is not the case in Figure 5.2(b) where the two die stack communicates with die 3 through the interposer. The interconnections not only pass through the vias but have to traverse laterally between the voltage and ground planes in the interposer before completing the communication path. The current loop here is formed between the vias (voltage, ground, signal), the interconnection in the interposer (represented as TLine for transmission line) and the voltage and ground planes. Therefore the signal integrity of this path is dictated by the part of the PDN that completes the current loop which includes the voltage and ground planes. The current loop worsens in Figure 5.2(c) where the signal interconnection traverses both the interposer and PCB and therefore the PDN in the die, interposer and PCB will affect signal integrity since they affect the current path. Since a smaller current loop will always minimize discontinuities, one can expect that the maximum channel speed and Signal Integrity can be achieved from Figure 5.2(a) followed by Figure 5.2(b) and then Figure 5.2(c).

Let's look at the power distribution in more detail in Figure 5.2(a). As mentioned previously, the parasitic that generates power supply noise is inductance which can be mitigated using capacitance. Resistance of the power distribution provides a DC level shift due to the ohmic drop. If so, in Figure 5.2(a), Die 1 will experience more inductance (since it is farthest from the PCB) as compared to Die 2 with the least inductance present for Die 3. Therefore, when the I/O circuits switch simultaneously, maximum power supply noise will occur in Die 1, followed by Die 2 and then Die 3. Though this seems intuitive, computing the inductances and capacitances in a PDN with 3D die stacks can be complex and it is therefore appropriate to understand the behavior of these PDNs based on detailed analysis. We will therefore rely on two detailed analysis available in the literature to better understand the PDN impedance and power supply noise.

5.2.1 *Power Distribution Impedance*

In the paper by [Kim et al., 2010], analysis of the PDN impedance for DRAMS stacked on a GPU (Graphical Processing Unit) using silicon interposer is discussed. We will use this example to better understand the impedance seen by each die in the die stack. The structure of the stacked die is shown in Figure 5.3 where three DRAMs are stacked on a silicon interposer which in turn is stacked on the GPU. The PDN in each of the DRAMs and GPU consists of a two layer on-chip power grid made up of orthogonal Power/Ground (P/G) lines as shown in Figure 5.3.

The dies are connected together using P/G through silicon vias (TSVs). The silicon interposer consists of a two layer orthogonal P/G grid as well which is then connected through TSVs to the Backside Re-distribution Layer (BS-RDL) on the GPU which then connects to the on-chip PDN in the GPU using TSVs. The BS-RDL consists of co-planar lines. Let's consider the case when the PDN providing power to the I/O circuits consists of sizes $4\text{mm} \times 1\text{mm}$ (DRAM), $10\text{mm} \times 1\text{mm}$ (GPU)

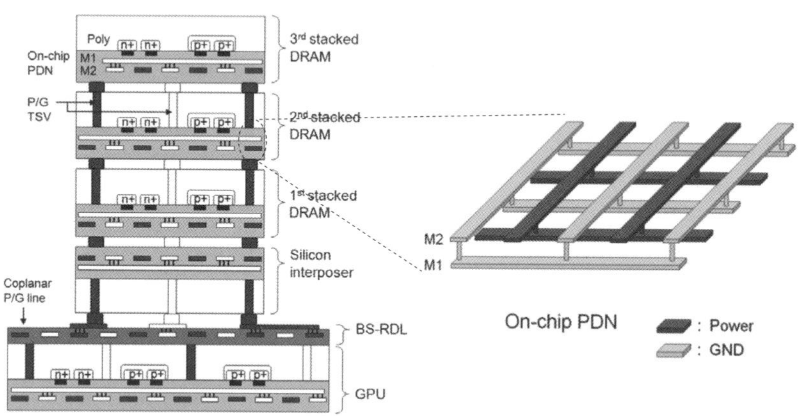

Figure 5.3: 3D Integration with DRAMs, interposer, BS-RDL and GPU showing Power Distribution Network [Kim et al., 2010].

and $4\text{mm} \times 2\text{mm}$ (interposer), as described in the paper. Though the details of the modeling method are described in the paper, we will focus mainly on the response, rather than on the intricacies of the modeling method itself.

Let's start by describing a target impedance (Z_{target}) for the PDN as 0.21Ω based on the voltage tolerance (ΔV) and current drawn by the I/O circuits. The input impedance seen from each of the DRAM die that includes the on-chip power grid, a pair of P/G TSVs in each of the stacked die and interposer, P/G lines in the interposer, BS-RDL and on-chip power grid in the GPU is shown in Figure 5.4(a). The impedance starts by being capacitive (negative slope) up to around 3GHz, reaches a minimum and then becomes inductive (positive slope) until 30GHz. Resonances and anti-resonances can be seen in the response due to the interaction between the inductances and capacitances of the power distribution network. The initial capacitive slope (region I) can be attributed to the capacitance in the on-chip and interposer power grid, shown as C_{DRAM}, $C_{interposer}$ and C_{GPU} in Figure 5.4(a). Similarly, regions II, III and IV contributions in Figure 5.4(a) are due to the inductance/resistance of the BS-RDL ($L_{BS\text{-}RDL}$, $R_{BS\text{-}RDL}$), capacitance of the DRAM/interposer (C_{DRAM}, $C_{interposer}$) and inductance of the DRAM/interposer/TSV (L_{DRAM}, $L_{interposer}$, L_{TSV}). The response in region V becomes interesting due to the resonances and anti-resonances caused by the interaction between the capacitances and inductances in the network. For example, the interaction between capacitance of the DRAM (C_{DRAM}) and inductance of the TSV (L_{TSV}) causes nulls and peaks in the response, as shown in Figure 5.4(a). The remaining interactions, also called modes, are also illustrated and indicated as "a-f" in the figure. The most important result of this analysis is the impedance profile seen by the three DRAM dies, as shown in Figure 5.4(a), where the impedance seen by each DRAM die has been plotted. In regions I and II, the results for the three stacked DRAMs are identical with a small deviation occurring in region III. However, in regions IV and V, larger deviations begin to occur between the impedances seen by DRAM dies 1, 2 and 3. Due to the positive slope of the impedance in these regions, the parasitic that dominates the response is the inductance formed between the power/ground (P/G) loops. Since DRAM 3 has higher impedance than

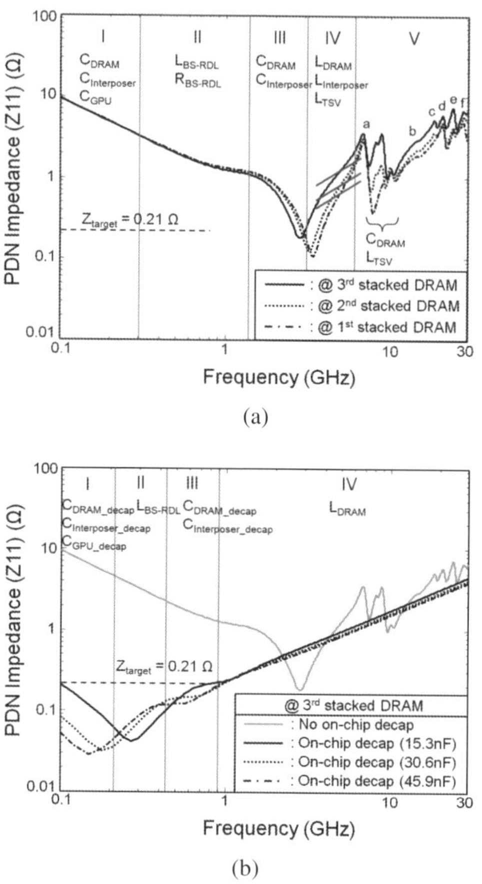

(a)

(b)

Figure 5.4: PDN input impedance seen by each DRAM die in the stack (a) without and (b) with on-chip capacitance [Kim et al., 2010].

DRAM 2 which in turn has higher impedance than DRAM 1, the dies higher in the stack will experience a larger P/G loop inductance, as expected, similar to the intuitive reasoning earlier.

One way to decrease the PDN impedance is by using decoupling capacitors [Swaminathan et al., 2007]. Consider the stack in Figure 5.3 where the GPU now contains capacitors in the on-chip power grid. Assuming 5%, 10% and 15% of the PDN area in the GPU contains decoupling capacitance, which translates to 15.3nF, 30.6nF and 45.9nF of total capacitance in this example, the resulting PDN impedance for

DRAM 3 in the die stack is shown in Figure 5.4(b). Clearly, the on-chip capacitance has reduced the impedance below the target impedance up to 1GHz, with the first resonance above 1GHz shifting to a lower frequency as the capacitance is increased. However, beyond 1GHz, the impedance is still inductive, with an increase in capacitance doing little to decrease the impedance. Though it is true that the P/G inductance will reduce as the number of P/G TSVs are increased (inductances in parallel), the loop inductance can still be a major problem as the dies are stacked. Since, the number of I/O circuits switching per unit area will increase for 3D integration as compared to having dies side by side (2D integration), one would expect that the inductive slope beyond 1GHz can cause excessive power supply noise, when circuits switch in this frequency range. This currently is a concern for 3D IC designers.

5.2.2 *Power Supply Noise*

Though related to each other, let's now look at power supply noise instead of PDN impedance. Unlike impedance, power supply noise is a transient phenomenon which occurs when circuits switch simultaneously and causes the power and ground terminals of the transistors to fluctuate with time. As explained in [Swaminathan et al., 2007], an increase in power supply voltage for an integrated circuit (IC) is a reliability limiter (transistors will cease to function) while a decrease in power supply voltage represents a performance limiter (transistors will run slower). An interesting paper by [Huang et al., 2012] considers the future of stacked ICs by addressing power supply noise in the 45nm node. Though the paper compares physical modeling with spice simulation results, we will once again focus on the results rather than the modeling approach itself. The fundamental difference between physical and spice modeling is that the former solves the partial differential equations (PDE) directly using appropriate boundary conditions as compared to the latter which uses circuit simulations. Since, the starting point is a circuit model for both, one would expect the two results to be the same if the PDE describes the circuit behavior appropriately. Figure 5.5(a) shows stacked dies on a package, with the dies connected to each other through TSVs and micro

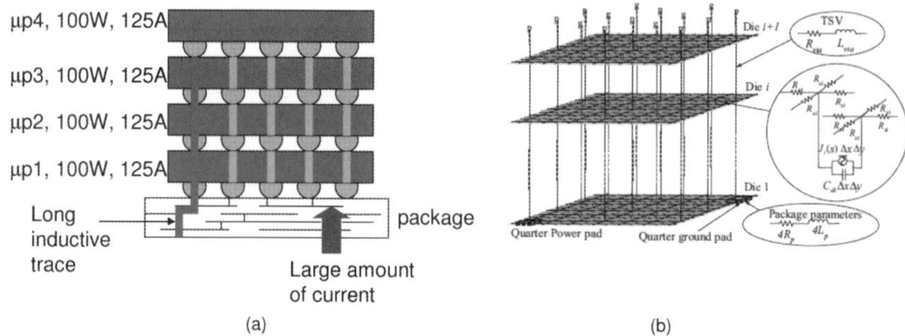

Figure 5.5: 3D Integration (a) stacked logic dies and (b) circuit model [Huang et al., 2012].

bumps. The die stack is powered from the bottom of the package. Since each die consumes 125A current, a large amount of current flows through the bottom of the package. The long inductive trace of the TSVs shown in the figure is the major parasitic that induces power supply noise. Since, each die consumes 100W of power, these are logic dies (represented as μP for microprocessor). Assuming the P/G connections at the bottom of the package are held at constant voltage, the goal is to estimate the peak power supply noise at the transistors in each of the stacked dies. This computation is possible through detailed modeling of the on-chip power grid of each die, TSVs, package interconnections and the on-chip capacitance for each die arising both due to the idle transistors and the added capacitance in the on-chip power grid. The power grid model of each die consisting of the two topmost metal layers is shown in Figure 5.5(b) along with the TSV model used for interconnecting the die models to each other. Since, the top metal layers of a die do not generally scale from one generation to the next, the physical dimensions used for the on-chip power grid correspond to 65nm technology in this example. In Figure 5.5(b), the TSVs are modeled as a series R-L circuit (R_{Via} in series with L_{Via}) while the on-chip power grid is modeled as a resistive network consisting of resistance R_{Si} between two nodes. The on-chip capacitance C_{di} is per unit cell with a cell area of $\Delta x \Delta y$. The current source $J_i(s)$ in each cell is also shown in the figure. The package resistance and inductance R_p and L_p correspond to a quadrant of the package. Dividing a chip into cells, connecting cells to form a power grid and connecting the power grids with TSV models leads to a complex network that can be analyzed in a circuit simulator (Spice) [Huang et al., 2012].

To enable circuit simulation, let's define the parameters of the model from [Huang et al., 2012]. Each die is partitioned into five functional blocks with each block switching simultaneously. The current drawn by the functional blocks translates into a current density of 100A/cm^2. The on-chip power grid resistance R_{si} is 0.22Ω with 43 P/G wires between P/G pads on the die. Decoupling capacitance covers 20% of the area of the die with an equivalent oxide thickness of 0.65nm. In Figure 5.5(b) the pad size is assigned as 1/6th of the pad pitch with 49 nodes in the on-chip power grid connected to a pad. The dies are stacked using solder bumps with 100μm diameter, and TSVs with 50μm diameter and 200μm height is used with a loop inductance of 0.12nH (0.06nH each for the power and ground trace). The package inductance L_p is assumed as 0.5nH. The supply current waveform for a functional block switching on and off is approximated by a ramp function with a rise time of 0.7ns.

Consider first the example of a single die switching in the die stack similar to a microprocessor communicating with memory, where most of the current drawn is from the microprocessor. A total of ten dies are modeled in the stack, as shown in Figure 5.6(a), where die i contains the switching circuits. The peak power supply noise computed in die i when the position of the die is varied in the stack (i = 1 to 10), is shown in Figure 5.6(b). Clearly, as the die position i in the stack is increased, power supply noise increases as well. This can be attributed to the increased loop inductance of the PDN, as the position of the die in the stack increases. In Figure 5.6(b), consider the linear part of the curve. Assuming the noise is purely inductive, the voltage drop across the power supply terminals of the transistors can be written as:

$$\Delta V_4 = L\frac{dI}{dt}$$

$$\Delta V_7 = (L + 3\Delta L)\frac{dI}{dt}$$

$$(5.3)$$

where ΔV_4 and ΔV_7 are the peak voltage droops (reduction in power supply voltage) on die 4 and die 7, respectively, L is the inductance to die 4, ΔL is the additional inductance between two adjacent dies and dI/dt is the transient current drawn by the switching circuits.

Since, only one die switches at any time, dI/dt is identical in the above two equations. From Figure 5.6(b), $\Delta V_4 = 105$mV and $\Delta V_7 = 130$mV, leading to $\Delta L/L = 0.08$. Hence, as the position of the die in the stack increases by one, an additional 8% inductance is added to the PDN, resulting in increased power supply noise. This gives an estimate on the amount of added inductance per stack, assuming that the voltage drop is linear. As is well known, with excessive inductance, transient current dI/dt decreases due to a slow down of the drivers, causing saturation. From Figure 5.6(b), with the top die switching in a 10 die stack, ~155mV of power supply noise is generated, which appears to be manageable since it represents only 14% of a 1.1V power supply typically used in 45nm technology. By adding additional decoupling capacitors in the dies, the noise can be controlled to within 5–10% of the power supply voltage, since the capacitors in the remaining nine dies can contribute towards reducing the noise in the switching die as well.

Let's now consider a different scenario where logic dies with similar footprint and functional blocks communicate with each other vertically. To enable communication between the dies, the functional blocks within

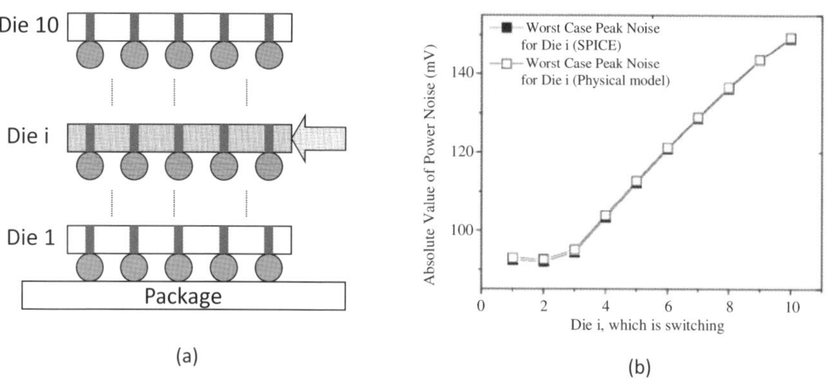

Figure 5.6: (a) Die i in a 10 die stack and (b) Power supply noise at die i due to circuits switching in die i [Huang et al., 2012].

the same foot print in each die will all switch simultaneously, causing the simultaneous switching of multiple die. This scenario is illustrated in Figure 5.7(a) where "n" dies with "n" varying from 1 to 10 are stacked on each other with all dies in the stack switching. The resulting noise is shown in Figure 5.7(b) for the top die and bottom die in the stack. From the figure, for a ten die stack, using the noise voltage at the bottom die of 400mV and at the top die of 800mV, $\Delta L/L = 0.11$, using an analysis similar to (5.3). Therefore, in the ten die stack, comparing the bottom and top die, the PDN inductance almost doubles. The slightly larger value of 0.11 as compared to 0.08 can be attributed to multiple dies switching simultaneously, resulting in less capacitance available for decoupling. For the bottom die, the noise saturates to 400mV while the noise increases linearly to 800mV for the top die in the ten die stack, which represents 73% of the 1.1V power supply voltage. This obviously is a major problem since most of the decoupling capacitor resources are exhausted (idle circuits are unavailable to provide the extra capacitance) to reduce the noise. From the discussions in this section, it is clear that multiple dies switching in a stack can cause excessive power supply noise in the system.

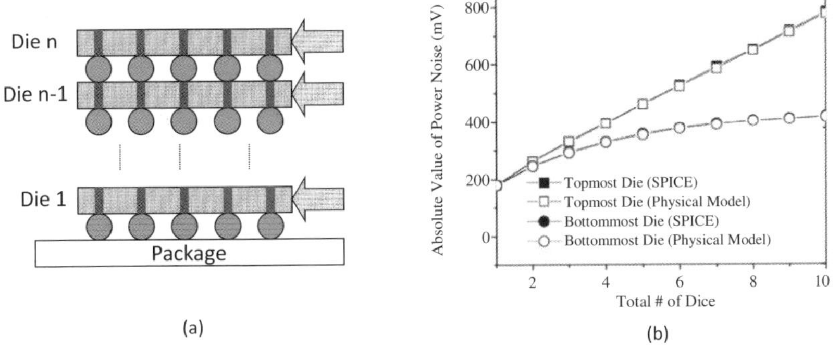

Figure 5.7: (a) Die stack with all dies switching and (b) power supply noise with variation in the die stack [Huang et al., 2012].

The two examples relating to PDN impedance and power supply noise correspond to the switching scenario shown in Figure 5.2(a) where the dies on the stack communicate with each other. *Based on the analysis the following can be ascertained for a 3D IC stack: 1) As the position of the die in the stack increases, so does the inductance (and resistance) of the power distribution, 2) The power supply noise increases as the position of the die in the stack increases, 3) Power supply noise can be excessive when many circuits switch simultaneously and 4) Decoupling capacitors are required to minimize power supply noise.*

However, it is important to realize that Figure 5.2(a) is just one embodiment of 3D integration. With thermal management still a major problem for removing heat in 3D ICs, trying to enable communication between stacked dies for high power dies as in Figure 5.5(a) can be very challenging. Hence, alternate solutions as shown in Figures 5.2(b) and (c) are required where the backside of the logic die (Die 3) is available for removing heat in the former and limited stacking (two logic dies) is used in the latter for better thermal and noise management. In both Figures 5.2(b) and (c), interconnects in the interposer and PCB are used for interconnecting dies together. With both the interposer and PCB containing voltage and ground planes, it is therefore important to understand their effect on the power supply noise. Let us therefore start by looking at the current paths during switching, and their role in affecting power supply noise.

5.3   Current Paths in IC and Package

Let's start with the scenario in Figure 5.2(a). Consider a push-pull driver I/O circuit on Die 1 communicating with a receiver circuit in Die 2. A simplified representation of the communication path is shown in Figure 5.8.

The on-chip power grid for both Die 1 and Die 2 are shown along with the microbumps connecting the two dies together using TSVs. The driver is powered from the power (P) and ground (G) rails of the on-chip power grid. The TSVs connect to the power grid through the interlayer metal dielectric (IMD) vias. Three TSVs are shown namely power (P), signal (S) and ground (G). The on-chip capacitance C_{di} across the driver and inductance of the TSV L_{TSV} are also shown in the figure (resistances not shown). The capacitance C is a cumulative sum of the capacitance between TSVs, between microbumps and the receiver capacitance (at the

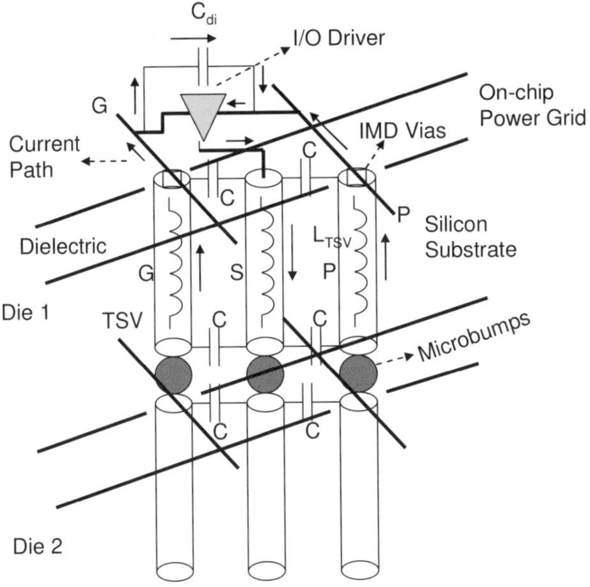

Figure 5.8: Communication path between Die 1 and Die 2.

far end). Consider a low-to-high switching event for the driver where the current is drawn from the power TSV to charge the signal TSV, as shown in the figure. This current flows from the power TSV, through the PMOS transistor of the driver (not shown) into the signal TSV and completes the loop through the local capacitances between the power and signal TSV. Due to the proximity of the ground TSV, the signal current induces a return current on the ground TSV as shown, which completes the loop through the decoupling capacitor, since this represents the low impedance path (nmos transistor is open during this transition). It is important to note that the local capacitances shown in the figure help in completing the loop and therefore neither the package nor PCB contributes to the noise. Assuming the capacitance is insufficient to provide the low impedance path, the current will then loop through the package and PCB causing increased power supply noise. Hence, the frequency of the switching activity is critical in determining the path of least impedance, which can be gathered from the PDN impedance plot similar to the one shown in Figure 5.4. Ultimately, the inductance and capacitance in the current loop will determine the power supply noise with resistance in the network providing the DC level shift and damping of the noise waveform. In this example, the decoupling capacitor has been used to provide the low impedance path to complete the current loop, thereby minimizing power supply noise.

Let's next turn to the scenarios in Figures 5.2(b) and (c) where the interconnections traverse the interposer and PCB containing voltage (also called power plane) and ground planes in them. The planes serve as a conduit of charge from the VRM and decoupling capacitors to the IC. A communication path between a push-pull driver and receiver circuit in either the interposer or PCB is shown in Figure 5.9, where the VRM on one side of the board powers the two planes. The planes provide charge to the IC which flows through the vias and other interconnects (not shown) to the driver and receiver circuits. The parasitics along the current path are lumped together as a resistor and inductor in series, connected to the driver power supply (TxPWR) and ground terminal, as shown in Figure 5.9. As is well known, due to the longer lengths in the interposer and PCB, the signal interconnections behave as transmission lines as frequency increases, as shown in the figure. The stack-up in the package (or PCB) consists of signal, power and ground layers. Depending on the application, the number of layers and their assignments may change and therefore Figure 5.9 only represents an example of a stack-up where the signal layer is above the power and

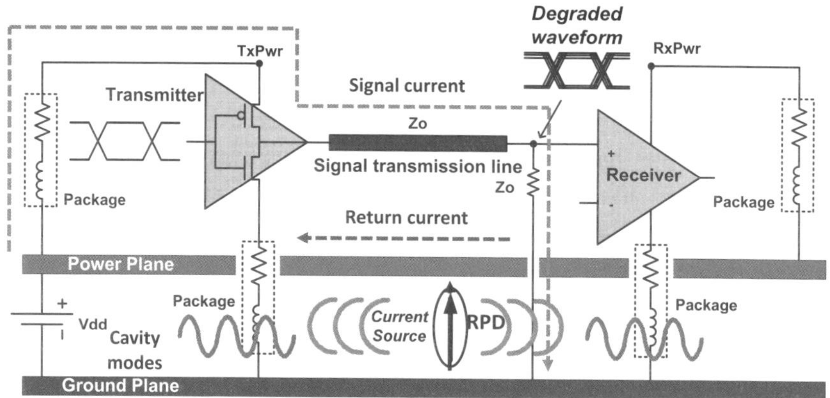

Figure 5.9: Die to die communication in interposer or PCB.

ground planes. When the driver switches, the charging and discharging of the signal transmission lines induce return currents on the power and ground planes [Swaminathan et al., 2010], as illustrated in Figure 5.9. Since, the return current follows the path of least impedance, it flows on the reference plane closest to the signal transmission line. In Figure 5.9, since the power plane is closer to the signal line, during low-to-high switching, the forward charging current in the signal line induces a return current in the opposite direction on the power (or voltage) plane. The forward current in the signal line and the return current on the power plane need to connect to each other, to complete the current loop. The three currents in the signal line, power and ground plane have to all come together without any high impedance discontinuity along its path, to minimize voltage fluctuations between the two planes. A notorious discontinuity (also called return path discontinuity or RPD) is shown in Figure 5.9, where the current jumps from the ground plane to the power plane through the interlayer capacitance, to complete the current loop. This current source excites the plane cavity formed between the power and ground planes, causing standing wave resonances. This disturbance causes signal integrity problems by reducing the eye height and increasing the jitter of the channel, illustrated as a corrupted received eye diagram in Figure 5.9. A common method to improve the quality of the received waveforms is by using decoupling capacitors between the power and ground planes. If these capacitors are placed at the location of the discontinuity, they can be used to provide a low impedance path for the current, thereby mitigating the RPD effect. In a complex package the cause for RPDs can be many (due to split planes, termination schemes, layer assignments to name a few) and since their position can be arbitrary, the need for low impedance return current paths has led to the development of new technologies such as thin dielectrics, embedded capacitance layers, high frequency decoupling capacitors, and others [Novak, 2000], [Balaraman et al., 2004], [Muthana et al., 2005], [Hobbs et al., 2001] and [Alvarez et al., 2000] at the package and board levels. In addition, the use of power and ground planes as part of the target impedance concept is making the packages and boards more complex, leading to the need for sophisticated tools and methodologies to account for all the discontinuities and mitigate them in a design. In the next section let's look deeper into the relationship between Signal and Power Integrity and determine the challenges that need to be addressed for high speed signaling outside the IC.

5.4 Signal and Power Integrity – Does One Affect the Other?

As illustrated in Figure 5.9, RPDs cause voltage fluctuations between the power and ground planes, which in turn induces signal integrity problems, leading to excess jitter and reduction in eye height of the received waveform. But how does this actually happen? To illustrate the interplay between signal and power integrity in the package and PCB, we use a simple test vehicle to describe the phenomenon [Swaminathan et al., 2010].

Vias in a package or board is a major source of return path discontinuities due to a change in the reference plane of the signal line due to via transitions. An interconnect path with via discontinuity can induce power supply noise in the PDN which is proportional to the PDN impedance at the via discontinuity. Since the PDN impedance is a function of frequency, the magnitude of noise induced in the PDN depends on the frequency of the signal propagating through the interconnection. Since digital signals are comprised of multiple frequencies with significant energy content at harmonic frequencies, the induced noise and jitter on the signal depends both on the PDN impedance and the harmonic content of the signal. The behavior of a signal through an interconnection can be quantified by understanding the network parameters of the interconnection and the harmonic content of the signal. The resulting jitter and noise on the signal can be estimated by, 1) determining the impact of PDN Impedance at the RPD on the insertion loss of the signal in the frequency domain and 2) relating the signal insertion loss and coupling to jitter and eye height of the signal in the time domain. These effects are quantified in this section using a test vehicle.

A test vehicle designed and fabricated to observe the impact of PDN impedance on the insertion loss of the signal and its effect on voltage amplitude and jitter is shown in Figure 5.10. The test vehicle consists of

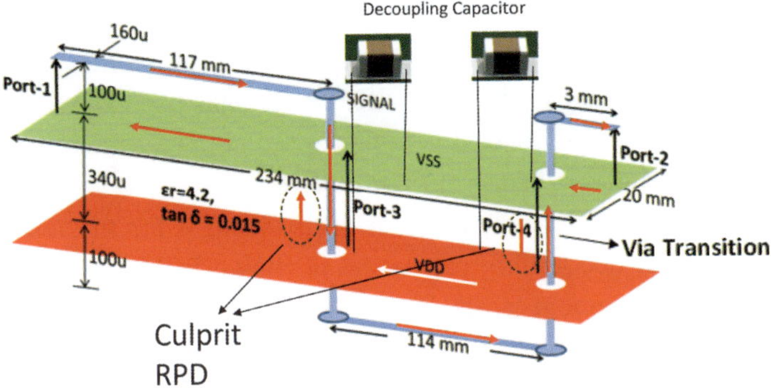

Figure 5.10: Test vehicle to illustrate return current and RPD effect.

four metal layers with microstrip line on the top and bottom layers and ground and voltage planes on the second and third layer, respectively, fabricated using PCB technology. The test vehicle contains two via transitions with each transition causing a return path discontinuity at the via location, as shown in Figure 5.10. Capacitor pads are provided near each via transition, so that suitable capacitors can be soldered to reduce the PDN impedance at the RPD to evaluate its effect on the signal waveform, as shown in Figure 5.10. The bypass (or decoupling) capacitors used have $C = 4700pF$, $ESL = 0.3nH$ and $ESR = 0.25Ohm$, where ESL and ESR are the equivalent series inductance and resistance, respectively. Details of the test vehicle with dimensions, port assignments and their placement are provided in Figure 5.10.

5.4.1 *Impact of PDN Impedance on Signal Insertion Loss*

As indicated earlier, the return current on the planes induce standing wave resonances between the planes in the frequency domain, when the current flows through a RPD. The resulting noise voltage induced between the planes is equal to the product of the return current and the PDN impedance at the discontinuity. As the PDN impedance is a function of frequency, the noise voltage also depends on the frequency of the current. At the anti-resonance frequency of the PDN, the impedance increases, resulting in a large noise voltage being induced between the planes, resulting in a smaller amount of energy propagating through the signal line from port 1 to port 2 in Figure 5.10. This manifests itself as an increase in the insertion loss of the signal, which can be measured using a Vector Network Analyzer (VNA). This effect has been captured in this section using the test vehicle. It is important to note that the reference for ports 1 and 2 is the VSS plane in Figure 5.10.

The measured PDN impedance at ports 3 and 4 for the test vehicle with and without decoupling capacitors is shown in Figure 5.11. In the figure, the measurement ports 3 and 4 are placed in the vicinity of the RPD and hence represent the plane impedance at the discontinuity. The difference in the position of the ports (port 4 closer to the edge than port 3) reflects in the frequency of the anti-resonance (positive peak), resulting in more anti-resonances in Figure 5.11(b) as compared to Figure 5.11(a).

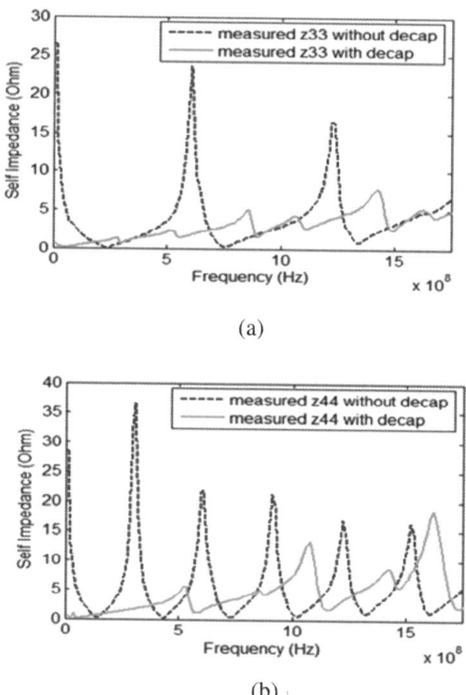

(a)

(b)

Figure 5.11: Measured PDN impedance with and without decoupling capacitors at (a) Port 3 and (b) Port 4 [Swaminathan et al., 2010].

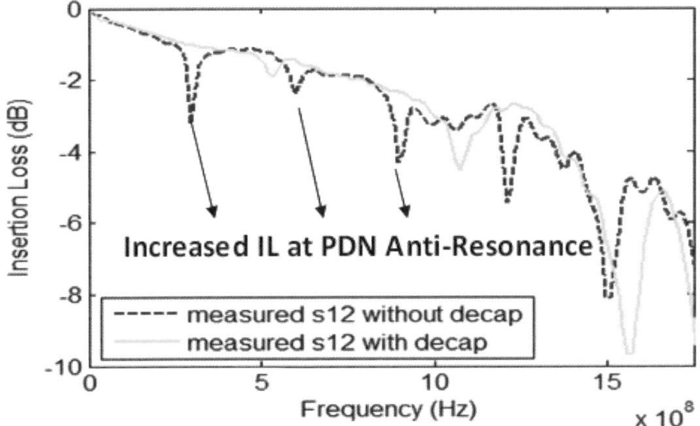

Figure 5.12: Comparison of measured insertion loss between Port 1 and 2 with and without decoupling capacitors [Swaminathan et al., 2010].

The combined effect of both of these anti-resonances reduces the insertion loss of the signal line, as shown in Figure 5.12, where the increase in the insertion loss (IL) coincides with the frequency of anti-resonance between the voltage and ground planes. The placement of the capacitor at the RPD reduces the plane impedance in Figure 5.11 which reflects in an improved insertion loss on the signal line up to 1GHz, as shown in Figure 5.12.

Beyond 1GHz, the impedance of the capacitor increases, making the capacitor less useful in providing a low impedance return path for the current. Clearly, there is a strong relationship between the PDN impedance at an anti-resonance frequency and the corresponding insertion loss of the signal at the same frequency. This interaction occurs due to the RPD. Elimination of the RPD requires a continuous return path which can be accomplished by using decoupling capacitors at the RPD location. Since decoupling capacitors are non-ideal, the goal is to ensure that the impedance of the decoupling capacitor is less than the PDN impedance at the anti-resonance frequency. The smallest impedance can be achieved by selecting a capacitor that resonates at the anti-resonance frequency of the PDN, resulting in the smallest insertion loss of the signal at that frequency.

5.4.2　*Impact of Insertion Loss and PDN Impedance on Jitter and Amplitude of Clock Signal*

The frequency spectrum of a clock signal consists of the fundamental and harmonics at odd multiples of the clock frequency. Therefore, a significant amount of energy of the clock signal is stored at these frequencies. For example, a 600MHz clock has significant harmonic components present at 600MHz, 1800MHz …, as shown in Figure 5.13. If the harmonics of the clock coincide with the frequencies at which the signal line insertion loss is large, they undergo a large attenuation causing reduction in the amplitude and increase in rise/fall time of the clock signal.

However, since the clock signal is periodic, the increased insertion loss on the signal line at the PDN anti-resonance frequencies does not increase jitter of the clock signal. Hence, the RPD's effect is a reduction in the voltage amplitude with little impact on the jitter of the clock signal. This effect has been illustrated in this section using a 0.8V, 600-MHz clock signal, propagated from Port 1 to Port 2 on the microstrip line in the test vehicle so that its fundamental frequency coincides with the increased insertion loss at 600-MHz, as shown in Figure 5.12.

The attenuation of the fundamental clock frequency due to the insertion loss causes a reduction in the amplitude of the clock signal with amplitude of 590mV, as shown in Figure 5.14(a). With the addition of

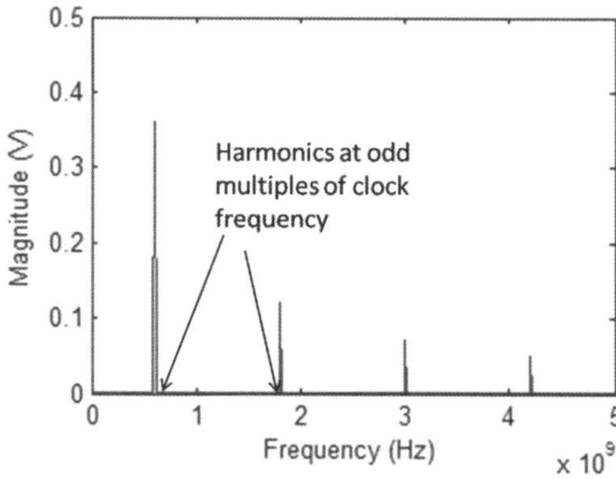

Figure 5.13: Spectrum of 0.8V 600 MHz Clock Signal [Swaminathan et al., 2010].

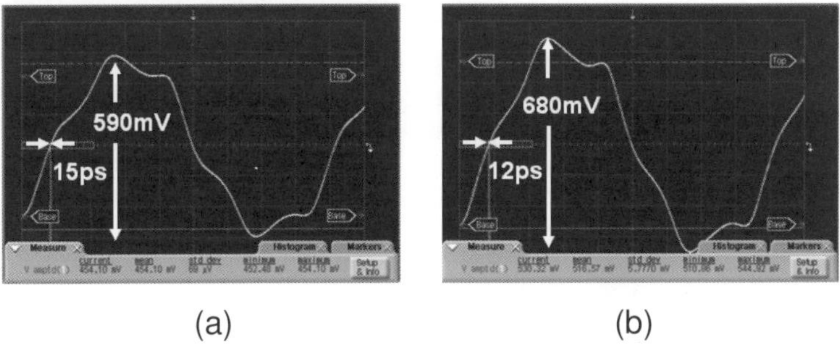

(a)   (b)

Figure 5.14: Comparison of 600 MHz clock waveforms at Port-2 (a) without and (b) with decoupling capacitors [Swaminathan et al., 2010].

the decoupling capacitor at the RPD, the insertion loss of the signal line improves, resulting in an increase in the clock amplitude to 680mV, as shown in Figure 5.14(b). However, the improvement in jitter is insignificant from 15ps to 12ps, which can be attributed to the source waveform uncertainty rather than the signal line insertion loss, since the energy content of the clock signal is concentrated at specific frequencies.

5.4.3 *Impact of Insertion Loss and PDN Impedance on Jitter and Noise of PRBS Signal*

The frequency spectrum of a Pseudo Random Bit Stream (PRBS) consists of harmonics distributed across multiple frequencies based on the data pattern. The envelope of the spectrum is a "sinc squared" function with nulls at multiples of the bit rate of the PRBS. As an example, the spectrum of a 600Mbps PRBS consists of nulls at multiples of 600MHz and has significant harmonic content at 900MHz, 1500MHz …, as shown in Figure 5.15. If significant harmonics of the PRBS signal coincide with large insertion loss peaks of the signal line as shown in Figure 5.12, they undergo large attenuation reducing the amplitude of the PRBS pulses and increasing their rise/fall times. However, since most of the energy of the PRBS signal is distributed across multiple frequencies, attenuation at a few discrete frequencies does not cause a large reduction in the amplitude of the PRBS signal. However, due to the randomness of the data pattern, the switching

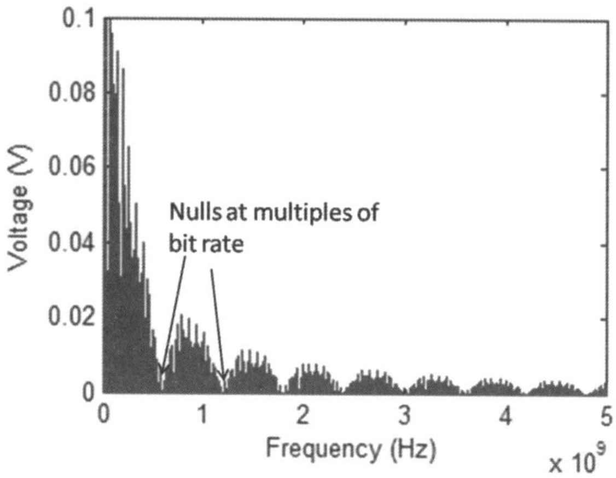

Figure 5.15: Spectrum of 0.8V, 600-Mbps PRBS signal [Swaminathan et al., 2010].

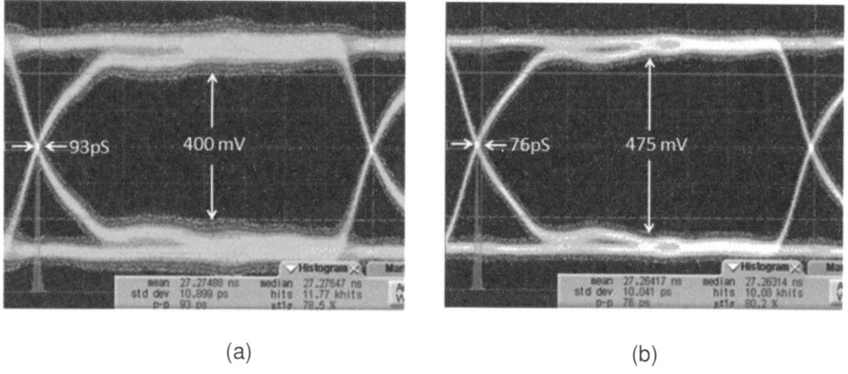

Figure 5.16: Measured eye diagram at port-2. (a) without decoupling capacitors (b) with decoupling capacitors [Swaminathan et al., 2010].

sequence does change the noise on the PDN, leading to a change in the rise/fall time of the signal, leading to an uncertainty in the timing of the signal edge. This effect manifests itself as increased jitter on the PRBS signal. To illustrate this effect, a 600-Mbps PRBS signal is applied to the test vehicle at Port 1 with a measurement of the received signal at Port 2.

A comparison of the eye diagram obtained at Port 2 with and without decoupling capacitors is shown in Figure 5.16. The addition of the decoupling capacitor reduces the PDN impedance, thereby improving the signal insertion loss, as shown in Figure 5.12. Hence, the eye diagram with decoupling capacitors shows an eye height of 475mV as compared to 400mV with the decoupling capacitor removed; a result that is smaller than the improvement for the clock signal in Figure 5.14. However, the addition of the decoupling capacitor improves the jitter significantly from 93ps to 76ps, a large effect which was absent with the clock signal.

In conclusion, RPDs generate noise in the PDN which affects the signal amplitude and jitter depending on the nature of the signal being transmitted. Hence, signal and power integrity are related to each other and therefore cannot be separated.

In Chapter 4, the cavity resonances between the power and ground planes when using a silicon interposer were mitigated due to the loss of silicon. However, irrespective of whether silicon, glass or any other material is used for the interposer, from a design standpoint, the RPD effect needs to be first quantified in the frequency domain by looking at the relationship between the PDN impedance and signal insertion loss, as described in this chapter. After applying suitable design methods on the PDN such as removing RPDs, choosing the right interposer or adding decoupling capacitors at appropriate locations to improve the insertion loss of the signal, the effect of the design change on the time domain response has to be evaluated. The time domain response can change in the presence of non-linear drivers and terminations, which needs to be evaluated as well. For a complex package or PCB, this can obviously be a very laborious process (requires detailed modeling and simulation) and hence the design of a PDN can be very time consuming and often times leads to expensive solutions where either many layers are used or the package or PCB is packed with decoupling capacitors to eliminate RPD effects.

To summarize, the following conclusions can be reached outside the IC in the interposer, package and PCB, namely: 1) Voltage and ground planes are invariably used to distribute power, 2) Return path discontinuities excite the cavity between the planes causing standing wave resonances, 3) The anti-resonances in the power distribution degrade signal insertion loss, 4) The interplay between power distribution and signal transmission can result in degradation of signal quality and 5) Decoupling capacitors are used to reduce the impact of RPDs.

5.5 Challenges for Addressing Power Distribution in 3D ICs and Interposers

The discontinuities along the return path that affect signal integrity are many, as illustrated in Figure 5.17 where a subset of the discontinuities experienced by the signal lines are shown for one layer of a package. Of course, not every discontinuity can be fixed using capacitors since other means need to be used as well such as changing stack-up, re-routing the lines around discontinuities, removing via transitions to name a few.

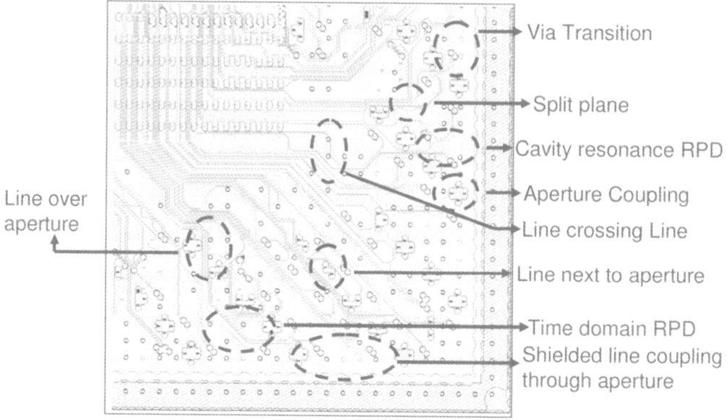

Figure 5.17: Typical Return Path Discontinuities in a package.

Our community has addressed these challenges until now using two fundamental approaches namely, 1) relying on signal and power integrity tools to model and asses the impact of the problem and 2) reducing the power distribution impedance at the RPDs, to improve performance. Needless to say, over the last decade, several signal and power integrity tools have emerged that has helped us immensely to tackle the RPD identification problem. To reduce PDN impedance designers have relied on the use of capacitors. Depending on the frequency range, capacitors have been used in the die, package and PCB, to manage power integrity. This has led to the PDN impedance reducing to the order of milli-ohms today, as compared to two decades back, as illustrated in Figure 5.18.

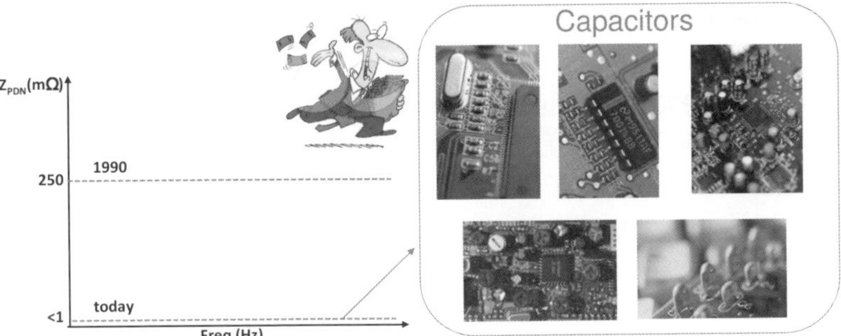

Figure 5.18: PDN impedance and role of capacitors.

Today, the capacitors we use in a system are enormous and often times require larger real estate than the ICs they service. With the electronics industry being very cost conscious, the luxury of relying on capacitors to continuously reduce the PDN impedance from one generation to the next is becoming increasingly difficult. Simply put, adding capacitors is akin to throwing away money as depicted in Figure 5.18 and this represents one overwhelming challenge we face today. This could very well be an opportunity for research. Some ideas are discussed in the next chapter which hopefully acts as a catalyst to transform the way we address the challenges we face, with a focus on signaling between ICs. These ideas can be applied for communication between ICs in a 3D stack, interposer or PCB.

5.6 Thermal Management and its Effect on Power Distribution

Any electronic system requires thermal management solutions to ensure that the temperature of the transistors do not exceed the maximum junction temperature allowed which is anywhere between 85 to 100 deg C for CMOS circuits. Several publications and books are available in the open literature discussing thermal management solutions. Our goal in this section is not to repeat these ideas but to focus on the thermal impact on power distribution. Our focus therefore is to evaluate the impact of steady state thermal gradients on DC drop in the context of systems containing stacked dies. For completeness numerical methods for computing the DC drop and thermal gradients simultaneously to analyze 3D systems are provided. It is important to note that computing the temperature and voltage gradients are both system level problems requiring system level modeling approaches where the impact of the PCB, package, interposer and interconnections become critical. Our objective therefore is to compute the temperature and voltage gradients up to the IC and not inside the IC. The packaging side of the problem is therefore addressed in this section.

5.6.1 *What is Joule Heating*

As we know, all metals have finite conductivity associated with them. When electric current flows through these conductors it releases heat. This process is called Joule heating. It was first studied by James Prescott Joule in 1841 where he concluded that the heat produced Q was proportional to the product of the square of the current I and resistance of the conductor R [Wikipedia] given by:

$$Q \propto I^2 \times R \tag{5.4}$$

The relationship in (5.4) is known as Joule's first law.

5.6.2 *Modeling Electrical-Thermal Effects in the Steady State*

In steady state, the governing equation for computing the voltage distribution across a conductor can be expressed as [Xie et al., 2011]:

$$\nabla \cdot \left(\frac{1}{\rho(x,y,z,T)} \nabla \phi(x,y,z) \right) = 0 \tag{5.5}$$

where, $\rho(x,y,z,T)$ and $\phi(x,y,z)$ represent the temperature-dependent electrical resistivity and voltage distribution.

Similarly, for steady-state thermal analysis, the governing heat equations for solid medium and fluid flow can be expressed as [Xie et al., 2011]:

$$\nabla \cdot \left[k(x, y, z) \nabla T(x, y, z) \right] = -P(x, y, z) \tag{5.6a}$$

$$\sigma c_p \vec{v}(x, y, z) \cdot \nabla T(x, y, z) = \nabla \cdot (k_f \nabla T(x, y, z)) \tag{5.6b}$$

where, $k(x, y, z)$ and $T(x, y, z)$ represent the thermal conductivity of solid medium and temperature distribution, respectively; σ, c_p and $\vec{v}(x, y, z)$ represent the density, heat capacity and velocity distribution of the fluid, respectively and k_f is the thermal conductivity of the fluid [Dragoni et al., 2002], [Hettiarachchi et al., 2008]. The details on the rationale for looking at fluid mediums for cooling stacked ICs will be discussed later in this chapter.

In (5.6a) $P(x, y, z)$ is the total heat source excitation including the heat dissipation from the chip and Joule heating generated by ohmic losses due to current flowing through the conductors in the power distribution network (PDN). The Joule heating effect can be expressed as:

$$P_{Joule}(x, y, z) = \vec{J} \cdot \vec{E}(x, y, z) \tag{5.7}$$

where, $\vec{J}$ is the current density and $\vec{E}(x, y, z)$ is the electrical field distribution in the PDN. Since the electrical resistivity is temperature-dependent, it is described by:

$$\rho = \rho_0 [1 + \alpha(T - T_0)] \tag{5.8}$$

where, ρ_0 is the electrical resistivity at T_0 which is 20 °C, and α is the temperature coefficient of the electrical resistance. The coupled set of equations therefore arise from (5.7) and (5.8) where electric current flowing through conductors produces heat through Joule heating which in turn changes the electrical resistivity of the conductor, thereby causing a change in its voltage drop. Similar to Chapter 4, the electrical resistivity variation with temperature for three conductors is shown in Figure 5.19 where the resistivity increases linearly with temperature. It is important to note that since silicon is a semi-conductor, its resistivity also changes as discussed in Chapter 4.

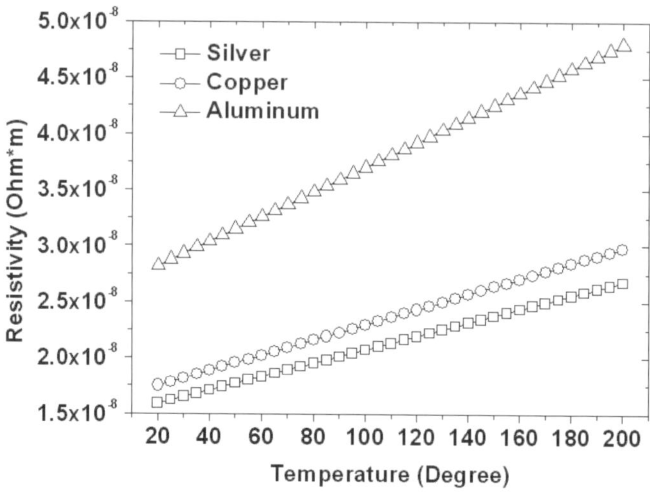

Figure 5.19: Temperature-dependent resistivity of silver, copper and aluminum [Xie et al., 2011].

Due to the temperature-dependent electrical resistivity $\rho(x, y, z, T)$ and Joule heating generated in the conductors, the electrical and thermal characteristics couple to each other and form a system of equations, as shown in Figure 5.20 [Xie et al., 2011]. To obtain accurate voltage distribution with conduction, convection and Joule heating effects, it is therefore required to solve the electrical-thermal equations (5.5)–(5.6), simultaneously. An iterative electrical-thermal co-simulation method therefore has to be developed that covers the following important procedures:

(i) Uses input information from package layout parameters, material properties, excitations, and boundary conditions for steady state electrical and thermal analysis.

(ii) Enables voltage distribution simulation for computing the voltage, current, and power distribution profiles in the PDN.

(iii) Calculates heat sources from Joule heating in the power distribution.

(iv) Uses the updated Joule heating excitation for thermal simulation with heat removal methods such as conduction, air convection and fluidic cooling.

(v) Updates the electrical resistivity of conductors based on the temperature in the PDN thereby accounting for the thermal effect on voltage drop.

(vi) Determines the convergence of temperature and voltage distributions. The final thermal and voltage distributions are displayed if convergence is reached.

These procedures are described in the context of Finite Volume Formulation in the next section.

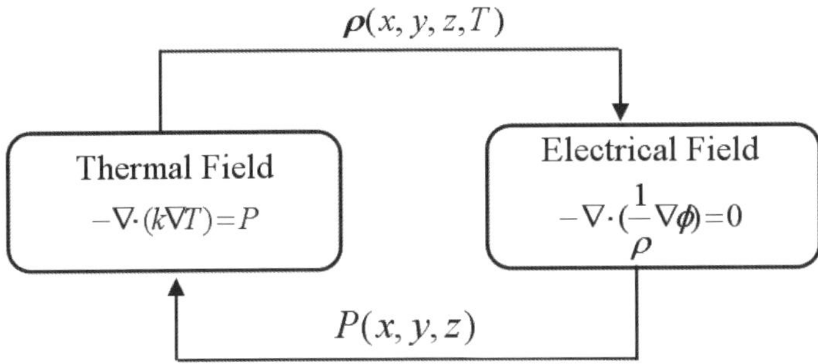

Figure 5.20: Interaction between electrical and thermal fields. Courtesy: [Xie et al., 2011].

5.6.3 *Finite Volume Formulation*

A simple procedure is provided here for simulating the electrical and thermal response by including Joule heating [Xie et al., 2011]. For establishing a simulation procedure, (5.5)–(5.6) are solved in an iterative manner using the same mesh for computing both the voltage and temperature distributions. Since a 3D system can contain geometries with different sizes (TSVs, C4s, planes etc) a non-uniform mesh is required to reduce the number of unknowns and to capture all the features. Here a non-uniform rectangular mesh is used to grid the volume of the structure. The formulations are explained using a 2D non-uniform grid for simplicity, which can be readily extended to 3D geometries.

5.6.3.1 *Voltage Distribution Equation*

The formulation for solving the voltage distribution equation (5.5) is performed using temperature-dependent resistivity. In Figure 5.21, $\phi_{i,j}$ represents the voltage at grid point (i, j) which is surrounded by four nodes. In the figure Δx_1 , Δx_2 , Δy_1 and Δy_2 are the nodal distances between node (i, j) and its adjacent nodes in x and y directions, respectively. The four surrounding cells of node (i, j) have different temperature T_1 , T_2 , T_3 and T_4 , which can be obtained from thermal simulation. In order to apply the finite volume method, node (i, j) is surrounded by a finite volume cell (dashed line) in Figure 5.21. The intersection points between the dashed cell and other four cells are the center points of each cell. By integrating (5.5) over the dashed cell and applying the divergence theorem, results in:

$$\int_{\substack{dashed \\ line}} \frac{1}{\rho(x, y, z, T)} \nabla \phi(x, y, z) \cdot \hat{n} dl = 0 \tag{5.9}$$

where, $\hat{n}$ is the outward pointing unit normal vector at the boundary of the dashed cell. By applying the finite difference approximation to the derivative of the potential ϕ in (5.9) and assuming temperature dependent resistivity in the adjacent cells, the finite volume scheme for computing the voltage distribution can be written as:

$$\left(\frac{\Delta y_1}{\rho(T_1)\Delta x_1} + \frac{\Delta y_2}{\rho(T_4)\Delta x_1} \right)\left(\phi_{i,j} - \phi_{i-1,j} \right)$$

$$+ \left(\frac{\Delta y_1}{\rho(T_2)\Delta x_2} + \frac{\Delta y_2}{\rho(T_3)\Delta x_2} \right)\left(\phi_{i,j} - \phi_{i+1,j} \right)$$

$$+ \left(\frac{\Delta x_1}{\rho(T_1)\Delta y_1} + \frac{\Delta x_2}{\rho(T_2)\Delta y_1} \right)\left(\phi_{i,j} - \phi_{i,j-1} \right)$$

$$+ \left(\frac{\Delta x_1}{\rho(T_4)\Delta y_2} + \frac{\Delta x_2}{\rho(T_3)\Delta y_2} \right)\left(\phi_{i,j} - \phi_{i,j+1} \right) = 0 \tag{5.10}$$

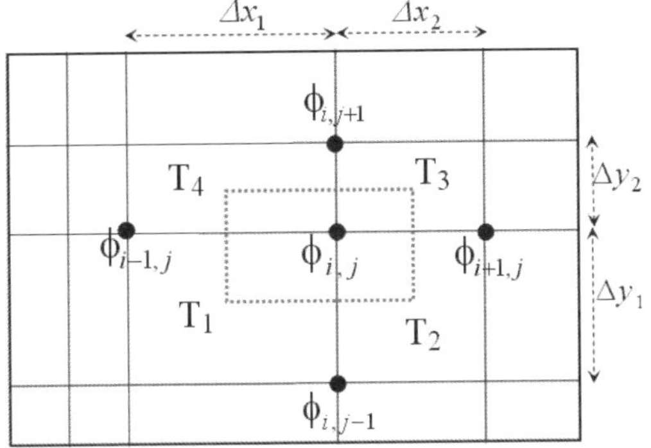

Figure 5.21: 2D rectangular grid for computing voltage distribution [Xie et al., 2011].

5.6.3.2 *Heat Equation for Solid Medium with Convection Boundary Condition*

For thermal simulation, the thermal conductivity 'k' is considered to be constant. For solving the heat equation (5.6a) with only conduction, since it has the same form as (5.5), the same finite volume formulation as in the previous section can be applied resulting in:

$$\frac{T_{i,j}-T_{i-1,j}}{\dfrac{\Delta x_1}{kd}}+\frac{T_{i,j}-T_{i+1,j}}{\dfrac{\Delta x_2}{kd}}+\frac{T_{i,j}-T_{i,j-1}}{\dfrac{\Delta y_1}{kw}}+\frac{T_{i,j}-T_{i,j+1}}{\dfrac{\Delta y_2}{kw}}=P_{total} \qquad (5.11)$$

where, $P_{total}=\displaystyle\iint\limits_{\substack{dashed\\cell}}-P(x,y,z)dS$ is the total heat source in the dashed cell in Figure 5.21.

Most electronic systems use air cooling using natural or forced methods (such as a fan). Therefore, implementing the convection boundary condition becomes important as well which is given by:

$$k\left.\frac{\partial T}{\partial n}\right|_{convection}=-h_c(T-T_a) \qquad (5.12)$$

In (5.12) T_a and h_c represent the ambient temperature and convection coefficient, respectively. The same finite volume formulation procedure described for the solid medium [Ozisik, 1994] can also be applied at the

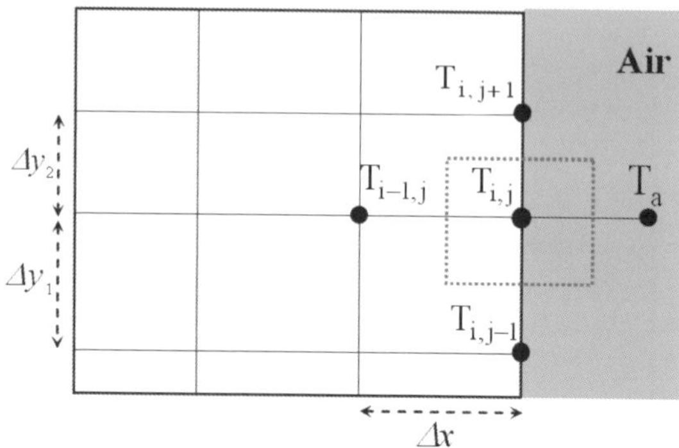

Figure 5.22: Convection boundary with non-uniform mesh [Xie et al., 2011].

convection boundary using a non-uniform mesh, as shown in Figure 5.22 where node (i, j) at the convection boundary is surrounded by a finite volume cell (dashed line). By integrating (5.6a) over the dashed cell and applying the divergence theorem, results in:

$$\int_{\substack{dashed \\ line}} k(x, y, z)\nabla T(x, y, z) \cdot \hat{n} dl = \iint_{\substack{dashed \\ cell}} -P(x, y, z)dS \qquad (5.13)$$

By applying the finite difference approximation to the first order derivative of $T(x, y, z)$ in (5.13) and incorporating (5.12), the finite volume scheme for solving the heat equation with convection boundary condition at node (i, j) can be expressed as:

$$\frac{T_{i,j} - T_a}{\dfrac{1}{h_c d}} + \frac{T_{i,j} - T_{i-1,j}}{\dfrac{\Delta x}{kd}} + \frac{T_{i,j} - T_{i,j-1}}{\dfrac{\Delta y_1}{k\Delta x/2}} + \frac{T_{i,j} - T_{i,j+1}}{\dfrac{\Delta y_2}{k\Delta x/2}} = P_{total} \qquad (5.14)$$

where $d = (\Delta y_1 + \Delta y_2)/2$.

5.6.3.3   *Heat Equation for Fluid Flow*

With stacking of dies, removing heat can be a major problem, especially for ICs used in high performance applications where power dissipation levels can be large. Unlike mobile hand held applications where memory is stacked on logic, in several other applications such as in laptop computers and servers, several logic dies need to be stacked on each other with each die dissipating upwards of 50W. In such situations, an attractive technology for removing heat is through water cooling. We therefore provide the formulation for including fluidic cooling into the thermal management solution using micro-channels.

Since the cross-sectional dimension of the micro-channel is much smaller than its length, the flow velocity along the longitudinal direction in such structures is much larger than along its lateral direction. It can therefore be assumed that the water only flows in the longitudinal direction and therefore the flow velocity is constant. The average flow velocity 'v' along the y direction has been used for simulating the fluid flow in this chapter. As a result, (5.6b) can be written as:

$$\sigma c_p v \frac{\partial T(x, y, z)}{\partial y} = \nabla \cdot (k_f \nabla T(x, y, z)) . \tag{5.15}$$

By integrating (5.15) over the dashed cell in Figure 5.23 and applying the divergence theorem, (5.15) becomes:

$$\int_{S1+S2} \sigma c_p v T \, \hat{y} \cdot \hat{n} dl = \int_{dashed\ line} k_f \nabla T \cdot \hat{n} dl \tag{5.16}$$

where, *S1* and *S2* are the upper and bottom boundaries of the dashed cell in Figure 5.23.

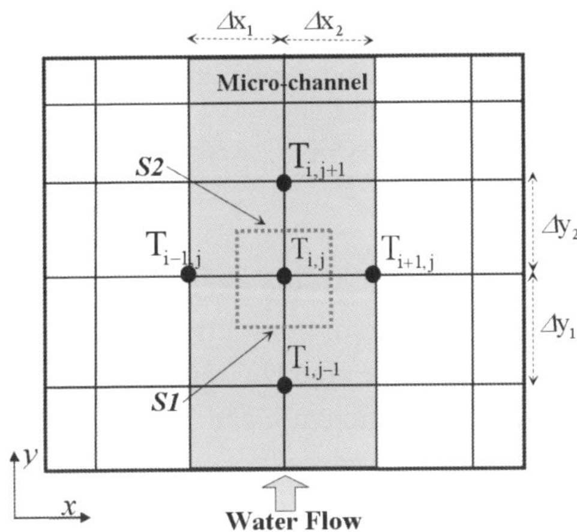

Figure 5.23: Non-uniform mesh for simulating micro-channel in IC chip [Xie et al., 2011].

For the right hand side of (5.16) the earlier formulation can be used. For the left hand side, since the central finite difference scheme can generate instability in certain cases [Ozisik, 1994], the backward difference approximation is used. The finite volume scheme for solving the heat equation for fluid flow can now be expressed as:

$$\frac{T_{i,j}-T_{i-1,j}}{\dfrac{\Delta x_1}{kd}}+\frac{T_{i,j}-T_{i+1,j}}{\dfrac{\Delta x_2}{kd}}+\frac{T_{i,j}-T_{i,j-1}}{\dfrac{\Delta y_1}{kw}}+\frac{T_{i,j}-T_{i,j+1}}{\dfrac{\Delta y_2}{kw}}=\sigma c_p v(T_{i,j}-T_{i,j-1})$$

$$(5.17)$$

where $w=(\Delta x_1+\Delta x_2)/2$, $d=(\Delta y_1+\Delta y_2)/2$.

Since the average flow velocity along the longitudinal direction is used in the model, the heat transfer coefficient h needs to be applied at the boundary of the micro-channel to model the heat transfer between the solid medium and the fluid flow. The effect of this boundary condition is important, since eliminating it can cause incorrect chip temperatures [Radmehr et al., 2004]. For water flow in micro-channels of ICs, the Reynolds number is usually less than 2300 and the flow is laminar [Kays et al., 2004]. The Reynolds number provides a ratio of inertial forces to viscous forces during fluidic flow with a large number depicting turbulent flow and a small number representing laminar flow. For fully developed laminar flow inside rectangular micro-channels with constant heat flux, the Nusselt number can be expressed as [Shah et al., 1978]:

$$Nu = 8.235\left(1-\frac{2.0421}{\alpha}+\frac{3.0853}{\alpha^2}-\frac{2.4765}{\alpha^3}+\frac{1.0578}{\alpha^4}-\frac{0.1861}{\alpha^5}\right) \quad (5.18)$$

where, α is the aspect ratio of the rectangular micro-channel. The Nusselt number provides a ratio of convective to conductive heat transfer across a boundary with a large number representing active convection. The average heat transfer coefficient h can be obtained analytically from the Nusselt number and expressed as:

$$h = Nu \cdot k / D_h \quad (5.19)$$

where, D_h is the hydraulic diameter of the micro-channel [Radmehr et al., 2004]. The same formulation for air convection boundary used in (5.14) can be used to model the water convection boundary between the solid medium and water flow.

5.6.3.4 *Solving the Coupled Equations*

The thermal equations consisting of heat flow in conducting and fluidic medium along with convection lead to a coupled set of equations that can be solved to compute the temperature in each cell. The temperature in each cell can be used to modify the conductivity which can then be used to solve the voltage equations resulting in the voltage distribution across all cells. The resulting heat produced serves as sources in each cell for solving the coupled thermal equations. Clearly, an iterative technique can be developed to compute the thermal and voltage distributions until convergence is reached. Since this problem needs to be solved at the system's level (as described in the examples), computationally efficient methods such as domain decomposition and multi-grid techniques are required which enable the use of an optimum mesh for each sub-domain, combining the various sub-domains together by invoking the necessary boundary conditions at the interface and the use of mesh refinement techniques that speed up computations. These computational methods are considered outside the scope of this chapter and are not discussed here. Instead, we will focus on examples to illustrate the importance of Joule heating and temperature effects.

5.6.4 *Examples*

5.6.4.1 *Joule Heating Effects using Air and Micro-Fluidic Cooling*

As is obvious, thermal management can be a major problem for stacked chips since heat cannot escape easily. With the stacking of chips two major thermal issues emerge namely i) the heat flux increases based on the number of chips stacked (as an example two chips each with a heat flux of $50W/cm^2$ stacked on each other leads to a cumulative heat flux of

$100W/cm^2$ as opposed to a heat flux of $50W/cm^2$ when they are next to each other) and ii) due to the presence of a large number of TSVs, thermal coupling between chips arises where heat flows from a hotter to a cooler chip thereby creating hot spots and increased junction temperatures due to the absence of a low thermal resistance path to the heat sink. Since the typical junction to ambient thermal resistance of conventional air-cooled heat sinks is around 0.5C/W [Dang et al., 2010], as the power dissipation of the chips increases, air-cooling based technologies may no longer be viable for cooling stacked chips. Micro-channels provide an interesting alternative for stacked chips since they provide a highly integrated solution within the silicon chip without the need for large heat sinks and with a junction to ambient thermal resistance of ~$0.09C\text{-}cm^2/W$ they provide the ability to remove large heat fluxes of the order of $790W/cm^2$ [Dang et al., 2010]. A schematic of future stacked dies using micro-fluidic cooling is shown in Figure 5.24 where each silicon die has an integrated micro-channel, through silicon copper vias for the electrical connections, through silicon hollow vias for fluidic flow, solder bumps (electrical I/Os) and micro-scale polymer pipes (fluidic I/Os). In an arrangement as in Figure 5.24, the electrical TSVs can be used for power delivery and signal connections while the fluidic TSVs can be used to connect to the micro-channel and for thermal management. A micro-pump on the board connects to the fluidic I/Os to pump and circulate DI water. It has been shown in [Dang et al., 2010] that microprocessors with similar power dissipation using fluidic micro-channels can reduce the junction temperature by 50% as compared to using air-cooling with heat sink. In this section, we compare joule heating effects using air and fluidic cooling using the modeling methods described earlier.

Consider the 3D integrated system shown in Figure 5.25 which consists of two sets of stacked dies, nine micro-channels in each silicon die, hundreds of TSVs connecting the dies together, C4 bump technology for assembling the dies on each other and to the glass ceramic substrate, which serves as the package [Xie et al., 2011]. The board is not included in this analysis. The package has five metal layers including two signal layers, two power plane layers and one ground plane layer, as shown in

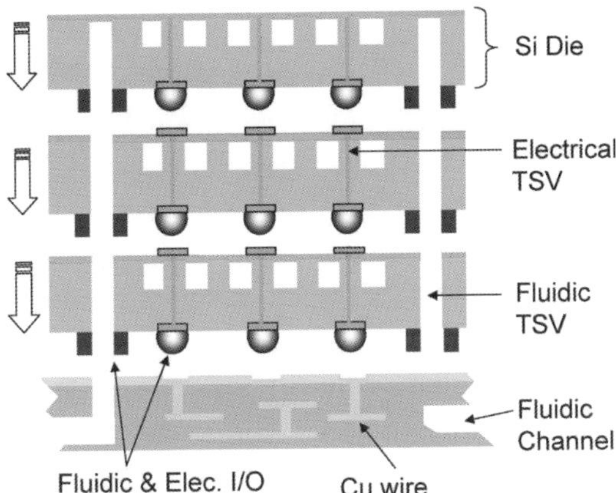

Figure 5.24: Stacked chips with micro-channels and fluidic I/Os [Dang et al., 2010].

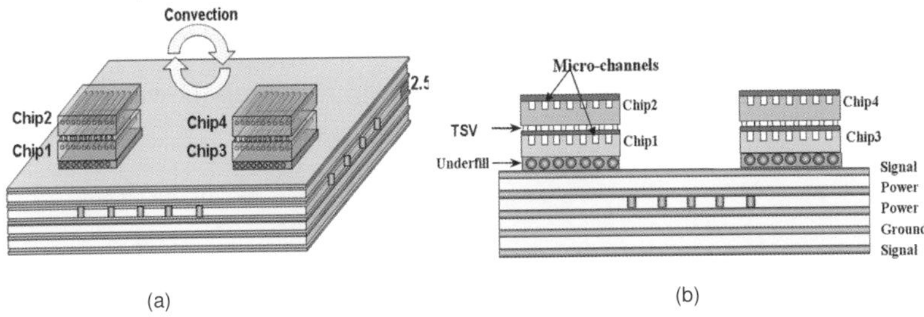

Figure 5.25: 3D Integrated System (a) system configuration and (b) cross section [Xie et al., 2011].

Figure 5.25(b). The two power plane layers are shorted together with multiple-vias in the glass ceramic substrate for reducing ohmic (IR) drop. An ideal 2.5V voltage source is placed at the corner of the package as shown in Figure 5.25(a). Obviously, from a power delivery and signal integrity standpoint, the stack-up in the package is not optimum. However, since our focus is on steady state thermal and electrical analysis, this stack-up should work fine. The package is of size 20cm × 20cm with each chip of size 1.1cm × 1.1cm.

In the system, the power consumption for Chip1, Chip2, Chip3, and Chip4 are 100W, 100W, 50W and 50W, respectively. A uniform power map is used for each die. Though this is a very optimistic assumption, it does provide a base for understanding the impact of joule heating and evaluating the importance of fluidic cooling in 3D systems. The modeling method described is capable of incorporating non-uniform power maps as well. The power dissipation translates to roughly 200W/cm^2 and 100W/cm^2 for the first and second stack, respectively. Both air-cooling and micro-fluidic cooling are used here, where chilled water is used in Figure 5.25(b) for micro-channel cooling. In each silicon die, nine micro-channels with cross-section of 0.6mm $\times$ 0.2mm are used as shown in Figure 5.26 along with the TSV arrangement configuration shown. In Figure 5.26, the thermal TSVs carry the fluid while the remaining TSVs are for power distribution. The physical dimensions and material parameters for this example are summarized in Table 5.1.

Air convection with heat transfer coefficient of $5\ W/(m^2 K)$ was applied to both the top and bottom surfaces of the package. For fluidic cooling all four dies were supplied with the same water flow rate. The input water temperature at the inlet of the micro-channels was set to 22C.

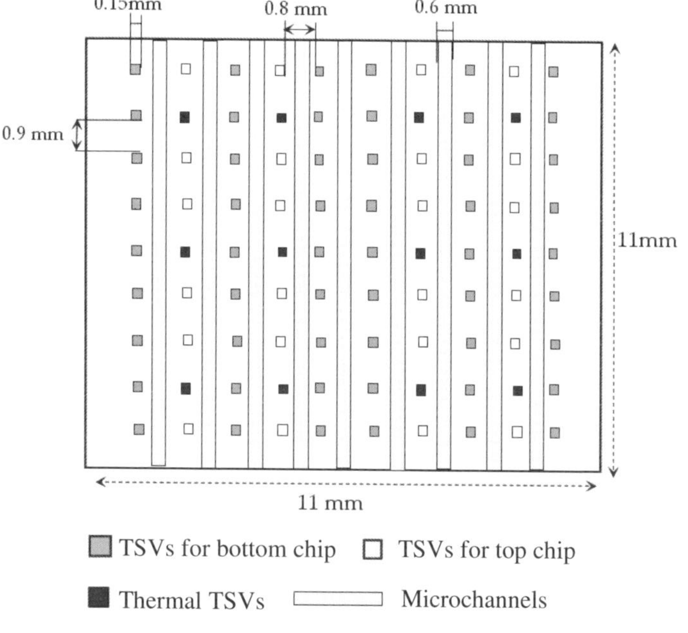

Figure 5.26: Microchannel and TSV configuration.

Table 5.1: Physical and Material parameters.

	Material Thickness (mm)	Thermal Conductivity (W/mK)
Glass-ceramic	0.35	5
Copper	0.036	400
Chip	0.5	110
Underfill	0.2	4.3
C4	0.2	60
TIM	0.2	2.4
TSV (Tungsten)	0.5	174
Micro-channel	0.2	0.6

To compare with air-cooling using heat sink, a thermal interface material (TIM) was used with the heat sink assumed to be ideal with a constant room temperature of $25\,^{\circ}C$.

In the simulation, a 3D non-uniform rectangular grid was used with $\sim 166K$ unknowns for the thermal simulation. For the voltage distribution computations, since only conductor cells are considered as unknowns in the simulation, only $110K$ unknowns were required. The simulation took 5 iterations to converge with a total simulation time of 401.4 seconds, indicating that system level simulations are indeed possible to evaluate trends using reasonable computational resources.

With water flow rate of 104ml/min for each die, the simulated temperature using micro-fluidic cooling and traditional heat sink are shown in Figures 5.27(a) and (b). In the figure, the horizontal axis is the iteration number. As the number of iterations increase the results converge where in Figures 5.27(a) and (b), the results converge in the 5th iteration with final temperatures of $167.6\,^{\circ}C$, $156.8\,^{\circ}C$, $97.5\,^{\circ}C$ and $91.9\,^{\circ}C$ for dies (chips) 1,2,3 and 4 respectively using heat sink technology. With micro-fluidic cooling the temperatures reduce significantly to $97.5\,^{\circ}C$, $101.5\,^{\circ}C$, $60.3\,^{\circ}C$ and $61.8\,^{\circ}C$ for the four dies which illustrates that micro-fluidic cooling technology can greatly reduce chip temperatures especially for dissipating large heat fluxes. The simulated voltage drop at the dies using micro-fluidic cooling and heat sink technology are shown in Figures 5.27(c) and (d). The horizontal axis once again represents the iteration number where the results converge at the 5th iteration. The vertical axis represents the voltage seen by the chip. The voltage drop is therefore the difference between the ideal supply voltage of 2.5V and the voltage seen by the chip. For example, for chip1, the voltage drop is $2.5 - 2.4213 = 78.8$mV (initial) at room temperature. In the figure, the initial voltage drops for Chip1, Chip2, Chip3 and Chip4 are 78.8mV, 83.2mV, 60.9mV and 63.2mV, respectively. Using heat sink, the final voltage drops (5th iteration) for Chip1, Chip2, Chip3 and Chip4 are 102.5mV, 109.6mV, 75.8mV, and 78.7mV, respectively indicating that the thermal effects increase voltage drop by 30%, 32%, 24% and 25%, for the respective dies. However, with micro-fluidic cooling, the thermal effects (as compared to initial room temperature) increase the voltage drop for Chip1, Chip2, Chip3 and Chip4 by only 20%, 20%, 18% and 18%, respectively, indicating that fluidic cooling can help with power distribution by controlling voltage drops. The temperature of the chips can be further reduced by increasing the flow rate of the fluid [Xie et al., 2011].

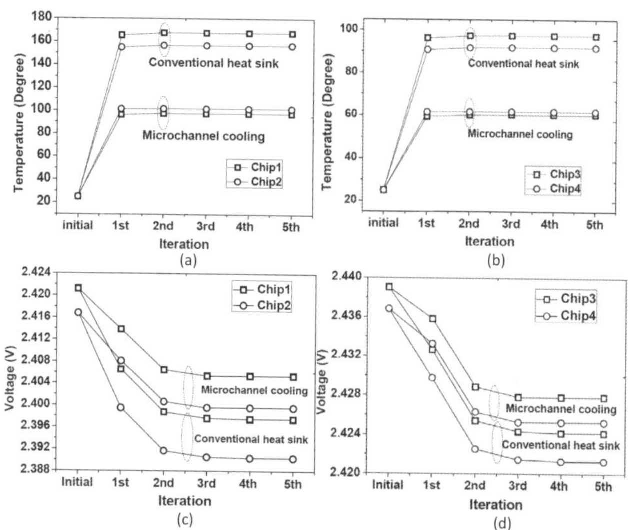

Figure 5.27: Temperature of (a) Chip1, Chip 2, (b) Chip3, Chip4 and Voltage of (c) Chip1 and Chip2 and (d) Chip3, Chip4 with electrical-thermal iterations [Xie et al., 2011].

5.6.4.2 *Impact of Silicon and Glass Interposer*

In Figure 5.25, the stacked dies are assembled directly onto a package such as the glass ceramic substrate. This is one embodiment of 3D stacking. Another option being pursued by industry is to introduce an interposer between the stacked dies and the package, as shown in Figure 5.28. The purpose of the interposer is to provide the necessary redistribution between the fine pitch micro-bumps at the bottom of the 3D stack and the package, as described in Chapter 1. This embodiment also provides better testability and depending on the type of interposer used, can also provide a better CTE (coefficient of thermal expansion) match to the stacked dies. In Chapter 4, silicon and glass interposer based solutions were compared for signal and power integrity. Another important criterion for comparison is thermal management. In this section we compare the thermal performance of silicon and glass interposers in their ability to spread and dissipate heat by incorporating them into the system shown in Figure 5.25(a) with a cross section as in Figure 5.28. The comparison is based on air-cooling only with a heat sink, as shown in Figure 5.28. The physical and material properties in this example are identical to Table 5.1 except for the presence of the interposer. The power consumption for Chip1, Chip2, Chip3, and Chip4 are 70W, 10W, 40W and 10W, respectively. A uniform power map was used for all the dies. Air convection is applied to the top and bottom surfaces of the package. The thermal conductivities of glass and silicon

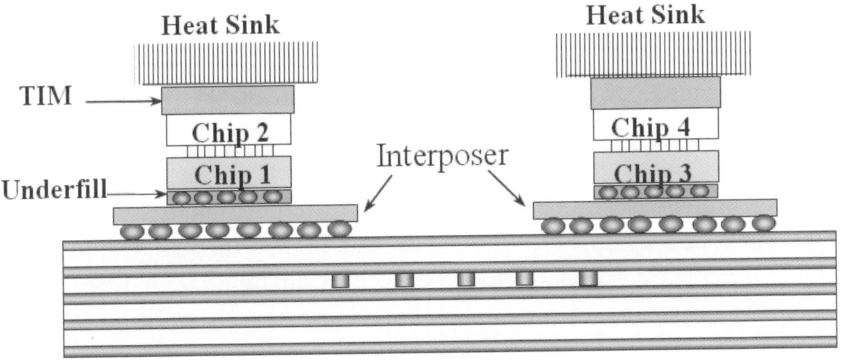

Figure 5.28: 3D system with interposer [Xie et al., 2011].

interposers are 1.14W/(mK) and 148W/(mK), respectively. The thickness of the interposer is 0.3mm.

The simulated temperature of Chip1 and Chip3 with glass and silicon interposers is shown in Figure 5.29 as a function of the convection coefficient. A few interesting observations are possible from the figure namely, i) the silicon interposer reduces the chip temperature as compared to the glass interposer solution. This is expected since silicon has a higher thermal conductivity which enables the spreading of heat laterally; ii) as the convection coefficient increases the difference in chip temperature between silicon and glass interposer solutions increases as well.

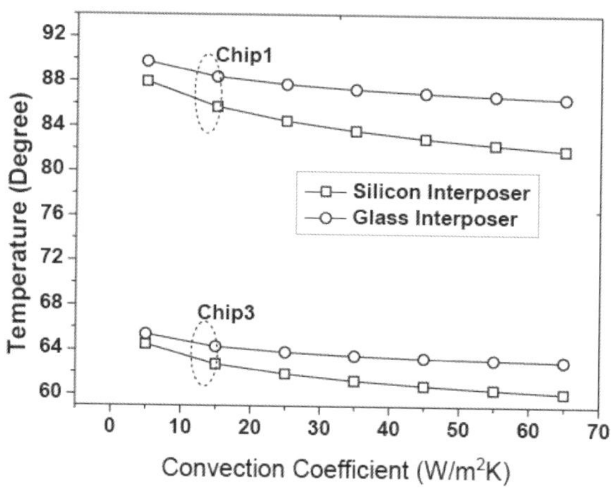

Figure 5.29: Chip 1 and Chip 3 temperature as a function of convection coefficient for silicon and glass interposer based solutions [Xie et al., 2011].

For example, with convection coefficient of $35\ W/(m^2 K)$, the temperature of Chip1 is $83.6\,^{\circ}C$ using silicon interposer, which is about $3.7\ ^{\circ}C$ lower than that of using glass interposer. When the convection coefficient increases to $65\,W/(m^2 K)$, this difference increases to $\sim 5\,^{\circ}C$. This can once again be attributed to the higher thermal conductivity of the silicon interposer which results in a smaller thermal resistance from the die to the surrounding air medium. Hence, heat spreads out through the silicon interposer which is then removed through convection.

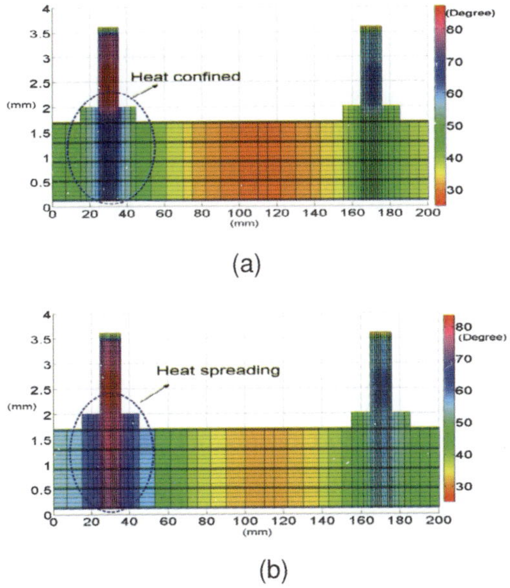

Figure 5.30: Temperature distribution (a) glass interposer and (b) silicon interposer based solutions [Xie et al., 2011].

This effect is shown in Figure 5.30 where the temperature profile is plotted across the cross section of the system using a convection coefficient of 35 $W/(m^2 K)$. From the figure it can be observed that the glass interposer confines heat a lot more than the silicon interposer which acts as a good heat spreader.

5.7 Summary

As in 2D ICs, power distribution continues to be a major challenge and can be a show stopper for 3D integration. Designing power distribution networks for future ICs (2D, 2.5D and 3D) has two major problems that need to be addressed namely, a) eliminating return path discontinuities (RPD) in the package and PCB and b) reducing power supply noise. Both these phenomena affect the speed at which ICs can communicate with each other since they can lead to the closure of the eye and generate excessive jitter in the transmitted waveforms. As has been shown, stacking of ICs leads to increased inductance as the position of the die in the stack increases. Therefore, one would expect that the dies higher up in the stack would generate more power supply noise than the dies lower in the stack. This has been shown in this chapter through detailed modeling and simulations. To address the two problems identified in this chapter, decoupling capacitors are generally used as the design solution. Instead, an alternate power distribution scheme is proposed in the next chapter that addresses both the RPD and power supply noise issues.

Another very important performance limiter for 3D integration is thermal management where the heat flux can increase substantially. Fluidic cooling using micro-channels is an attractive solution for high power chips (though the cost effectiveness is questionable) which can be used not only to decrease chip temperatures but also to reduce joule heating effects. The numerical methods provided in this chapter have been used in Chapter 1 to determine the parameters that need to be passed between dies for generating design exchange formats. Also, the effect of silicon and glass interposer on chip temperatures have been discussed briefly showing the improved thermal performance possible using silicon interposer, resulting in the conclusion drawn in Table 1.4 of Chapter 1.

References

1. [Swaminathan et al., 2007] Madhavan Swaminathan and Ege Engin, "Power Integrity Modeling and Design for Semconductors and Systems", Prentice Hall, 2007.
2. [Novak et al., 2007] Istvan Novak and Jason R. Miller, "Frequency Domain Characterization of Power Distribution Networks", Artech House, 2007.
3. [Kim et al., 2010] Kiyeong Kim, Woojin Lee, Jaemin Kim, Taigon Song, Joohee Kim, Jun So Pak and Joungho Kim, "Analysis of Power Distribution Network in TSV-based 3D-IC", IEEE Conference on Electrical Performance of Electronic Packaging and Systems (EPEPS), pp. 177–180, 2010.
4. [Huang et al., 2012], Gang Huang, Muhannad Bakir, Azad Naeemi and James Meindl, "Power Delivery for 3-D Chip Stacks: Physical Modeling and Design Implication", IEEE Trans. on Components, Packaging and Manufacturing Technology, Early IEEE Xplore access, 2012.

5. [Swaminathan et al., 2010] M. Swaminathan, D. Chung, S. Grivet-Talocia, K. Bharath, V. Laddha, and J. Xie, "Designing and Modeling for Power Integrity," IEEE Trans. Electromagn. Compat., Vol. 52 , No. 2, pp. 288–310, May 2010.

6. [Novak, 2000] Istvan Novak, "Lossy Power Distribution Networks With Thin Dielectric Layers and/or Thin Conductive Layers," IEEE Trans. Adv. Packag., Vol. 23, No. 3, pp. 353–360, Aug. 2000.

7. [Balaraman et al., 2004] D. Balaraman, J. Choi, V. Patel, P. M. Raj, I. R. Abothu, S. Bhattacharya, L. Wan, M. Swaminathan, and R. Tummala, "Simultaneous Switching Noise Suppression Using Hydrothermal Barium Titanate Thin Film Capacitors," in Proc. Electronic Components and Technology Conference, pp. 282–288, June 2004.

8. [Muthana et al., 2005] P. Muthana, M. Swaminathan, E. Engin, P. Markondeya Raj, and R. Tummala, "Mid frequency decoupling using embedded decoupling capacitors," in Proc. Electr. Perform. Electron. Packag., pp. 271–274, Oct. 2005.

9. [Hobbs et al., 2001] J M. Hobbs, H. Windlass, V. Sundaram, S. Chun, G. E. White, M.Swaminathan, and R. Tummala, "Simultaneous Switching Noise Suppression for High Speed Systems Using Embedded Decoupling," in Proc. Electron. Compon. Technol. Conf., pp. 339–343, May 2001.

10. [Alvarez et al., 2000] E. Diaz-Alvarez, J. P. Krusius, and F. Kroeger, "Modeling and simulation of integrated capacitors for high frequency chip power decoupling," IEEE Trans. Components and Packaging Technologies, Vol. 23, No. 4, pp. 611–619, Dec. 2000.

11. [Wikipedia] http://en.wikipedia.org/wiki/Joule_heating

12. [Xie et al., 2011] Jianyong Xie, M. Swaminathan, "Electrical-thermal co-simulation of 3D integrated systems with micro-fluidic cooling and Joule heating effects," IEEE Transactions on Components, Packaging and Manufacturing Technology, Vol. 1, No. 2, pp. 234–246, Jan. 2011.

13. [Dragoni et al., 2002] M. Dragoni, F. D'Onza and A. Tallarico, "Temperature distribution inside and around a lava Tube," Journal of Volcanology and Geothermal Research, Vol. 115, No. 1–2, pp. 43–51, Sept. 2002.

14. [Hettiarachchi et al., 2008] H.D. M. Hettiarachchi, M. Golubovic, W. M. Worek and W.J. Minkowycz, "Three-dimensional laminar slip-flow and heat transfer in a rectangular microchannel with constant wall temperature," International Journal of Heat and Mass Transfer, Vol. 51, pp. 5088–5096, May 2008.

15. [Ozisik, 1994] M. N. Ozisik, Finite Difference Methods in Heat Transfer, CRC Press, 1994.

16. [Radmehr et al., 2004] A. Radmehr and S.V. Patankar, "A flow network analysis of a liquid cooling system that incorporates microchannel heat sinks," The 9th Intersociety Conference on Thermal and Thermomechanical Phenomena in Electronic Systems (ITHERM), Vol. 1, pp. 714–721, Jun. 2004.

17. [Kays et al., 2004] W. Kays, M. Crawford and B. Weigand, Convective Heat and Mass Transfer, McGraw Hill *Higher Education, 2004.*

18. [Shah et al., 1978] R.K. Shah and A.L. London, Laminar Flow Forced Convection in Ducts, Advance Heat Transfer (Suppl. I), Academic Press, 1978.

19. [Dang et al., 2010] B. Dang, M. S. Bakir, D. C. Sekar, C. R. King and J. D. Meindl, "Integrated microfluidic cooling and interconnects for 2D and 3D chips," IEEE Trans. Advanced Packaging, Vol. 33, no. 1, pp. 79–87, Feb. 2010.

Chapter 6

Alternate Methods for Power Distribution

Power distribution poses several challenges as described in the previous chapter where return path discontinuities (RPDs) and power supply noise causes increased jitter and closure of the eye. As a community we have relied on decoupling capacitors to fix most of these problems. With an objective of increasing the channel speed and communication bandwidth between ICs, the electronics industry is moving towards massively parallel I/Os between ICs by developing new technologies such as 2.5D and 3D integration to support it. This makes the power distribution in the interposer, package and printed circuit board even more complex.

In this chapter, new methods are discussed for addressing the challenges described in the previous chapter with a focus on *I/O power distribution*. This pertains to the part of the problem where the circuits communicating between chips need to be powered from a power supply on the printed circuit board (PCB).

This chapter starts by introducing power transmission lines (PTL) that helps eliminate RPDs in the package and PCB. To minimize power supply noise at the transistor terminals, alternate signaling schemes are introduced that enhance the quality of the transmitted and received waveforms. The application of these methods to a field programmable gate array (FPGA) chip is described to demonstrate practicality. Finally, the use of these techniques for 3D integration is demonstrated through modeling and simulation.

295

6.1 Introducing Power Transmission Lines

In Chapter 5, the current path for a signal line referenced to a voltage plane was discussed. As described in the previous chapter, the creation of RPDs between the voltage and ground planes can cause excessive power supply noise in the system. Since the discontinuities occur in the current return path, they can be controlled by reducing the power distribution network (PDN) impedance using capacitors or changing the stack-up to eliminate them. Let's look into eliminating these discontinuities using an alternate power distribution scheme shown in Figure 6.1 where transmission lines are used to replace the voltage (power) plane [Engin et al., 2008]. This is depicted as power transmission line or PTL in Figure 6.1. Both the PTL and signal transmission line are referenced to a common ground plane. Following the return current path in Figure 6.1, the loop current is continuous and therefore the RPD can be eliminated. The impedance of the PTL can now be determined based on the desired eye height at the receiver side using the turn-on impedance of the I/O driver and the impedance of the signal transmission line. The termination resistor at the source end helps absorb reflections due to any mismatch in the communication path. Since the characteristic impedance of the signal transmission line is typically 50ohms, the PTL impedance falls in the range of 10–25ohms, as shown in this chapter. This scheme has several advantages as listed below:

(i)   The power distribution impedance is of the order of ohms and not milli-ohms. Hence, this can be construed as a high impedance power distribution network (PDN).

(ii) Since the RPD is removed, no decoupling capacitors are required for mitigating RPDs and therefore the number of capacitors required to implement this scheme in a real design can be reduced.

(iii) Depending on the routing density, the PTL can be routed on the same layer as the signal line, thereby reducing layer count and

(iv) As will be shown later, designing the PDN using PTL is considerably easier than using voltage planes and hence the design cycle can be reduced as well.

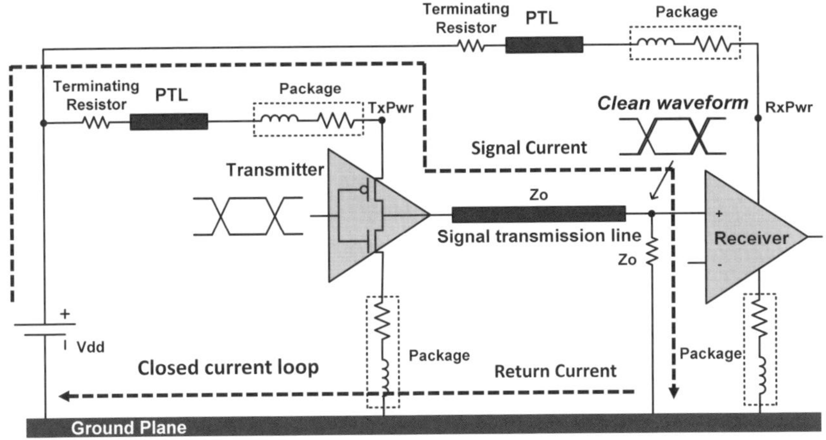

Figure 6.1: Die to die communication using power transmission line (PTL).

As shown in Figure 6.1, by eliminating the RPD between the power and ground plane, an improved waveform can be received. All these advantages seem too good to be true! Let's therefore delve deeper into PTLs to understand them better.

6.1.1 *Understanding Power Transmission Lines*

Consider Figure 6.2 where a transmitter and receiver are communicating with each other. The transmitter and receiver circuits are on two separate ICs and therefore the signal line shown as a transmission line with impedance $Zo = 50\Omega$ is in the package. The far end of the signal line is terminated in 50Ω. Consider the case when the signal is being charged with the PMOS transistor of the driver in the transmitter switching to ON state. Assuming the transistor on-resistance is 25Ω and the supply voltage is 2.5V, to obtain a 1.25V signal swing at the receiver end (data_rx in Figure 6.2) in the steady state requires a source termination resistance $Z = 25\Omega$. The PTL is next introduced into the circuit with an impedance of 25Ω where the source resistor is used to absorb all reflections from the power supply junction (shown as node TxPwr) when the driver switches states. Both the PTL and signal line are referenced to a common ground plane in the figure.

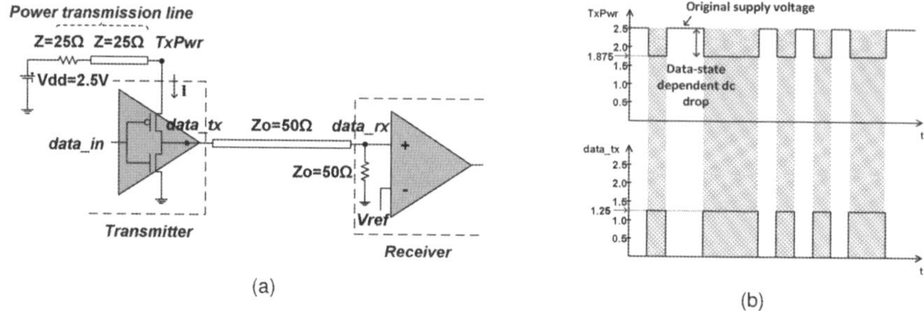

Figure 6.2: Single ended signaling using PTL (a) Schematic and (b) Waveform.

For a voltage-mode driver with far end termination, charging the signal transmission line will cause current to flow through the source termination resistor causing a DC drop across it, resulting in the power supply voltage at the driver to be 1.875V ($V_{TxPwr} = 2.5 \times (25 + 50)/(25 + 25 + 50)$). On the other hand, discharging the signal line will cut off the flow of current from the power supply resulting in a supply voltage of 2.5V at TxPwr node of the driver. As a result, the DC drop across the source termination resistor becomes dynamic depending on the state of the output data. This is illustrated in Figure 6.2(b) where the DC voltage level at the power supply node (TxPwr) of the I/O driver alternates between 1.875V and 2.5V based on the switching pattern of the driver. It is interesting to note in Figure 6.2(b) that the DC variations of the power supply node TxPwr does not affect the amplitude of the data output (data_tx).

Figure 6.3 provides a comparison between a power plane (Chapter 5) and PTL based (Figure 6.1) designs. When a power plane is used in the I/O PDN, the turn-on impedance of the I/O driver is typically matched to the signal transmission line, which is then terminated using a resistor at the far end to minimize reflections. Hence, the output voltage swing is half of the supply voltage. This impedance matching scheme cannot be applied directly to the PTL-based I/O PDN due to the presence of the source termination resistor. Here the sum of the source resistance and turn-on impedance of the I/O driver is matched to the characteristic impedance of the signal transmission line. This matching ensures an eye

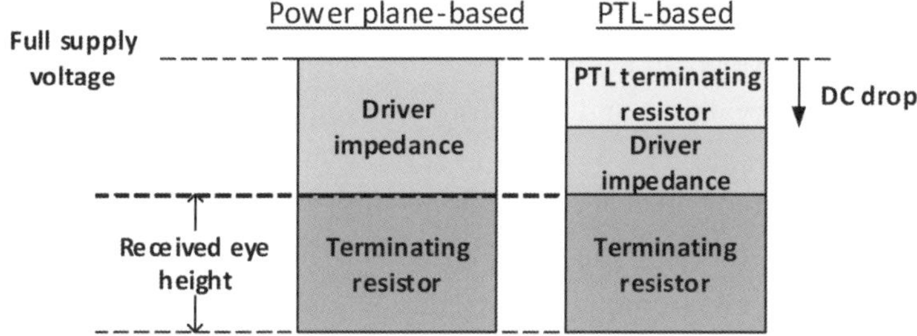

Figure 6.3: Comparison of voltage levels between power plane and PTL designs.

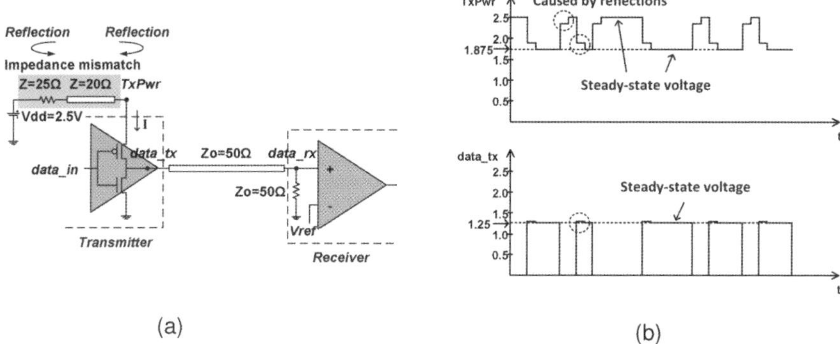

Figure 6.4: Impedance mismatch with PTL for single ended signaling (a) Schematic and (b) Waveform.

height that is half the supply voltage which also creates a DC drop across the source resistor, as shown in Figure 6.3.

The dynamic dc drop shown in Figure 6.2(b) creates problems when combined with any other impedance mismatch occurring in the circuit, especially coming from manufacturing variations. This can result in signal integrity issues. As an example, when the PTL and source termination resistor have a mismatch of 20% occurring due to manufacturing variations, multiple reflections are created in the circuit, as shown in Figure 6.4(b). These reflections induce a staircase-step waveform at the power supply node (TxPwr) and cause signal overshoot (circle) at the output node (data_tx) of the driver, [Huh et al., 2011].

For the use of PTLs in a PDN to be practical, a single PTL needs to feed multiple driver and receiver circuits. Else, the routing congestion can become very large and therefore the implementation is no longer practical. If a single PTL is used to service multiple drivers, the power supply nodes of the drivers need to be tied together, which results in varying current through the PTL based on the switching data pattern. This affects the DC voltage level at the common power supply node of the drivers and therefore the amplitude of the transmitted and received signal waveforms can further degrade. This is illustrated in Figure 6.5, where a PTL is used to feed two I/O drivers. Since the output data of the two drivers are independent of each other, there are three possible data patterns in terms of the number of 1s transmitted, namely 00, 01/10, and 11. The three possible combinations result in three different amounts of current through the PTL, which leads to three different DC voltage levels at the TxPwr node, and three different amplitudes of the waveform at the input and output of the signal transmission lines (data1_tx, data2_tx, data1_rx, and data2_rx). Hence, due to a combination of manufacturing variations and data transitions, the voltage at the driver output and power supply node will vary, leading to signal and power integrity problems. Clearly, this issue will get exacerbated as the number of drivers being served by a single PTL increases. Therefore, though the PTL-based power distribution scheme eliminates RPDs, it creates other problems that need to be addressed which can be listed as follows:

(i) The PTL circuit creates a voltage drop across the source termination resistor, which results in a change in voltage at the power supply terminals of the transistor, based on the switching pattern of the driver. This is called as dynamic DC drop in this chapter.

(ii) The performance of the PTL circuit is susceptible to manufacturing variations and

(iii) When a single PTL is used to service multiple drivers, the effects of (i) and (ii) are amplified, which degrade the quality of the transmitted waveforms.

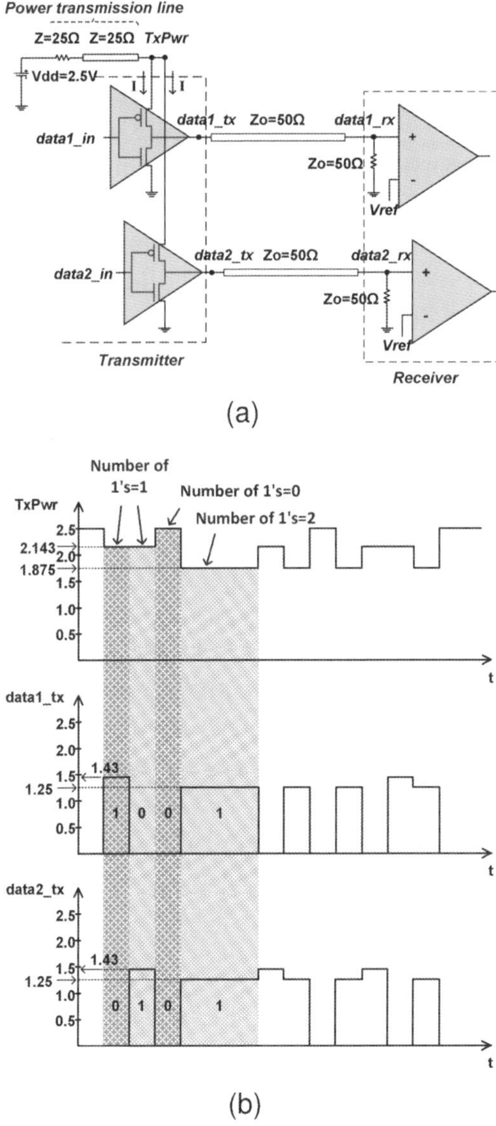

Figure 6.5: Two drivers sharing one PTL (a) Schematic and (b) Waveform.

To address these issues we introduce three signaling schemes namely,
1) Constant Current, 2) Pseudo-balanced and 3) Constant Voltage.

6.2 Constant Current Power Transmission Line (CCPTL)

6.2.1 *Basic Concept*

The constant current PTL (CCPTL) scheme resolves the above issues related to dynamic DC drop and mismatch between the PTL and source termination resistor due to manufacturing variations [Huh et al., 2011], [Huh et al., 2011a] and [Huh et al., 2011b]. The varying DC drop in Figure 6.2 is due to the current flowing through the PTL only during the high state of the output data. To maintain the DC voltage level constant, a current path is required during the low state of the data as well. In the CCPTL scheme, an additional current path from the power supply terminal to ground is created using a data pattern detector and dummy path, as shown in Figure 6.6(a). The data pattern detector detects the state of the input data and determines whether to connect or disconnect the dummy path to the PTL. For a single driver, the data pattern detector is the inverted data input signal to the driver. The dummy path is a resistive path, whose impedance is matched with that of the signal path so as to induce the same amount of current during both the low and high state of the data. It can be implemented with transistors whose width and length are optimized to yield the desired resistance. As a result, the DC voltage level at the power supply node (TxPwr) of the driver stays constant regardless of the output data state, as shown in Figure 6.6(b). The DC drop due to the source termination resistor can be compensated either by reducing the turn-on impedance of the output driver or by

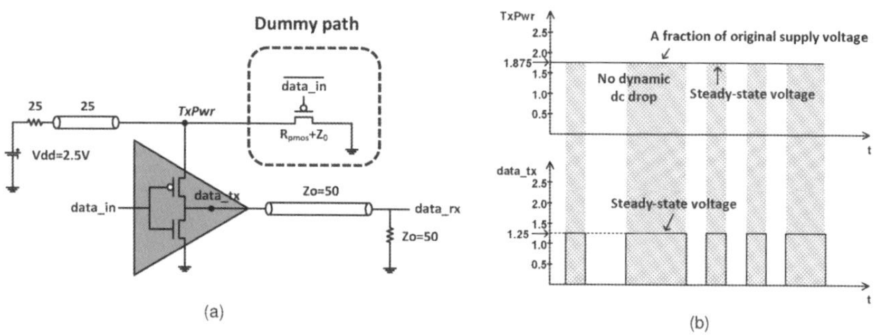

Figure 6.6: Single ended signaling using CCPTL (a) Schematic and (b) waveform.

increasing the supply voltage. Between the two compensation methods, the latter has been used for implementation in this chapter, which will be discussed later.

The constant current through PTL eliminates the repeated charging and discharging of the PTL. Since the PTL is always kept charged, the transmission line's response to applied voltage is resistive rather than reactive, even though the current path contains inductances and capacitances. This removes the possibility of any impedance mismatch in the circuit. Consequently, the mismatch effect will not appear on the signal line even when there is an impedance mismatch between the source termination resistor and PTL. Even with a 20% impedance mismatch, the DC voltage level at the power supply node (TxPwr) and the amplitude of the signal at the output of the driver (data_tx) maintain their expected levels, as shown in Figure 6.6(b).

The CCPTL scheme is quite easy to implement since for a single driver the dummy path is a PMOS transistor whose input is an inversion of the data input (data_in in Figure 6.6(a)). The on resistance of the pass transistor is made to equal the sum of the on resistance of the PMOS transistor of the driver and termination resistance at the far end of the signal line. In Figure 6.6(a), the on resistance of the PMOS transistor is set to $25\Omega + 50\Omega = 75\Omega$. This ensures that in the low state, the same amount of current as in the high state is drawn from the power supply, thereby resulting in a constant voltage at the TxPwr node of 1.875V.

The CCPTL scheme can be extended to support multiple I/O drivers. This requires a data pattern detector and multiple dummy paths. When the power supply node of multiple drivers is tied together, the amount of current through the PTL once again varies based on the data pattern. Therefore, the number of dummy paths needs to equal the number of possible data patterns. Figure 6.7(a) shows the CCPTL scheme used for 2-bit transmission. Two bits generate four different patterns, which are 00, 01, 10, and 11. In terms of the number of high states, three patterns are possible, as indicated in the table in Figure 6.7(b). The maximum PDN current is drawn by the drivers when all the drivers are turned on with all the output data being high, while no current flows when all the drivers are turned off. The data pattern detector detects the number of

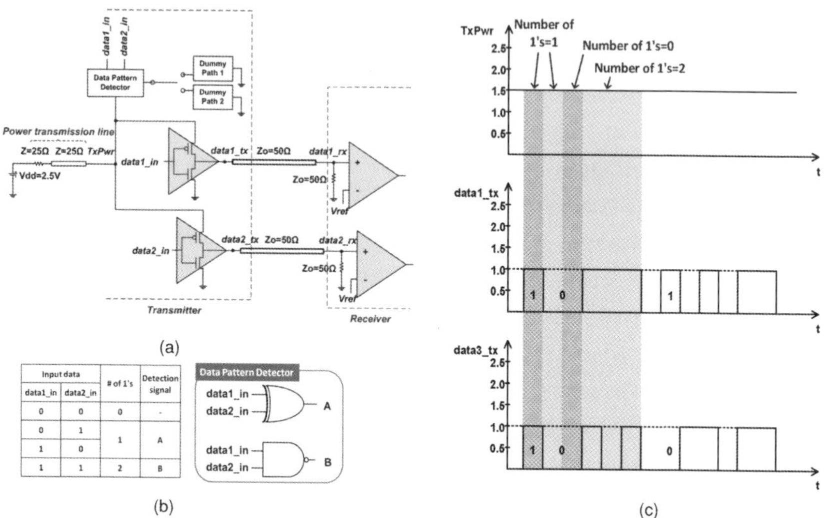

Figure 6.7: Two I/Os being powered by one PTL (a) Schematic, (b) Data Pattern Detector and (c) Waveforms.

turned-off drivers and enables the corresponding dummy paths to carry current from the PDN. As a result, the total amount of current drawn from the PDN by either the drivers or the dummy paths will be kept constant regardless of the data pattern. Figure 6.7(c) shows the resulting simulated waveforms with the voltage at the power supply pin (TxPwr) being constant at 1.5V, and the amplitudes of the signal at the output of the signal line being 1V. These voltages can be calculated using a voltage divider network based on steady state analysis, where the transmission lines are added later into the circuit, similar to the case for a single driver covered earlier.

6.2.2 CCPTL Test Vehicle

Two simple test vehicles using off the shelf chips have been used to illustrate the functioning of the CCPTL scheme and quantifying its performance. The first test vehicle consists of a board with a power plane while the other replaces the power plane using a PTL. A 16-pin QFN-packaged SiGe differential driver is used for both test vehicles as the driver IC. Since the driver has a differential output pin pair, one output pin is connected to a 66-mm long signal transmission line and terminated inside of the oscilloscope (during measurements) with 50Ω. The other output pin is directly connected to a 50Ω resistor to function as the dummy path outside the chip. During the high state of the output data, current flows along the signal transmission line, while the dummy path draws the same amount of current during the low state of the output data. Therefore, constant current flows through the PDN in both test vehicles at all time. The difference in signal integrity between the two test vehicles should provide insight into the advantages gained using the CCPTL scheme.

The top view of the two boards is shown in Figures 6.8(a) and (b). The width and length of the power and ground planes are 96.5mm and 63.5mm, respectively. For the PTL test vehicle, a 25ohm transmission line is used as the PTL, with a 25ohm source termination resistor. The ground plane serves as a reference conductor for the PTL and signal line. The only difference between the two test vehicles is the method used to provide power to the driver. The stack-up details of the test boards are shown in Figures 6.8(c) and (d). Copper and FR4 are used for the metal

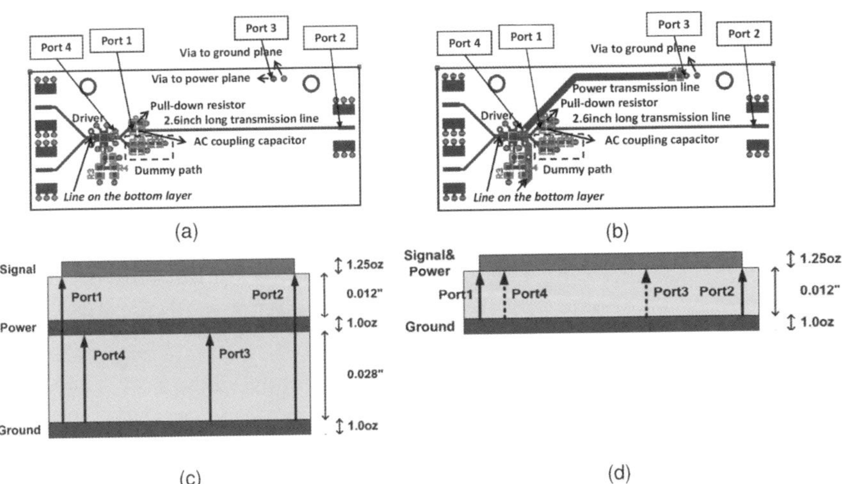

Figure 6.8: Layout of (a) Plane and (b) PTL test vehicles and Stack-up of (c) Plane and (d) PTL test vehicles.

and dielectric layers. The conductivity of copper is 5.8×10^7S/m. The dielectric constant and loss tangent of FR4 are 4.6 and 0.025, respectively. The board with the power plane consists of 4 layers (4th layer not shown in Figure 6.8(c)), which are signal-power-ground-signal, while the PTL board consists of only 3 layers (3rd layer not shown in Figure 6.8(d)), which are signal/power-ground-signal, with both signal and power transmission line routed on the top layer. It is important to note that in both boards, the signal transmission line and other connections are placed on the top signal layer, while the bottom signal layer is nearly empty with only a couple of lines used to connect control and power supply pins. The lines placed on the bottom layer are also shown in Figures 6.8(a) and (b).

As mentioned earlier, planes separated from each other (with different DC potentials) support cavity modes which induces resonances between the planes. A detailed discussion on cavity modes and plane resonances are available in [Swaminathan et al., 2007] and therefore will not be covered here. Based on the behavior of planes, the board stack-up in Figure 6.8(c) can support cavity modes between the power and ground plane. The resonant frequency f_{mn} of this structure can be calculated [Swaminathan et al., 2007] as:

$$f_{mn} = \frac{c}{2\pi\sqrt{\mu_r \varepsilon_r}} \sqrt{\left(\frac{m\pi}{a}\right)^2 + \left(\frac{n\pi}{b}\right)^2} \tag{6.1}$$

where c is the speed of light in air, a and b are the width and length of the plane structure, m and n are integers corresponding to the propagating modes, and ε_r and μ_r are the relative permittivity and permeability of the material between the power and ground planes. Given the dimensions of the planes in the plane test vehicle, the frequencies of the first four resonant modes are shown in Table 6.1.

From Table 6.1, the first resonant frequency occurs at 724MHz, where at this frequency the length of the plane is $\lambda/2$ long, where λ is the wavelength in the medium. Using Sphinx [Sphinx, 2010], an electromagnetic solver based on the multi-layered finite difference method (M-FDM) [Swaminathan et al., 2007], the two test vehicles can be modeled in the frequency domain to compare the insertion losses of

Table 6.1: Resonant frequencies of first four modes.

a(mm) - width	b(mm) - length	m	n	f_{mn}(GHz)
96.5	63.5	1	0	0.724
		0	1	1.1
		2	0	1.45
		2	1	1.82

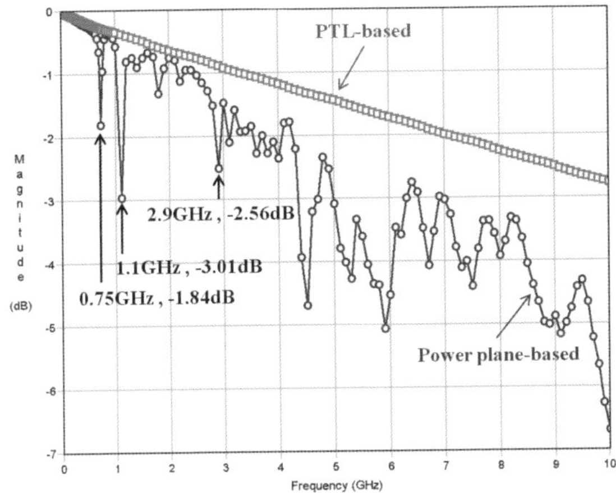

Figure 6.9: Insertion loss of signal line for plane and PTL based test vehicles.

the signal transmission lines between Port1 and Port2, as defined in Figure 6.8. Four ports are defined for each structure, two for the signal line (Port1 and Port2 at input and output) and the other two for the PDN, as in Figure 6.8. The reference used for all the ports is the ground plane. Ports represent either excitation or measurement points in the figure.

The insertion loss of the signal transmission lines for the two test vehicles are compared in Figure 6.9. The insertion loss of the PTL test vehicle has a smooth negative slope with -2.77dB insertion loss at 10GHz. This behavior is expected for a matched microstrip line with the main contributor to insertion loss coming from the conductor and dielectric loss of the board. On the other hand, the insertion loss of the signal line in the power plane test vehicle has multiple resonances. The dips in the insertion loss are consistent with the resonant frequencies in Table 6.1, similar to the behavior described in Chapter 5. The resonances cause rapid increase in the insertion loss reaching -6.69dB at 10GHz, as shown in Figure 6.9.

Consider next the test environment and measurement set-up, shown in Figure 6.10, which also shows the port locations where the power supply and signal transmission lines, are connected. An Agilent 81133A signal generator is used to generate a 2^7-1 PRBS pattern at the desired frequency. A supply voltage of 2.5V is used for the power plane test vehicle, while 3.98V is used for the PTL test vehicle, to maintain the same voltage at the TxPwr node of the driver.

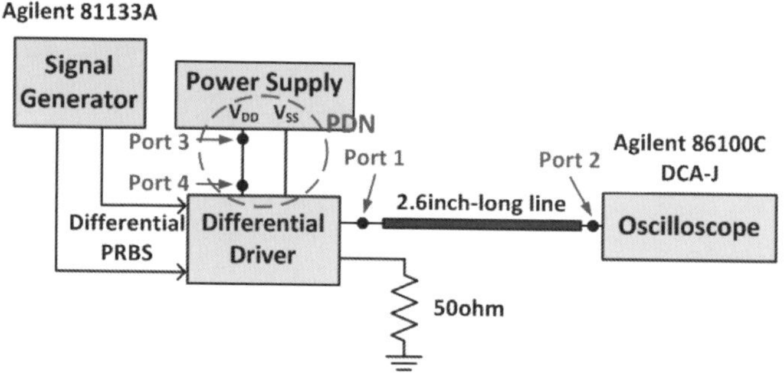

Figure 6.10: Measurement set-up for Test Vehicle.

The drivers of both test vehicles were excited at 1500Mbps. The fundamental frequency of the 1500Mbps PRBS pattern coincides with the resonant frequency of the signal line in the power plane test vehicle. Eye diagrams were measured at the output of the transmission line (Port2) after the signal travels a distance of 66mm along the transmission line, using an Agilent 86100C oscilloscope. To maintain consistency, 10K samples were used to construct the eye diagram. Two types of jitters were measured using the oscilloscope, namely the root mean square (RMS) jitter and the peak-to-peak (p-p) jitter. The RMS jitter utilizes all the 10K samples for calculation. Assuming that the mean is zero, the RMS jitter quantifies the standard deviation of the jitter distribution. This number is more meaningful compared to the p-p jitter which represents the distance between the two farthest data points. However, since the jitter is mainly caused by power supply noise here, its distribution can be non-Gaussian. In such a case, the peak-to-peak jitter becomes more useful. Hence, both jitters are presented as a function of data rate later in this section.

Figures 6.11(a) and (b) show the measured eye diagrams for a 1500Mbps PRBS at the receiver side in the power plane and PTL test vehicles, respectively. The output of the signal generator has an initial p-p jitter of 19.1psec. The p-p jitter increases to 36psec in the power-plane test vehicle, while it increases to 27psec in the PTL test vehicle. The difference between the p-p jitters is 9psec, which is a reduction of 25% for the PTL test vehicle. The eye height is ~463mV in the power plane test vehicle, and ~523mV in the PTL test vehicle, which shows an improvement of 13.0%. When 3.98V is used as the supply voltage for the PTL test vehicle, the DC voltage level at the power supply pin of the chip equals 2.498V, which matches the supply voltage provided to the driver in the power plane test vehicle. The increased voltage level raises the power consumption of PTL as compared to the plane test vehicle. As 3.98V is 59.2% larger than 2.5V, the power consumption is 59.2% higher as well with PTL, since the current drawn is the same for both the test vehicles. The power consumption for PTL can be reduced by reducing the turn-on impedance of the I/O driver by custom-designing the CCPTL scheme, or by eliminating the source termination of the PTL. Since the source termination of the PTL is used to prevent multiple reflections as mentioned earlier, it can be removed if the load-termination resistor is matched well to the characteristic impedance of the signal transmission line. In Figures 6.11(c) and (d), the received eye diagram of the plane and PTL test vehicles are shown without the source termination. Without the source-termination resistor, no compensation for the DC drop is required and therefore 2.5V can be used as the supply voltage for the PTL test vehicle as well. The difference between the p-p jitters in the two test vehicles without source termination is 9psec, which is a reduction of 25% for the PTL test vehicle. The eye heights are ~463mV and ~546mV in the plane and PTL test vehicles respectively, which shows an improvement of 17.9%.

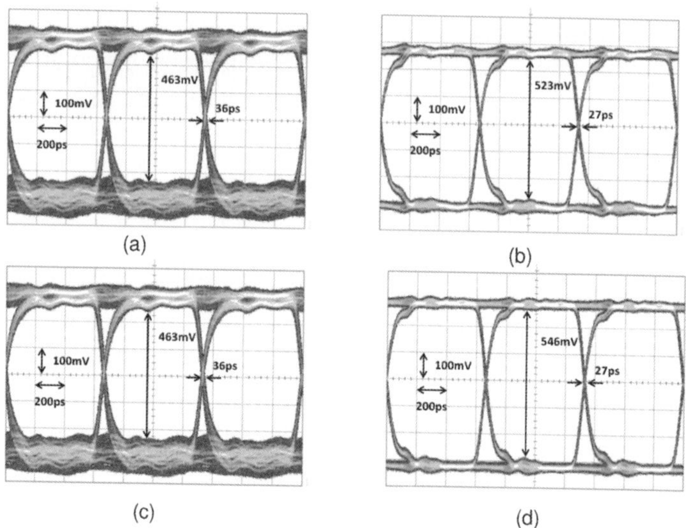

Figure 6.11: Measured eye diagrams for 1500Mbps PRBS with source termination (a) plane and (b) PTL and without source termination (c) plane and (d) PTL.

The RMS and p-p jitters at various data rates are shown in Figures 6.12(a) and (b), respectively. The jitter is plotted along the y-axis, while the data rate is plotted along the x-axis, with a range from 500Mbps to 3000Mbps. The line with triangular marker corresponds to jitter from the signal generator (Agilent 81133A). The lines with square (SE_PTL) and circular markers (SE_Plane) indicate the jitters of the PTL and plane test vehicles, respectively.

Based on the results, CCPTL provides an interesting alternative for power distribution that has the following advantages:

(i)   Removes RPDs, thereby improving signal integrity

(ii)   Removes the dynamic DC drop on the power supply by ensuring constant current through the PTL

(iii)   Removes the need for decoupling capacitors for mitigating RPDs, thereby reducing cost

(iv)   Enables the removal of the source termination resistor, if the far end of the signal line is terminated and

(v)   Provides a path for reducing layers if the PTL can be routed on the same layer as the signal line.

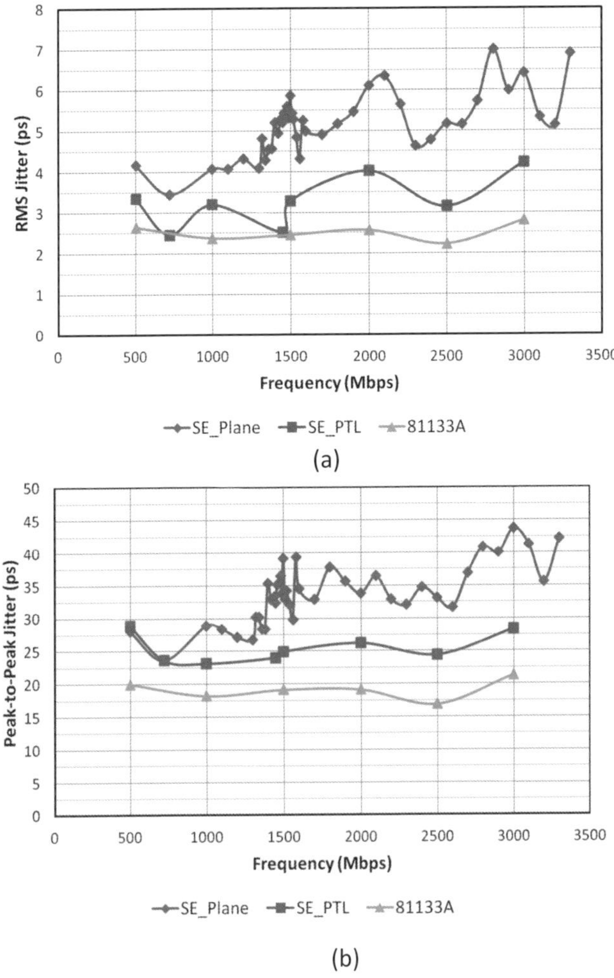

Figure 6.12: Jitter variation with data rate (a) RMS jitter and (b) p-p jitter.

So, what is the problem with CCPTL implementation? The issue is with POWER, since the PTL is always kept charged with constant current flowing through it. This drawback will be addressed in the following sections.

It is important to note that the CCPTL concept can be readily extended to differential signaling, since the two complementary signals draw constant current from the PDN by construction. Since differential signals are also prone to RPDs, the PTL will help improve eye diagrams for differential signaling as well. The focus of this chapter is primarily on improving signal integrity for single-ended drivers, and therefore differential signaling is not considered here.

6.3 Pseudo Balanced Power Transmission Line (PBPTL)

Since the current through the PDN is constant during both the low and high transitions for CCPTL, one would expect the power to double as compared to drawing current only during the high state. The power analysis will be addressed in detail later, while an alternate method for signaling is addressed in this section. The PBPTL scheme addresses the increased power consumption in CCPTL scheme by encoding the input data pattern.

In a balanced signaling scheme, encoding is performed prior to data transmission where N bits of data are mapped onto $(N + \log_2 N)$ bits or less to provide 2^N data patterns with an equal number of 1s and 0s at all times [Carusone et al., 2001]. According to [Tallini et al., 1998], the length of extra bits required for balanced coding is approximately $(0.5 \cdot \log N + 1)$ for N-bit data. This balanced signaling scheme minimizes the variation of the total driving current through the PDN by controlling the number of high and low states in the output data string. As the total driving current is maintained constant, total di/dt can be minimized, and therefore power supply noise can be reduced as well [Sim, 2007], [Oh et al., 2008]. These ideas will be used to implement the PBPTL scheme.

To transmit 4-bit information, 6-bit symbols with three 1s and three 0s are required to maintain an equal number of 1s and 0s in the encoded data word. This results in 50% overhead in terms of off-chip PCB trace and I/O pin count. The disparity between the total number of 1s and 0s is fixed at zero. To reduce such overhead, the pseudo balanced signaling scheme can be used which employs two types of disparity between the total number of 1s and 0s [Huh et al., 2011d] in the pattern. Let M be the length of the encoded data word, which is pseudo-balanced. The length M is determined to satisfy the following equation:

$$\binom{M}{M/2}+\binom{M}{M/2-1}\geq 2^N \quad M \quad even$$

$$\binom{M}{(M-1)/2}+\binom{M}{(M+1)/2}\geq 2^N \quad M \quad odd$$

$$(6.2)$$

where ( ) represents combination.

In balanced signaling scheme, only the first term on the left hand side of (6.2) is used to determine M, while two terms are included in pseudo-balanced signaling. If the minimum M to satisfy (6.2) is an even number, the difference between the number of 1s and 0s in an M-bit pseudo-balanced data word becomes either 0 or 2. If the minimum M is an odd number, the difference between the number of 1s and 0s becomes 1. For example, when the original data word consists of 4 bits ($N = 4$), the balanced coding scheme requires at least two additional bits, which makes $M = 6$. Using the pseudo-balanced coding scheme, (6.2) is satisfied with $M = 5$. Thus, the encoded string becomes 5-bits long, which is less than the required length for the balanced data word. The 5-bit pseudo-balanced data word can be categorized into two types; where the number of 1s and 0s are either (2,3) or (3,2). To maintain the total drive current constant, the counts of 1s and 0s need to be maintained constant. To achieve this, a balancing bit is added to the encoded string. When the number of 1s and 0s are (2,3), the balancing bit becomes 1, while it becomes 0 in the other case. As a result, the number of 1s and 0s become (3,3) at all times. The balancing bit is not transmitted, but terminated at the transmitter after serving as a parity bit, which causes the balancing of the current through the PDN. Only 5 bits are transmitted and arrive at the receiver, where the number of 1s is either 2 or 3. The 5-bit symbol provides enough patterns to uniquely encode the original 4-bit information and for recovery at the receiver.

As an example consider Figure 6.13 where 4 bits are encoded into 6 bits with only 5 bits being transmitted to the receiver. The data sequence 0100 is encoded to 010011 while the data sequence 0101 is encoded to 101100, where both cause three drivers to turn on, as shown

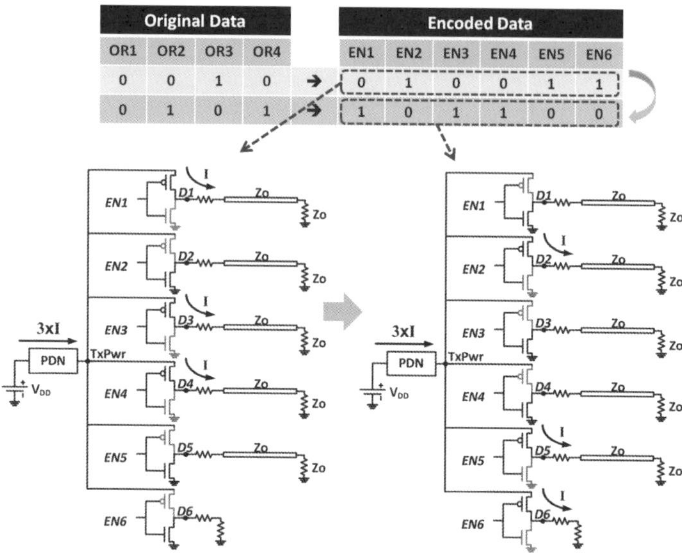

Figure 6.13: Encoding of data and the resulting driver switching [Huh, 2011c].

in Figure 6.13, resulting in constant current being drawn from the PDN. With a constant current of 3xI (I is the current from one driver) passing through the PTL, the power supply node of the driver remains constant. Since both the PTL and signal transmission lines are referenced to a common ground plane as before, there are no RPDs and hence the output waveform should have maximum signal integrity.

Two important issues are noteworthy here namely, 1) the encoding scheme eliminates the need for the dummy path used in the CCPTL scheme and 2) a single PTL feeds six drivers without the need for a data pattern generator as in CCPTL, since this is addressed through coding.

The encoding scheme for 4, 5 and 6 bits of data using both balanced and pseudo-balanced signaling are tabulated in Table 6.2. In the table, "M" represents the number of bits after encoding with balancing while "new M" is the data width using pseudo-balancing. The balancing bit is used in pseudo-balanced scheme to equalize the number of 1s and 0s in the pattern, but is never transmitted. Therefore, it is only used to draw constant current from the PDN. Table 6.2 also shows the overhead

Table 6.2: Comparison of Balanced and Pseudo-balanced Signaling [Huh et al., 2011c].

N	Number of patterns	M	(1s, 0s)	Overhead	new M	(1s, 0s)	With balancing bit	(1s,0s)	Number of patterns	Overhead
4	16	6	(3,3)	50	5	(2,3), (3,2)	→	(3,3)	20	25
5	32	7	(3,4)	40	6	(2,4), (3,3)	→	(3,4)	35	20
6	64	8	(4,4)	33.3	7	(3,4), (4,3)	→	(4,4)	70	16.7

involved with both balanced (M) and pseudo-balanced (new M) schemes with overhead decreasing (for both) with increase in data width. From the table, pseudo-balanced scheme has lower overhead and is therefore preferable and implemented in this section.

6.3.1 *PBPTL Test Vehicle with Off the Shelf Chips*

The PBPTL scheme is used here for the design of a board using off-the-shelf chips and compared with a board containing power plane. The layout of two fabricated boards is shown in Figure 6.14, using a cross section identical to Figures 6.8(c) and (d). For the test vehicle with the power plane, a pair of 237mm by 52mm planes (voltage and ground) with 0.7mm thick FR4 dielectric is used. For the PTL test vehicle, a 25ohm transmission line is used to power the drivers. A 20-pin TSSOP-packaged octal driver is mounted on each board. Only 6 drivers are used in this experiment. Each driver is connected to a 203-mm long signal transmission line. The signal transmission lines are both series- and parallel-terminated. A series termination of 200 ohm is used to limit the amount of current through the I/O driver and thus reduce the current load of the driver. Each signal transmission line is terminated using a 50Ω load through an SMA connector. A $0.1\mu F$ decoupling capacitor is placed between the power and ground pins of the device to respond to the sudden surge of current during transitions in both test vehicles.

Fourteen ports are defined for each structure: two ports for the PDN (port 1 and 2) and twelve ports for six signal transmission lines (ports 3, 5, 7, 9, 11, 13 at input and ports 4, 6, 8, 10, 12, 14 at output) with

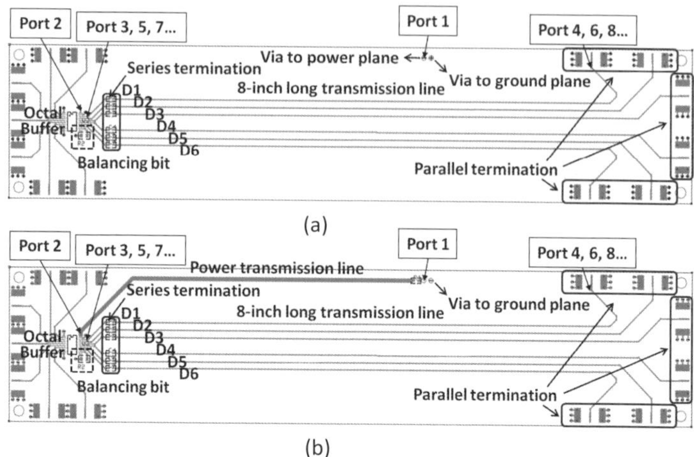

Figure 6.14: Layout of PBPTL Test Vehicle (a) with Power plane (TV1) and (b) with PTL (TV2).

reference to the ground plane, with a subset of the ports shown in Figure 6.14. These ports are defined for modeling purposes where the signal lines and PDN can be modeled together to extract the port responses, which can then be used in a circuit simulator for time domain simulation. Figure 6.15 shows the insertion loss between the input and output of one signal transmission line for test vehicles TV1 and TV2, before and after adding the decoupling capacitor to the PDN. The insertion loss of TV1 has multiple resonances, due to the resonances between the power and ground plane. The first dip appears at 313MHz, which coincides with the half-wavelength ($\lambda/2$) resonant frequency of the plane. When a 0.1μF decoupling capacitor with 2nH equivalent series inductance (ESL) is connected between the power and ground planes in TV1, the decoupling capacitor changes the response of the PDN in the frequency domain leading to a change in the insertion loss below 1GHz, as shown in the zoomed-in plot in Figure 6.15. Hence, the capacitor has a large effect on the insertion loss of the signal line. However, the insertion loss for the signal line in TV2 has a smooth negative slope. The decoupling capacitor added to the PTL does not affect the insertion loss of the signal line. It reaches -10.32dB at 10GHz, which is 2dB better than TV1. So, clearly the impact of the PDN in degrading the insertion loss of the signal line is

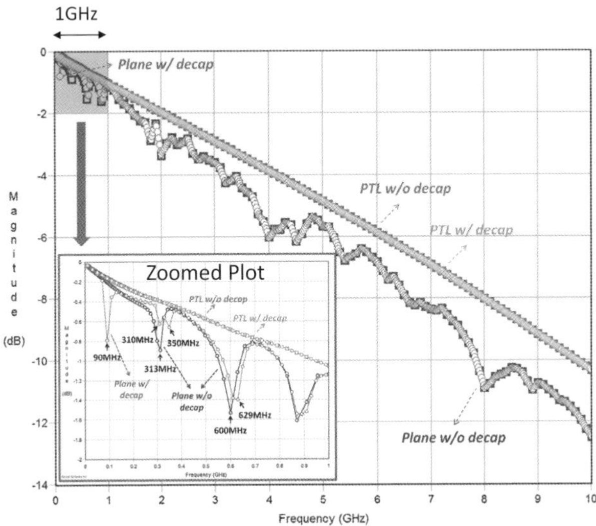

Figure 6.15: Comparison of insertion loss with and without decoupling capacitor for plane (TV1) and PTL (TV2) test vehicles [Huh, 2011c].

larger for TV1 than TV2, due to the presence of the RPD. An important point to note is the complexity involved in modeling the plane resonances and their effect on signal integrity. Unless the board is modeled in its entirety, plane resonances cannot be captured accurately. This requires complex Electronic Design Automation (EDA) tools. [Swaminathan et al., 2007]. However, PTLs are one-dimensional transmission line structures and therefore their modeling is much easier.

The measurement set-up used is shown in Figure 6.16 where an automated test equipment (ATE) system provides power and input data patterns to the two test vehicles. The power supply voltage is 5V, with a target data rate of 300Mbps. The source termination for the PTL in TV2 has been removed and therefore there is no DC drop at the power supply node of the driver. Eye diagrams are measured at the output of the 4th 203-mm long signal line, using an Agilent 86100C oscilloscope. All the signal lines are terminated at the far end using 50ohms.

The output waveforms are compared between TV1 and TV2 in Figure 6.17. For TV1, excessive ringing occurs, for both the negative and positive transitions of the output data, as shown in Figure 6.17(a). The

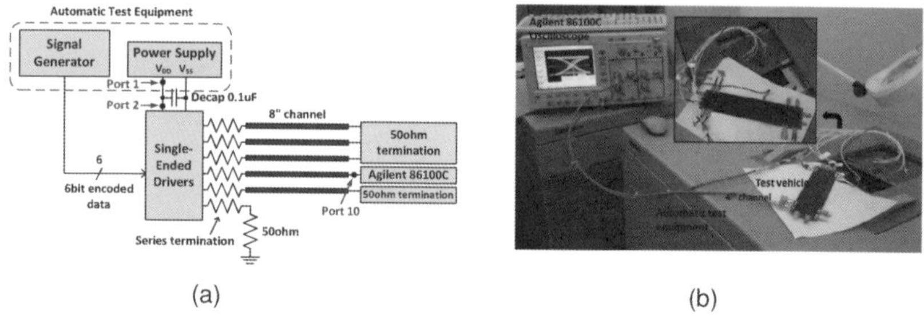

(a)   (b)

Figure 6.16: Measurement set-up (a) Block Diagram and (b) Photograph of test environment.

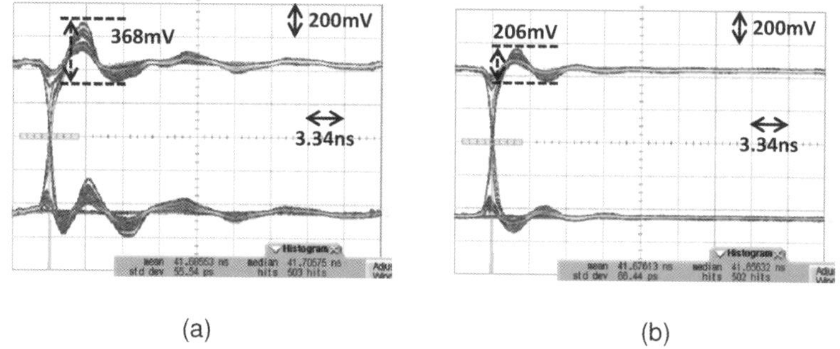

(a)   (b)

Figure 6.17: Comparison of output data waveform (a) TV1 and (b) TV2.

placement of the decoupling capacitor between the power and ground planes causes the ESL of the decoupling capacitor to be in parallel with the plane capacitance and the on resistance of the driver, which forms a parallel RLC circuit, resulting in ringing. The peak-to-peak value of the ringing is 368 mV in Figure 6.17(a).

Since the decoupling capacitor provides a current path at the RPD location in TV1 between the power and ground planes, the impact of the ESL on the waveform can be seen. The ringing continues for around 30nsec before it is damped. The ringing in the waveform for TV2 is less severe, as seen in Figure 6.17(b). In the PTL scheme, a closed current loop is achieved without the decoupling capacitor and therefore the role of the decoupling capacitor is to act as a charge reservoir responding to sudden surges of charge during data transitions as opposed to compensating for RPDs. This reduces the impact of the capacitor ESL on the signal waveform. This results in the peak-to-peak value of ringing to reduce to 206 mV, which is 44% smaller than that of TV1. In addition, the damping duration is reduced to less than 10 nsec.

The pseudo-balanced signaling scheme is next applied to both TV1 and TV2 test vehicles. The 6th bit is used as the balancing bit so that the 6th driver is connected to a resistor whose value equals the sum of the series and parallel terminations. It contributes to drawing constant current from the PDN, but does not increase the PCB trace count. The resulting eye diagrams at the output of the 4th channel on TV1 and TV2 are shown in Figure 6.18. Comparing Figures 6.18(a) and (b), the quality of the eye diagram improves significantly with a decrease of 151ps in p-p jitter (21.8%) and an increase of 203mV in eye height (34.3%) for TV2 as compared to TV1. The improvement in signal quality can be attributed to the elimination of RPD effects during signaling using power transmission line. The eye diagrams for all signal lines showed similar improvement using PTL as compared to power plane.

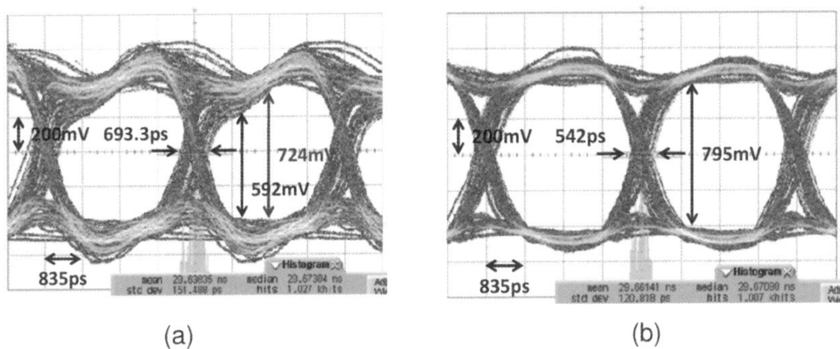

Figure 6.18: Pseudo-balanced signaling (a) TV1 with power plane and (b) TV2 with PTL.

6.3.2 *Design and Measurements of Custom Chip*

As is typically the case, custom designed chips always perform better compared to off the shelf chips since they can be fine tuned for obtaining the best signal and power integrity. A transmitter for the proposed

PBPTL scheme, custom-designed and fabricated using a 0.18μm CMOS process, is briefly described in this section. In Figure 6.19(a), a PRBS generator supplies 4-bit PRBS data to a 4b6b encoder. The encoder maps 4-bit information onto 6-bit symbols with an equal number of 1s and 0s. The resulting balanced 6-bit data are sampled at the rising edge of the clock signal by the edge-triggered D flip-flop. Therefore, the transition edges of the six signals are timing-aligned with the clock signal. The synchronized and balanced 6-bit data are used to drive the output buffers. Figures 6.19(b) and (c) show the output buffers used for the two test vehicles using power plane and PTL, respectively. Chip on Board (COB) assembly technology was used to directly mount the die on the board using wire bonding. The transmitter was tested with 81-mm long signal line on the FR4 board. One of the boards uses a pair of 140mm by 93mm planes with 0.7mm thick FR4 dielectric (TV1), while the other board has a 66-mm long source-terminated power transmission line with 0.7mm thick FR4 dielectric (TV2). For TV1, a supply voltage of 3.11V is used to power the driver circuits through power and ground planes while for TV2, a 25ohm transmission line is used to feed the same supply voltage of 3.11V to the driver circuits, with the PTL sharing a common ground plane with signal transmission lines. A supply voltage of 1.8V is used for the core circuits for both test vehicles. Each output buffer drives a 66-mm long transmission line.

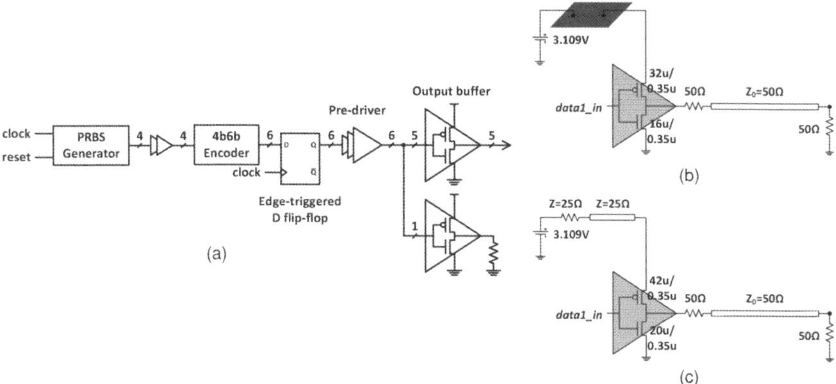

Figure 6.19: Schematic for (a) 4b/5b pseudo-balanced signaling, (b) output buffer using power plane for PDN (TV1) and (c) using PTL for PDN (TV2) [Huh et al., 2011c].

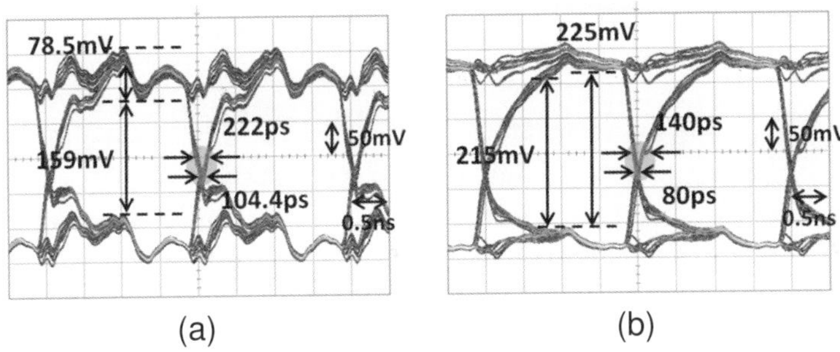

Figure 6.20: Output waveforms for custom chip at 500Mbps data rate (a) TV1 and (b) TV2 [Huh et al., 2011c].

The eye diagrams for TV1 and TV2 for 500Mbps data rate is shown in Figure 6.20. As can be seen, both the waveforms degrade due to the high speed. However, the signal quality is much better for TV2 than TV1. Comparing the two waveforms, TV1 has an eye height of 159mV as compared to 215mV for TV2 (35% improvement) while p-p jitter is at 104.4ps for TV1 as compared to 80ps for TV2 (23% improvement).

For completeness, Figure 6.21 shows the chip microphotograph. The active area of the pseudo-balanced scheme using power plane and PTL are shown in the figure occupying an area of $1.51 \times 10^{-2} mm^2$, which includes the PRBS generator, the encoder, the pre-drivers, and the output buffers. The area devoted to the encoder equals $3.29 \times 10^{-3} mm^2$, which is 22% of the transmitter area.

Figure 6.21: Chip layout showing pseudo-balanced implementation with plane and PTL power distribution.

The PBPTL scheme provides several advantages which are listed below:

(i)   The encoding scheme removes the need for dummy paths used in CCPTL and enables a single PTL to feed multiple drivers.

(ii) Compared to balanced-signaling, pseudo-balanced signaling reduces the number of signal lines in the printed circuit board.

(iii) Since capacitors are not used to mitigate RPDs, their ESL do not play a role in affecting signal integrity.

(iv) Compared to CCPTL, the power is reduced by 50%. This is discussed in detail in a later section.

6.4 Constant Voltage Power Transmission Line (CVPTL)

In this section, the power consumption is reduced more as compared to PBPTL. The configuration for this scheme is shown in Figure 6.22. For the constant voltage power transmission Line (CVPTL) scheme, depending on the input data, a data pattern detector is used to select a resistor path in the Power Distribution Network (PDN) to vary the current being drawn from the power supply. The resistor values (shown as R[00], R[01],[10], and R[11] in Figure 6.22) are carefully chosen to keep the power supply voltage seen by the drivers (V_{DD}) at a constant level regardless of driver states. For N drivers, there are a total of 2^N possible data combinations. However, the current that is drawn by the PDN is dictated not by the number of output combinations, but by the number of drivers that are in a "high" state at any given time. Consequently, for N drivers, $N + 1$ different values of source current can be drawn, and therefore $N + 1$ different resistor values are required in the PDN.

6.4.1 *Design Equations*

To appropriately vary the current through the PTL in Figure 6.22, the resistor network must be carefully designed. One important parameter is $R_{driver,k}$, which is the resistance looking into the power

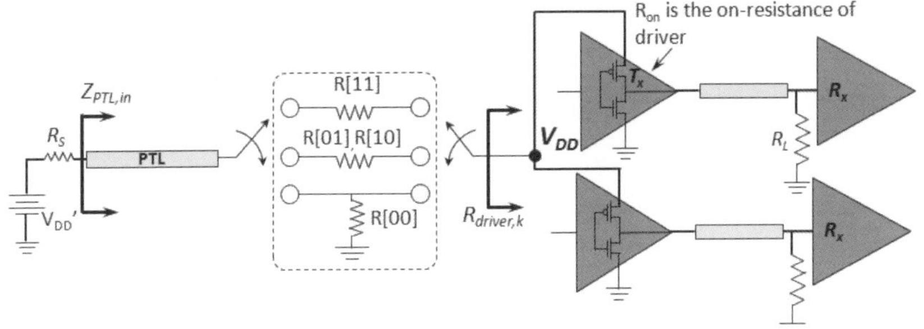

Figure 6.22: Constant Voltage Power Transmission Line (CVPTL).

supply node of the drivers, which is calculated as:

$$R_{driver,k} = \frac{R_{on} + R_L}{k} \tag{6.3}$$

The variable k denotes the number of ON drivers at any time with $k = 1,2,....N$ and R_L is the load resistance matched to the impedance of the signal transmission line. Consequently, the resistor value for each driver state can be determined [Telikepalli et al., 2013] using a voltage divider network as:

$$V_{DD} = V_{DD}' \frac{R_{driver,k}}{R_{driver,k} + R_s + R[ij]} \qquad i = 0,1; \; j = 0,1; i = j \neq 0; k = i + j \tag{6.4}$$

From (6.4), the resistor values R[ij] can be obtained as:

$$R[ij] = \frac{1}{V_{DD}}[V_{DD}'R_{driver,k} - V_{DD}(R_{driver,k} + R_s)] \tag{6.5}$$

where R_s is the source termination resistance. Equation (6.5) cannot be used when the output data are all zero since $R_{driver,0} \approx \infty$ under this scenario. When $k = 0$, using the voltage divider network:

$$V_{DD} = V_{DD}' \frac{R[00]}{R[00] + R_s} \tag{6.6}$$

R[00] can now be determined as:

$$R[00] = \frac{V_{DD}}{V'_{DD} - V_{DD}} R_s \qquad (6.7)$$

The independent variable in the above equations is k, the number of active drivers at any given time, and not the actual bit pattern itself. Therefore the data pattern detector selects the same resistor value when the data pattern is either "0101" or "1010". This means that the number of active drivers dictates the amount of current flow and not the actual bit pattern itself. The source termination resistor R_s in Figure 6.22 is used to prevent any reflections occurring on the line which can propagate across the line and back into the driver's power pin. In Figure 6.22, R_{on} is the 'on' resistance of the drivers where the same resistance is assumed for all drivers, V'_{DD} is the voltage provided by the power supply, V_{DD} is the desired supply voltage at the drivers, and R_L is the load resistance as seen by the driver's output pin. It is important to note that in this scheme the current that is needed to charge the signal transmission line is of appropriate strength.

6.4.1.1 *Example*

As an example consider Figure 6.22 where a PTL connects to two drivers. The PTL has an impedance of 25Ω and therefore $R_s = 25Ω$. The on-resistance of the driver is 50Ω, with each connected to a signal transmission line of impedance 50Ω and terminated in a load resistance $R_L = 50Ω$. Assume that $V'_{DD} = 5V$ and $V_{DD} = 2.5V$. From (6.3):

$$R_{driver,1} = 100Ω; R_{driver,2} = 50Ω$$

From (6.5):

$$R[01] = R[10] = 75Ω; R[11] = 25Ω$$

From (6.7)

$$R[00] = 25Ω.$$

6.4.2 *CVPTL Test Vehicle*

To determine the effectiveness of the CVPTL scheme, two test vehicles are shown in Figure 6.23, with each designed to transmit a 4-bit data sequence. Test vehicle TV1 in Figure 6.23(a) uses power and ground

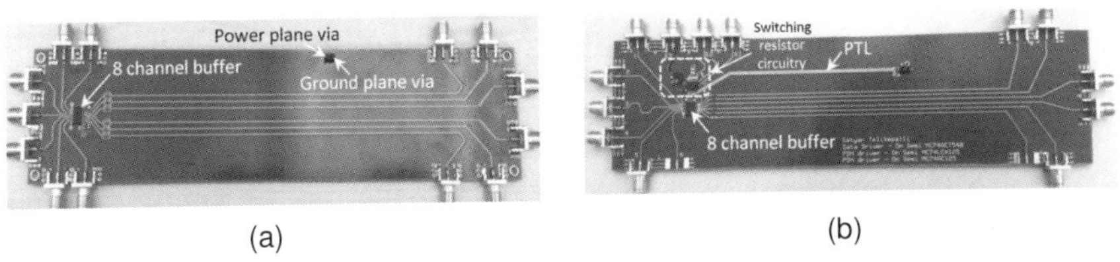

Figure 6.23: Test Vehicle (a) TV1 with power and ground planes and (b) TV2 with PTL and Constant Voltage signaling.

planes with a stack-up as in Figure 6.8(c) while TV2 in Figure 6.23(b) uses PTL routed on the same layer as the signal lines with a ground plane beneath it corresponding to the stack-up in Figure 6.8(d).

A commercially available 20-pin octal buffer/line driver is used for both test vehicles. The power and ground planes measure 240mm × 52mm in TV1. Test vehicle TV2 is implemented using a 25Ω power transmission line and 25Ω source termination resistor. At the output pin of each driver (with on resistance $R_{on} = 11\Omega$), a 450Ω series resistor is placed to limit the current required to charge the line and the current flowing into the oscilloscope. Each output connects to SMA connectors using a 205mm long 50Ω microstrip line, which is terminated in a 50Ω load resistor. In addition, a 0.1μF decoupling capacitor is connected between the power and ground pins of the driver to supply the high frequency charge for both boards. The switching resistor paths are implemented with extra drivers placed in series with the PDN. Therefore, enabling these extra drivers effectively serves to add series resistance into the power supply path of resistance 13Ω.

To implement four data channels, five unique resistor values are required based on the design equations. To simplify implementation, bus-inversion encoding scheme can be used [Cheng et al., 2001], [Stan et al., 1995]. This scheme requires an extra bit to be transmitted in addition to the 4-bit data sequence where the extra bit, or 'inversion' bit is set to '0' by default. When the number of ones in the 4-bit data sequence exceeds 2, the inversion bit is set to '1' and the data sequence is inverted. For example, when the data sequence to be transmitted is '1011', the actual transmitted sequence is "10100", where the first bit is the inversion bit and the remaining four bits are the compliment of the original data sequence. With this encoding scheme, instead of requiring five resistors in the PDN, only three are required since the encoding scheme will only transmit 0, 1, or 2 high states at any given time [Telikepalli et al., 2013]. To further reduce the number of resistors in the PDN, 4 bits can be encoded to 6 bits using the bus inversion scheme (shown under CVPTL) in the truth table in Table 6.4. In the table the 5th bit indicates inversion while the 6th bit is used to differentiate between the first and last row (0000 pattern). In Table 6.4, the data pattern has either 1 or 2 high states.

The drivers in both test vehicles use a supply voltage of 4.5V at the V_{DD} node. For the TV2 test vehicle $V'_{DD} = 5.3V$. From (6.3):

$$R_{driver,1} = \frac{11 + 450 + 50}{1} = 511\Omega; R_{driver,2} = \frac{11 + 450 + 50}{2} = 255.5\Omega.$$

From (6.5):

$$R[01] \approx 65.84\Omega; R[11] \approx 20.42\Omega.$$

Since 4 bits are encoded to 6 bits for CVPTL in Table 6.4, either one or two high states need to be transmitted. The resistor values required are therefore equivalent to transmitting 01 or 11 for two bit transmission and therefore the design equations (6.3) and (6.5) can be used. In Figure 6.22, switches in the PDN are implemented using buffers. Two buffers are required in the PDN for the test vehicle, one each for the one and two high states. Since the PDN buffers have an on resistance of 13Ω, the resistors required in series with the buffers have values of 52.84Ω and 7.42Ω for one and two high state transmission, respectively. In the test vehicle, the resistors were adjusted to 51Ω and 12Ω to achieve the best results.

Both test vehicles were tested with an HP 83000 Automated Test Equipment (ATE) with 256-bit pseudo-random bit sequences (PRBS) at

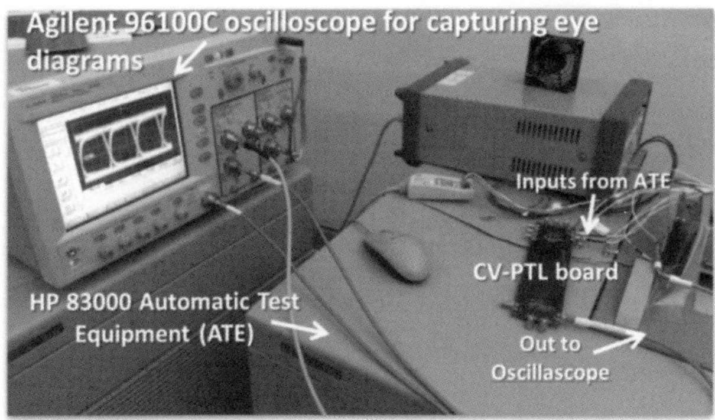

Figure 6.24: Measurement set-up.

data rates of 100Mbps, 200Mbps, and 300Mbps. The measurement setup is shown in Figure 6.24. The ATE was programmed to provide both the test vehicles with the appropriate PRBS input pattern. For TV2 test vehicle, the input was encoded using the bus inversion scheme, as described previously in this section.

6.4.3 *Measurements and Comparison*

The measured eye diagrams, and time-domain waveforms at the driver output and power supply node (V_{DD}) are shown in Figure 6.25.

The CVPTL test vehicle (TV2) shows significant improvement in performance as compared to TV1, as shown in Figure 6.25. At 300Mbps, the CVPTL scheme exhibits ~324ps of p-p jitter, while the test vehicle using power and ground planes exhibit ~418ps of jitter. In addition, the voltage fluctuation on the V_{DD} pin is much lower for CVPTL. The results for both test vehicles are summarized in Table 6.3. Comparing the two test vehicles TV2 provides between 22%–46% improvement in p-p jitter and 31%–44% improvement in power supply noise, which represent significant improvements.

As compared to CCPTL and PBPTL, CVPTL provides further reduction in power consumption, as will be discussed in the next section. This is made possible through inversion coding and by varying the current through the PDN rather than keeping it constant, as was done with both CCPTL and PBPTL schemes.

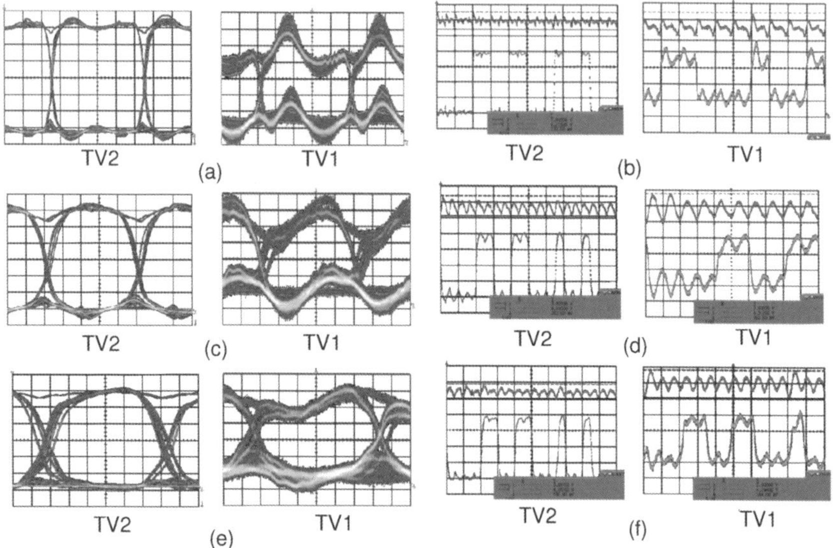

Figure 6.25: a) 100Mbps PRBS, (b) 100Mbps power supply noise (top) and driver output waveform (bottom), (c) 200Mbps PRBS, (d) 200Mbps power supply noise (top) and driver output waveform (bottom), (e) 300Mbps PRBS and (f) 300Mbps power supply noise (top) and driver output waveform (bottom).

Table 6.3: Measurement results and comparison between TV1 and TV2.

	TV1		TV2			
Data Rate	p-p jitter (ps)	ΔV_{DD} (mV)	p-p jitter (ps)	Jitter %Δ	ΔV_{DD} (mV)	ΔV_{DD} (mV) %Δ
100Mbps	**322**	**364**	**196**	-39%	**220**	-39.5%
200Mbps	**533**	**382**	**290**	-46%	**262**	-31.4%
300Mbps	**418**	**344**	**324**	-22.5%	**190**	-44.7%

6.5 Power Calculations

This section compares the power consumption for the four signaling schemes namely, 1) standard, 2) constant current, 3) pseudo-balanced and 4) constant voltage. The standard scheme refers to the use of the drivers in its current form without any modifications made to it. Since it has been shown in the previous sections that PTL provides better signal integrity due to improved power integrity, the focus in this section is on power calculations for the four signaling schemes.

As a simple example, consider a 4 bit data sequence transmitted through four output drivers. The truth table for the four bits is shown in Table 6.4, under the standard column. The columns labeled A, B, C and D show the total number of 'high' states transmitted during each cycle for standard, CVPTL, CCPTL and PBPTL schemes respectively, which represent the states that consume power. From the previous sections, irrespective of the number of 'high' states in the truth table, the number of PMOS transistors ON at any time is 4 for CCPTL and 3 for PBPTL. For the standard method, the number of 'high' states varies between 0-4. Using the inversion coding scheme for CVPTL, the number of high states is between 1-2. In Table 6.4, the first four bits for CVPTL represent the bits transmitted after inversion, high state for the 5th bit indicates inversion and a high state for the 6th bit is used to transmit all zeros without inversion. Consequently, using this scheme for a PRBS input, on average, only 1.625 drivers are in the high state as compared to 2 for standard, 4 for CCPTL and 3 for PBPTL. Hence, CVPTL results in 19% less drivers in the high state as compared to the standard design. In addition, CVPTL has 45% and 59% less drivers in the high state as compared to PBPTL and CCPTL, respectively.

Consider next the test vehicles described in the previous sections for standard, CCPTL, PBPTL and CVPTL schemes. Using the parameters shown in Table 6.5 the power can be calculated. The PTL for the CVPTL case has a characteristic impedance of 25Ω and is source terminated with a 25Ω resistor. A similar PTL structure is used for the CCPTL and PBPTL schemes except that two cases are considered namely, with and without source termination. To obtain 4.5V at the power pins of the drivers, and to compensate for the voltage drop across the PTL source termination resistor and the PDN drivers and resistors, the power supply voltage used for CVPTL is 5.3V, 5.39V for CCPTL and 4.5V for the standard design. With the desired 4.5V at the power pins of the drivers and 500Ω load resistor (450Ω output resistor to limit current and 50Ω load resistor), each driver consumes 8.81mA of current in the 'high'

                    Design and Modeling for 3D ICs and Interposers

Table 6.4: Truth Table and average high states for Standard (A), CVPTL (B), CCPTL (C) and PBPTL (D).

Standard				A	CVPTL						B	C	D
0	0	0	0	0	0	0	0	0	0	1	1	4	3
0	0	0	1	1	0	0	0	1	0	0	1	4	3
0	0	1	0	1	0	0	1	0	0	0	1	4	3
0	0	1	1	2	0	0	1	1	0	0	2	4	3
0	1	0	0	1	0	1	0	0	0	0	1	4	3
0	1	0	1	2	0	1	0	1	0	0	2	4	3
0	1	1	0	2	0	1	1	0	0	0	2	4	3
0	1	1	1	3	1	0	0	0	1	0	2	4	3
1	0	0	0	1	1	0	0	0	0	0	1	4	3
1	0	0	1	2	1	0	0	1	0	0	2	4	3
1	0	1	0	2	1	0	1	0	0	0	2	4	3
1	0	1	1	3	0	1	0	0	1	0	2	4	3
1	1	0	0	2	1	1	0	0	0	0	2	4	3
1	1	0	1	3	0	0	1	0	1	0	2	4	3
1	1	1	0	3	0	0	0	1	1	0	2	4	3
1	1	1	1	4	0	0	0	0	1	0	1	4	3
Average				2	Average						1.625	4	3

Table 6.5: Power Calculations.

	CV-PTL	CC-PTL (w/ R_S)	CC-PTL (w/o R_S)	PB-PTL (w/ R_S)	PB-PTL (w/o R_S)	Standard
Current for 1 Driver (mA)	8.81	8.81	8.81	8.81	8.81	8.81
Average "High" Drivers	1.625	4.0	4.0	3	3	2.0
Average Current (mA)	14.31	35.24	35.24	26.43	26.43	17.62
Supply Voltage V_{DD}' (V)	5.30	5.39	4.50	5.17	4.50	4.50
Total Power (mW)	75.86	189.94	158.58	136.64	118.94	79.29

state. Since, on average, there are 1.625 drivers activated for a PRBS input for CVPTL, the current required is 14.31mA. With a 5.3V supply voltage, the CVPTL circuit will dissipate on average 75.86mW of power. This compares to 79.29mW for standard design, 189.94mW for CCPTL (with source termination) and 136.64mW for PBPTL (with source termination). As expected, the power consumption for CCPTL and PBPTL without source termination reduces to 158.58mW and 118.94mW, respectively. Hence, the CVPTL scheme (with inversion coding) consumes less power on average than either the CCPTL, PBPTL or standard signaling case.

6.6 Application of Power Transmission Lines to FPGA

A brief illustration of the practicality of using PTL to power the output drivers in a large IC are described in this section. The Xilinx Spartan-6 LX45 FG(G) 484 with 338 I/O drivers is shown in Figure 6.26(a) mounted on a board. These drivers are connected to 50Ω signal transmission lines on the board with a far end termination of 50Ω. The PBPTL scheme with 12 to 18 coding (three groups with each containing 4 bits encoded to 6 bits as described earlier) was used which required a PTL with impedance 10Ω. A single PTL was used to feed all the drivers, as shown in Figure 6.26(b). The eye diagram at the far end of a signal line for a PRBS at 600MHz is shown in Figure 6.26(c), showing a p-p jitter of 100ps and eye height of 1.37V, for a 2.5V power supply [Telikepalli et al., 2013b]. Given the quality of the eye diagram, this is an example demonstrating that the new power distribution scheme can indeed be used for providing voltage and current to the I/O drivers in a real world application.

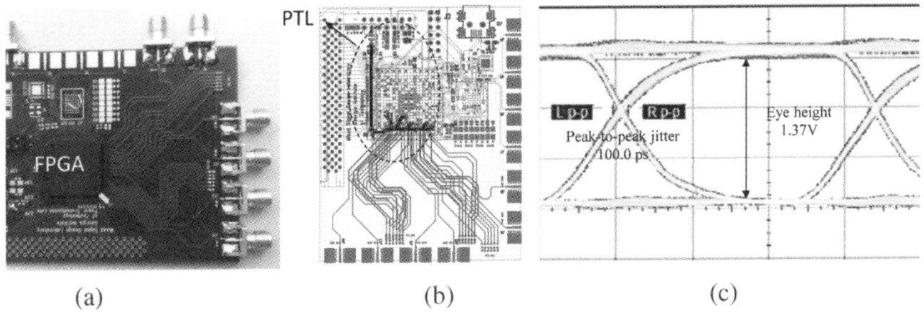

Figure 6.26: (a) FPGA IC on board, (b) Layout showing PTL as part of PDN and (c) Eye diagram at 600MHz.

6.7 Managing Signal and Power Integrity for 3D ICs

As described in Chapter 5, the inductance in the PDN increases as the position of the die in the stack increases. This can be compensated for by using additional capacitance within the die. As has been shown in Chapter 5, when all the dies in the stack switch simultaneously, the noise in the top die is much larger than in the bottom die. In this section the

CCPTL, PBPTL and CVPTL methods applied to 2D ICs are extended to 3D ICs where the performance of 3 stacked dies powered through standard (called conventional here) means using power/ground planes is compared with PTL based methods. The discussions for eye diagrams and power supply noise are restricted to the CVPTL scheme in this section while power comparisons are provided that include all methods.

The power distribution scheme for 3 stacked dies is shown in Figure 6.27(a) where power and ground planes are used in the PCB to connect the power supply to the ICs. In contrast, PTLs are used in the PCB to provide power to the ICs as shown in Figure 6.27(b) where the CVPTL scheme has been implemented. Simulations are used in this section to show that even though the inductance of the PDN increases from Die 1 to Die 3, the power supply fluctuation is reduced in Figure 6.27(b) and is much lower as compared to Figure 6.27(a) where standard methods are used. *To amplify the impact of noise, decoupling capacitors are not used in the simulation. In this section, the focus is on the I/O communication between dies with an objective of minimizing power supply noise at each die and improving the eye diagrams.*

Similar to Chapter 5, detailed models are constructed here to estimate the performance of each of the PDN architectures shown in Figure 6.27. Each die in Figure 6.27 is of size 1mm × 1mm and contains a TSV layer, PDN and digital logic. The TSV layer in each die consists of three TSV

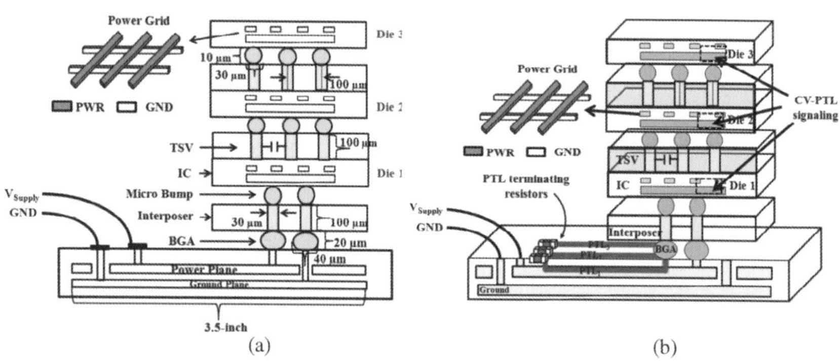

Figure 6.27: Power distribution for 3 stacked dies (a) conventional design using power/ground planes and (b) using CVPTL [Telikepalli et al., 2013a].

types namely, power, signal and ground. The interposer contains a power and ground TSV pair. All TSVs are surrounded by 0.1μm thick oxide layer. The micro bumps and BGA balls between layers are modeled as a resistor in series with an inductor. Resistance and inductance of BGA balls used are 2.4mΩ and 6.44pH, respectively, while those of micro bumps are 2.8mΩ and 0.8pH, respectively. The dimensions of TSV, BGA, and micro bumps are all shown in Figure 6.27. The PDN contains power and ground grid corresponding to the top metal layers in the IC, similar to the structures in Chapter 5. In each grid, the rails are laid out orthogonally in two separate layers separated by 40um with rail width of 40um. Each grid is divided into unit cells with individual cell size of 200μm × 200μm, as shown in Figure 6.28. Each cell is modeled as two separate resistor networks with resistance in each branch (R_b) equal to 51mΩ. The coupling capacitance between power and ground rails within a cell is approximated to be 4.2fF. The PCB for Figure 6.27(a) is a two layer board with a power and ground plane. The source is modeled as an ideal voltage source in the simulations.

The impedance profile of the power rail at each of the three IC dies is simulated for the structure in Figure 6.27(a) and shown in Figure 6.29(a). In addition, the impedance profile for just the interposer and die stack (not including the PCB package) is shown in Figure 6.29(b) to show the effect of the power and ground plane resonances on the PDN. The impedance profile in Figure 6.29(a) shows many fluctuations with

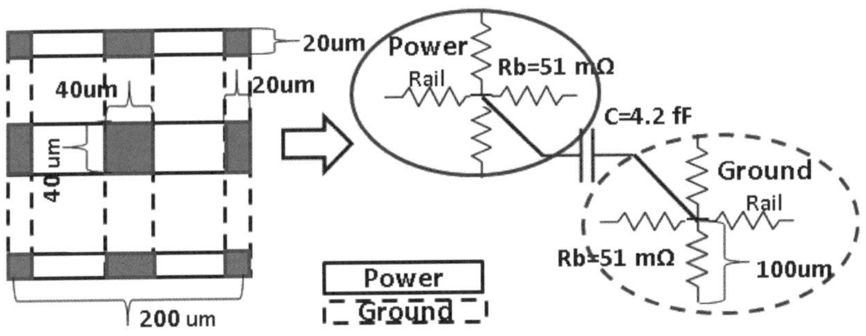

Figure 6.28: Die power grid unit cell and equivalent circuit model [Telikepalli et al., 2013a].

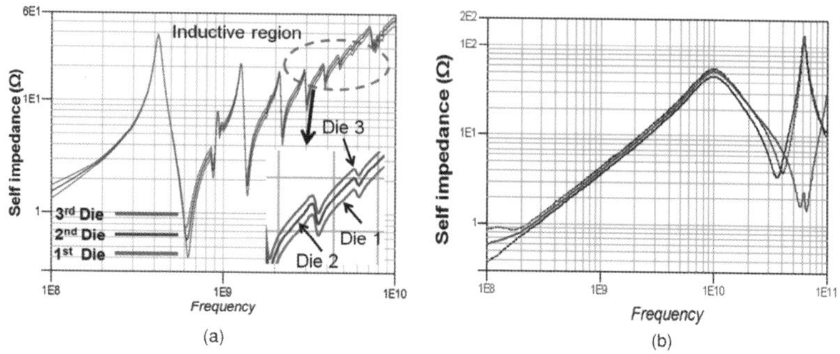

Figure 6.29: PDN impedance profile at each die (a) with PCB and (b) without PCB [Telikepalli et al., 2013a].

frequency. These are due to the cavity mode resonances of the power and ground plane cavity. When the switching frequency of the data driver corresponds to a cavity resonance frequency, this can cause large voltage fluctuations on the power supply node due to higher PDN impedance at that particular frequency. In addition, the impedance of the PDN at each die increases as the die height increases. More importantly, the impedance in the inductive region increases as the position of die in the stack increases, leading to the PDN of Die 3 having more inductance than either Die 1 or 2, similar to the trend shown in Chapter 5. As a result of the increased inductance, the power supply noise will increase as the position of the die in the stack increases. Comparing Figures 6.29(a) and (b), it can be seen that when the PCB package is not included, the impedance profile is smooth up to about 10GHz. This leads to the conclusion that the resonance peaks are primarily attributed to the power and ground planes in the PCB, and by modifying the PDN in the PCB, the effect of switching noise can be mitigated.

Figure 6.30 shows the resulting eye diagrams at the output of each driver for each die. A 1Gbps, 4-bit pseudo-random bit stream (PRBS) pattern is used as the input. Each eye diagram is measured at the output of the signal buffer and the supply noise is measured at the supply terminal of each buffer in the die. In addition, each buffer is terminated to 50Ω and is fed by an input signal on the same layer. In Figure 6.30, the eye diagram deteriorates as the position of the die in the stack

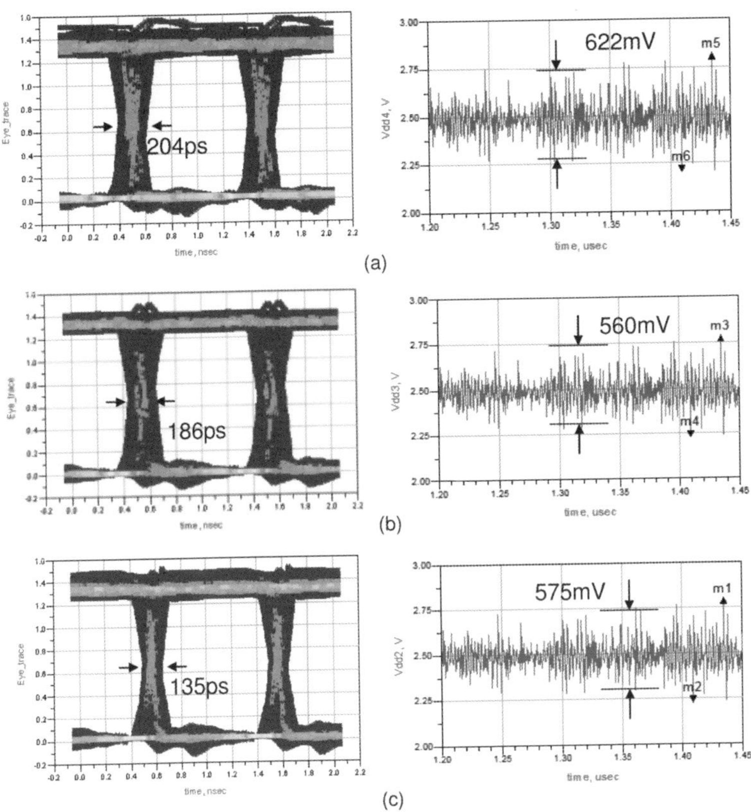

Figure 6.30: Eye diagram (left) and power supply noise (right) @ 1Gbps PRBS for conventional design (a) Die 3: Jitter = 204ps, ΔV_{DD} = 622mV, (b) Die 2: Jitter = 186ps, ΔV_{DD} = 560mV, (c) Die 1: Jitter = 135ps, ΔV_{DD} = 575mV [Telikepalli et al., 2013a].

increases, due to an increase in power supply noise. The p-p jitter from the eye diagrams is 204ps, 186ps, and 135ps, for Die 3 (top), Die 2 (middle), and Die 1 (bottom), respectively. Correspondingly, the peak-to-peak voltage variation (ΔV_{DD}) on the power supply terminal of the drivers is 622mV, 560mV, and 575mV, for Die 3, Die 2, and Die 1, respectively.

Though the CVPTL implementation is shown in Figure 6.27(b) [Telikepalli et al., 2013a], CCPTL and PBPTL methods can be used as well [Zhang et al., 2012]. The CVPTL scheme is simulated with 4-data bits. However, by utilizing the bus-inversion encoding scheme described in Section 6.5, the input data is encoded into 6-bits. The bus inversion scheme is a simple encoding method used to reduce the amount of variation in the driving current through the PDN by encoding the original data bits into data words with minimal change in the number of 1s and 0s that are transmitted, as described earlier. The resulting resistor values, based on the design equations are $R[01] = 28\Omega$ (one high state) and $R[11] = 2\Omega$ (two high states). These are calculated using the following parameters: $V_{DD}^{'} = 4V, V_{DD} = 2.5V, R_{s} = 25\Omega, R_{L} = 50\Omega, R_{on} = 39\Omega$.

The simulation models for the data drivers, TSVs, silicon interposer, micro bumps, BGA balls, and PDN are identical in Figures 6.27(a) and (b). The main difference is that on the PCB layer, instead of utilizing a power and ground plane, the PDN uses three power transmission lines (PTLs) connecting the power supply to the BGA balls located below the bottom layer. The PTL supplies power to the buffers in each die in the stack. In addition, each die includes the extra CVPTL circuitry for implementing the dynamic resistor paths. This implementation is shown in Figure 6.27(b). Unlike the standard scheme, the current is varied to enforce a constant voltage at the power supply terminals of the buffer in the CVPTL method. Hence, a PDN impedance as in Figure 6.29 provides little information on power supply noise and therefore is not shown here.

Figure 6.31 shows the resulting eye diagrams and power supply noise for each die using the CVPTL PDN scheme. The p-p jitter determined from the eye diagrams and the corresponding peak-to-peak power supply noise for CVPTL scheme are 93ps and 366mV, 88ps and 283mV, and 83ps and 207mV, for Die 3, Die 2 and Die 1, respectively. The reduction in jitter and power supply noise compared to the PDN scheme in Figure 6.27(a) is approximately 38.5% and 64%, 52.7% and 49.5%, and 54.4% and 41.2%, in the bottom, middle, and top die, respectively, which is substantial. Once again, it is important to note that no decoupling capacitors are used in both the schemes shown in Figure 6.27 to amplify the impact of power supply noise on eye diagrams. Due to smaller power supply noise, it is expected that the CVPTL scheme will require lower decoupling capacitance to reduce the noise further.

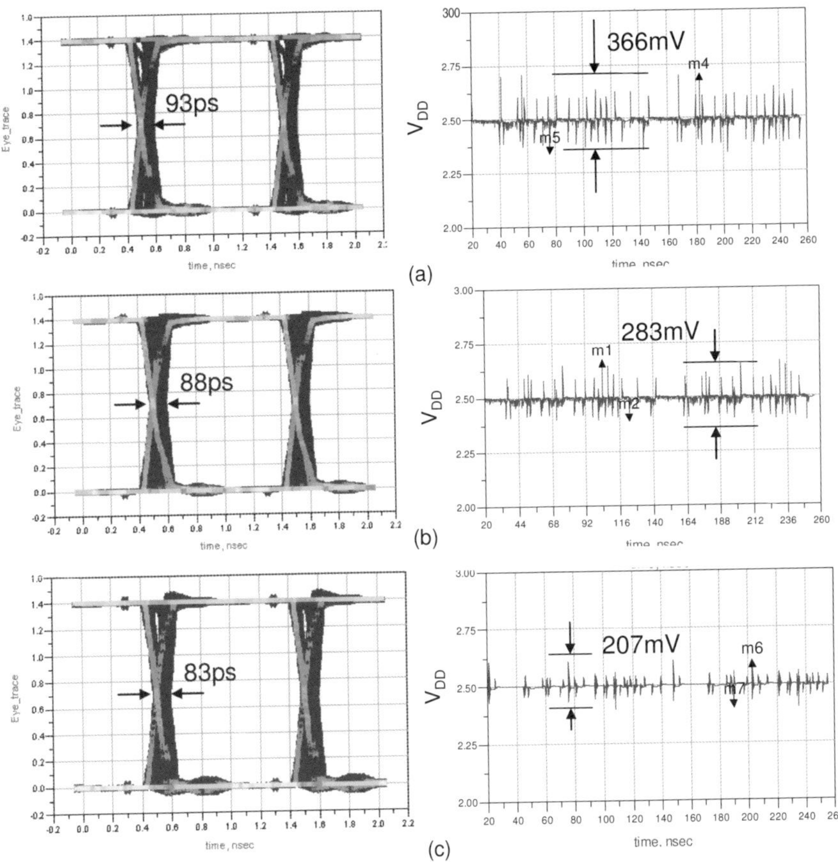

Figure 6.31: Eye diagram (left) and power supply noise (right) @ 1Gbps PRBS for CVPTL (a) Die 3: Jitter = 93ps, ΔV_{DD} = 366mV, (b) Die 2: Jitter = 88ps, ΔV_{DD} = 283mV, (c) Die 1: Jitter = 83ps, ΔV_{DD} = 207mV [Telikepalli et al., 2013a].

6.7.1 *Energy Calculations*

Unlike the power calculations shown earlier, the instantaneous power is simulated in this section for the three die stack shown in Figure 6.27 for a 256-bit PRBS input pattern at 1Gbps.

Figures 6.32(a) and (b) show the instantaneous simulated power consumed by the CVPTL and standard schemes, respectively. These simulations are also repeated for the CCPTL and PBPTL methods. The

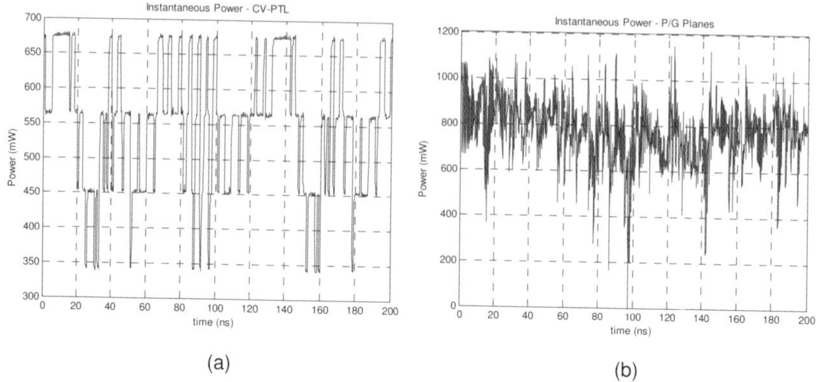

(a)   (b)

Figure 6.32: Instantaneous power consumed for 256-bit long PRBS signal @ 1Gbps (a) CVPTL and (b) Conventional [Telikepalli et al., 2013A].

total energy consumed by the circuit and energy/bit for the same 256-bit PRBS input at 1Gbps is summarized in Table 6.6. These results account for the losses in the resistor network and the increased supply voltage, and therefore all loss mechanisms are included here.

In Table 6.6, the CVPTL scheme uses a supply voltage of 4.0V (to compensate for the source termination resistor), while the PBPTL and conventional methods use 2.5V supply. As mentioned earlier, the PTL source termination resistor serves to prevent any reflections that may occur on the line that can propagate across the line and back into the driver's power pin. The source termination resistor can be removed in most scenarios if the load termination resistor is matched well to the signal transmission line. For the PBPTL circuit in Table 6.6 a 4b/6b encoding is used. From Table 6.6, the percentage increase in the energy per bit as compared to the CVPTL scheme is shown, which provides insight into the power levels involved with each method.

Table 6.6: Energy comparison between the four methods [Telikepalli et al., 2013A].

Scheme	Total Energy (μJ)	Energy/bit (pJ)	%Δ CVPTL
CVPTL	0.099	32.2	NA
PBPTL	0.124	40.4	+25
CCPTL	0.165	53.7	+66
Conventional	0.110	35.8	+11.0

6.8 Summary

In Chapter 5, two issues were identified for power distribution namely; return path discontinuities (RPD) and excessive power supply noise. To address these two problems alternate methods for power distribution have been proposed in this chapter using power transmission lines (PTL) in the package and PCB along with constant current, pseudo-balanced and constant voltage schemes. It has been shown through simulations and measurements on test boards that these methods can indeed be used to reduce power supply noise, reduce jitter and increase eye height, thereby improving eye diagrams significantly for both 2D and 3D ICs. Since the capacitors used to mitigate RPDs can be eliminated using the PTL based PDN, the number of capacitors required for implementation can be reduced as well, as compared to the traditional methods. In addition, since the power supply noise is reduced, fewer decoupling capacitors are required to further reduce the noise. Depending on the manner in which the PTLs are routed (Vdd routed on the same layer as signal lines), the layer count can be reduced.

Though substantial work is still required before such PDN schemes are adopted for 3D ICs (and 2D ICs), the results are indeed encouraging and provide alternatives. A very important item to note is that the PTL schemes described are based on a higher impedance power distribution network (as compared to the traditional methods used today) and has been applied in this chapter primarily for communication between ICs. The higher impedance required for the PDN is due to the need for matching the PTL impedance to the rest of the circuitry.

In general, components such as capacitors in the package and PCB raise cost and increase the space required for mounting them. In comparison, transistors (in general) come for free. The alternate methods described in this chapter for power distribution minimize the extra components required in the package and PCB for improving signal and power integrity by making use of these free transistors.

References

1. [Swaminathan et al., 2007] Madhavan Swaminathan and Ege Engin, "Power Integrity Modeling and Design for Semconductors and Systems", Prentice Hall, 2007.

2. [Engin et al., 2008] A. E. Engin and M. Swaminathan, "Power Transmission Lines: A New Interconnect Design to Eliminate Simultaneous Switching Noise" in Proc. Electron. Compon. Technol. Conf., pp. 1139–1143, May 2008.

3. [Huh et al., 2011] S. Huh, M. Swaminathan, and D. Keezer, "Constant Current Power Transmission Line based Power Delivery Network for Single-Ended Signaling," IEEE Trans. Electromagnetic Compatibility, Vol. PP, No. 99, pp. 1–15, 2011.

4. [Huh et al., 2011a] S. Huh, M. Swaminathan, and D. Keezer, "Constant Current Power Transmission Line based Power Delivery Network for Single-Ended Signaling with Reduced Simultaneous Switching Noise," in Proc. Workshop on Signal Propag. Interconn., pp. 47–50 , May 2011.

5. [Huh et al., 2011b] S. Huh, M. Swaminathan, and D. Keezer; "Low-Noise Power Delivery Network Design using Power Transmission Line for Mixed-Signal Testing," IEEE 17th IEEE International Mixed-Signal, Sensors, and Systems Test Workshop (IMS3TW), pp. 53–57, 2011.

6. [Sphinx, 2010] Sphinx V2.5, 2010; E-System Design. [Online]. Available: http://www.e-systemdesign.com/

7. [Huh, 2011c] S. Huh, "Design of Power Delivery Networks for Noise Suppression and Isolation using Power Transmission Lines", PhD Dissertation, School of Electrical and Computer Engineering, 2011.

8. [Carusone et al., 2001] A. Carusone, K. Farzan, and D. Johns, "Differential signaling with a reduced number of signal paths," IEEE Trans. on Circuits and System II, Vol. 48, No. 3, pp. 294–330, Mar. 2001.

9. [Tallini et al., 1998] L. G. Tallini and B. Bose, "Design of Balanced and Constant Weight Codes for VLSI Systems," IEEE Trans. on Computers, Vol. 47, pp. 556–572, May. 1998.

10. [Sim, 2007] J. Y. Sim, "Segmented group inversion coding for parallel links," IEEE Trans. on Circuits and System II, Vol. 54, No. 5, pp. 328–332, Apr. 2007.

11. [Oh et al., 2008] D. Oh, F. Ware, W. P. Kim, J.-H. Kim, J. Wilson, L. Luo, J. Kizer, R. Schmitt, C. Yuan and J. Eble, "Pseudo-differential signaling scheme based on 4b/6b multiwire code," in Proc. Electrical Performance of Electronic Packaging, pp. 29–32, Oct. 2008.

12. [Huh et al., 2011d] S. Huh, M. Swaminathan, and D. Keezer, "Pseudo-Balanced Signaling Using Power Transmission Line for Parallel Link," in Proc. Symposium on Electromagnetic Compatibility, pp. 871–876, Aug. 2011.

13. [Cheng et al., 2001] Wei-Chung Cheng and M. Pedram, "Memory bus encoding for low power: a tutorial," 2001 International Symposium on Quality Electronic Design, pp.199–204, 2001.

14. [Stan et al., 1995] M. Stan and W. Burleson, "Bus-invert coding for low-power I/O," IEEE Trans. Very Large Scale Integr. (VLSI) Syst., Vol. 3, pp. 49–58, Mar. 1995.

15. [Telikepalli et al., 2013] S. Telikepalli, M. Swaminathan and D. Keezer, "Minimizing Simultaneous Switching Noise at Reduced Power with Constant Voltage Power Transmission Lines for High-Speed Signaling," International Symposium on Quality Electronic Design (ISQED), pp. 714–718, 2013.

16. [Zhang et al., 2012] David C. Zhang, Madhavan Swaminathan and Suzanne Huh, "New power delivery scheme for 3D ICs to minimize simultaneous switching noise for high speed I/Os", Proceedings of the IEEE Electrical Performance of Electronic Packaging, pp. 87–90, 2012.

17. [Telikepalli et al., 2013a] S. Telikepalli, D. Zhang, M. Swaminathan and D. Keezer, "Constant Voltage Based Power Delivery Scheme for 3D ICs and Interposers", To be published in the IEEE Transactions on Components, Packaging and Manufacturing Technology, 2013.

18. [Telikepalli et al., 2013b] S. Telikepalli, S. K. Kim, S. J. Park, M. Swaminathan, Y. Han "Managing Signal and Power Integrity using Power Transmission Lines and Alternate Signaling Schemes", Proceedings of Latin American Symposium on Circuits and Systems (LASCAS), Cusco, Peru, Feb. 2013.

Appendix

A.1. Derivation of the Solution of the Diffusion Equation (2.1)

By inserting $\vec{J} = J_z(\rho,\varphi)\hat{z}$, (2.1) is simplified to the following equation in cylindrical coordinates:

$$\frac{1}{\rho}\frac{\partial}{\partial\rho}\left(\rho\frac{\partial J_z}{\partial\rho}\right) + \frac{1}{\rho^2}\frac{\partial^2 J_z}{\partial\varphi^2} + \alpha^2 J_z = 0. \qquad (A.1.1)$$

By using the separation of variables, i.e., $J_z(\rho,\varphi) = R(\rho)\Phi(\varphi)$, (A.1.1) is separated into the following two ordinary differential equations.

$$\rho^2 R''(\rho) + \rho R'(\rho) + \left(\alpha^2\rho^2 - v^2\right)R(\rho) = 0. \qquad (A.1.2)$$

$$\Phi''(\varphi) + v^2\Phi(\varphi) = 0. \qquad (A.1.3)$$

Since the current density distribution should be continuous over the conductor cross section, the solutions of (A.1.3) are periodic (harmonic) functions, and v should be an integer n. Substituting v^2 with n^2 converts (A.1.2) to the Bessel differential equation of order n. Therefore, the basis functions, which are the solutions of the diffusion equation, have the following form.

$$\cos\left(n(\varphi - \varphi_0)\right)J_n(\alpha\rho) \qquad n = 0, 1, 2, \cdots, \qquad (A.1.4)$$

where $J_n(\alpha\rho)$ is the nth order Bessel function or Kelvin function [Abramowitz et al., 1965], the asymptotic behavior of which is the exponential function of ρ.

A.2. Derivation of the Effective Areas

When $n = 0$, the effective area should satisfy the following:

$$\int_{S_i} \vec{w}_{i0} \cdot d\vec{S} = 1 \tag{A.2.1}$$

where $\vec{w}_{i0}$ is the SE-mode CMBF defined in (2.3) and S_i is the cross section of the *i*-th cylinder. Thus, the effective area can be found as

$$A_{i0} = \frac{2\pi\rho_i}{\alpha} J_1(\alpha\rho_i) \tag{A.2.2}$$

However, the integration of the higher-order basis functions ($n \geq 1$) should be zero because of the harmonic component in the φ direction. Thus, the normalization is redefined as follows.

$$\int_{S_{in}} \vec{w}_{in} \cdot d\vec{S} = \frac{1}{2n} \tag{A.2.3}$$

where $\vec{w}_{in}$ is the PE-mode CMBF defined in (2.4) or (2.5) and S_{in} is a part of the cross section occupied by a half period of the harmonic function, which is $\frac{1}{2n}$ of the entire cross sectional area. By inserting (2.4) or (2.5) into (A.2.3):

$$A_{in} = \frac{2^{2-n}\alpha^n\rho_i^{2+n}}{(2+n)n!} {}_1F_2\left(1+\frac{n}{2}; \left\{2+\frac{n}{2}, 1+n\right\}; -\frac{1}{4}\alpha^2\rho_i^2\right) \tag{A.2.4}$$

where ${}_1F_2$ is one of the forms of the generalized hypergeometric function.

A.3. Derivation of the Surface Charge Density Distribution (2.21)

As discussed in Section 2.4.1, we can assume the axial variation of charge density distribution is uniform and simplify Laplace's equation to the following two-dimensional form:

$$\frac{1}{\rho}\frac{\partial}{\partial\rho}\left(\rho\frac{\partial\phi}{\partial\rho}\right) + \frac{1}{\rho^2}\frac{\partial^2\phi}{\partial\varphi^2} = 0 \tag{A.3.1}$$

The linear combination of all the solutions of (A.3.1) results in the following general expression for potential:

$$\phi(\rho,\varphi) = C_1 \ln \rho + C_2 + \sum_{n=1}^{\infty}\{\rho^n (A_n \sin n\varphi + B_n \cos n\varphi)$$
$$+ \rho^{-n}(A'_n \sin n\varphi + B'_n \cos n\varphi)\}$$
$$(A.3.2)$$

where C_1, C_2, A_n, B_n, A'_n, and B'_n are all arbitrary constants. By applying (A.3.2) to the boundary condition of normal electric fields, the following expression for charge density distribution can be obtained:

$$\sigma = \hat{n}\cdot\varepsilon_0\vec{E}\Big|_{\substack{conductor\\surface}} = \hat{\rho}\cdot\varepsilon_0(-\nabla\phi(\rho,\varphi)) = \sigma_0 + \sum_{n=1}^{\infty}\{\sigma_q \sin n\varphi + \sigma_d \cos n\varphi\} \quad (A.3.3)$$

where σ_0, σ_d, and σ_q are undefined constants of surface charge density in C/m^2.

A.4. Partial Element Formula of Brick-Type Conductors

<u>Partial self inductance</u>
The partial self inductance of a brick element can be found from the following formula [Ruehli, 1972].

$$\frac{\pi}{2\mu l}Lp_{ii} = \frac{\omega^2}{24u}\left[\ln(\frac{1+A_2}{\omega}) - A_5\right] + \frac{1}{24u\omega}\left[\ln(\omega+A_2) - A_6\right]$$

$$+ \frac{\omega^2}{60u}(A_4 - A_3) + \frac{\omega^2}{24}\left[\ln(\frac{u+A_3}{\omega}) - A_7\right] + \frac{\omega^2}{60u}(\omega - A_2)$$

$$+ \frac{1}{20u}(A_2 - A_4) + \frac{u}{4}A_5 - \frac{u^2}{6\omega}\tan^{-1}\left(\frac{\omega}{uA_4}\right) + \frac{u}{4\omega}A_6 - \frac{\omega}{6}\tan^{-1}\left(\frac{u}{\omega A_4}\right)$$

$$+ \frac{A_7}{4} - \frac{1}{6\omega}\tan^{-1}\left(\frac{u\omega}{A_4}\right) + \frac{1}{24\omega^2}\left[\ln(u+A_1) - A_7\right] + \frac{u}{20\omega^2}(A_1 - A_4)$$

$$+ \frac{1}{60\omega^2 u}(1 - A_2) + \frac{1}{60u\omega^2}(A_4 - A_1) + \frac{u}{20}(A_3 - A_4)$$

$$+ \frac{u^3}{24\omega^2}\left[\ln(\frac{1+A_1}{u}) - A_5\right] + \frac{u^3}{24\omega}\left[\ln(\frac{\omega+A_3}{u}) - A_6\right]$$

$$+ \frac{u^3}{60\omega^2}\left[(A_4 - A_1) + (u - A_3)\right]$$

$$(A.4.1)$$

where

$$A_1 = (1+u^2)^{\frac{1}{2}}, \; A_2 = (1+\omega^2)^{\frac{1}{2}}, \; A_3 = (\omega^2 + u^2)^{\frac{1}{2}}, \; A_4 = (1+\omega^2 + u^2)^{\frac{1}{2}},$$

$$A_5 = \ln\left(\frac{1+A_4}{A_3}\right), \; A_6 = \ln\left(\frac{\omega + A_4}{A_1}\right), \; A_5 = \ln\left(\frac{u+A_4}{A_2}\right), \; u = \frac{l}{W}, \; \omega = \frac{T}{W},$$

and l, T, and W are the length, the thickness, and the width of a rectangular conductor segment, respectively. If the thickness T is negligible, a simple self inductance formula can be used instead [Ruehli, 1972].

Partial mutual inductance between parallel conductors
When two rectangular conductors are in parallel, the partial mutual inductance between the two conductors can be found by using the following weighted sum of the self inductances of 64 virtual conductor segments, as shown in the following formula [Zhong et al., 2003].

$$M = \frac{1}{W_0 T_0 W_1 T_1} \frac{1}{8} \sum_{i_0,i_1,j_0,j_1,k_0,k_1=0}^{1} (-1)^{i_0+i_1+j_0+j_1+k_0+k_1+1} \times A^2_{p_{i_0,j_0,k_0},q_{i_1,j_1,k_1}} L_{p_{i_0,j_0,k_0},q_{i_1,j_1,k_1}}$$

$$(A.4.2)$$

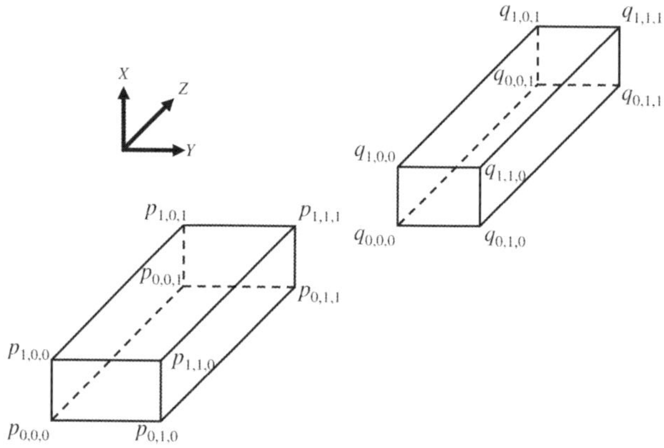

Figure A.4.1: Two parallel brick elements and their point indices to calculate partial mutual inductance [Zhong et al., 2003].

where $L_{p_{i_0,j_0,k_0},q_{i_1,j_1,k_1}}$ represents the self inductance of a brick element that has the two diagonal end points p_{i_0,j_0,k_0} and q_{i_1,j_1,k_1}. All the point indices are shown in Figure A.4.1. The partial self inductance formula can be found in the previous section of this appendix.

<u>Partial coefficient of potential between parallel conductors</u>
For computing the capacitive coupling between two parallel capacitive cells, the following formula can be used [Ruehli et al., 1973].

$$
\begin{aligned}
4\pi\varepsilon Pp_{i,j} = \frac{1}{f_a f_b s_a s_b} \sum_{k=1}^{4}\sum_{m=1}^{4} (-1)^{k+m} &\left[\frac{b_m^2 - C^2}{2} a_k \ln(a_k + \rho) \right. \\
&+ \frac{a_k^2 - C^2}{2} b_m \ln(b_m + \rho) - \frac{1}{6}\left(b_m^2 - 2C + a_k^2\right)\rho \\
&\left. - b_m C a_k \tan^{-1}\frac{a_k b_m}{\rho C} \right]
\end{aligned}
\tag{A.4.3}
$$

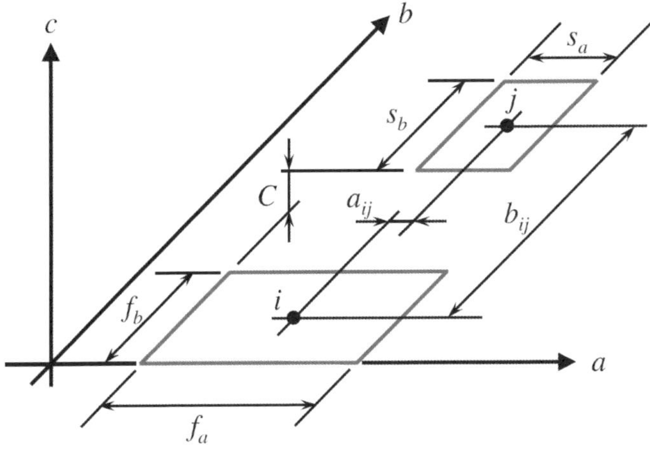

Figure A.4.2: Two parallel panel cell elements and their coordinate variables to calculate partial mutual coefficient of potential [Ruehli et al., 1973].

where

$$\rho = \sqrt{a_k^2 + b_m^2 + C^2}\,, \quad \{a,b\}_1 = \{a,b\}_{ij} - \frac{f_{\{a,b\}}}{2} - \frac{s_{\{a,b\}}}{2}\,,$$

$$\{a,b\}_2 = \{a,b\}_{ij} + \frac{f_{\{a,b\}}}{2} - \frac{s_{\{a,b\}}}{2}\,, \quad \{a,b\}_3 = \{a,b\}_{ij} + \frac{f_{\{a,b\}}}{2} + \frac{s_{\{a,b\}}}{2}\,,$$

$$\{a,b\}_4 = \{a,b\}_{ij} - \frac{f_{\{a,b\}}}{2} + \frac{s_{\{a,b\}}}{2}\,,$$

$\{a,b\}_{ij}$ s are the relative distances between cells in a and b directions, C is the relative vertical distance between cells, and $\{f,s\}_{\{a,b\}}$ s are the sizes of cells shown in Figure A.4.2.

A.5. Indefinite Integral for Axial Variables in Mutual Inductance between a Cylinder and a Plane

With the variables that are defined in Fig. 2.18, the distance between two points is formulated as follows.

$$R_{12} = \left| \vec{R}_2 - \vec{R}_1 \right| = \sqrt{z_2^2 + 2fz_2 + g} \tag{A.5.1}$$

where

$$f = \left[(D_x - x_1)\sin\alpha - (D_y - y_1)\cos\alpha\right]\sin\beta + (D_z - z_1)\cos\beta\,,$$

$$\begin{aligned}
g = {}& D_{21}^2 + \rho^2 + x_1^2 + y_1^2 + z_1^2 + 2\rho[-\cos\varphi(x_1\cos\alpha + y_1\sin\alpha) \\
& + \sin\varphi(x_1\sin\alpha - y_1\cos\alpha)\cos\beta - z_1\sin\varphi\sin\beta \\
& + (D_x\cos\alpha + D_y\sin\alpha)\cos\varphi + D_z\sin\beta\sin\varphi \\
& + (-D_x\sin\alpha + D_y\cos\alpha)\sin\varphi\cos\beta] - 2x_1D_x - 2y_1D_y - 2z_1D_z\,.
\end{aligned}$$

$\rho_{nx} = \rho_n\cos\varphi_n$, $\rho_{ny} = \rho_n\sin\varphi_n$, and α, β are the rotation angles of conductors based on Euler angles ($n = 1,2$). By using (A.5.1) in the integral over z_2, the following indefinite integral can be found.

$$I_z(\rho_2,\varphi_2,x_1,y_1) = \int_{-\frac{L_2}{2}}^{+\frac{L_2}{2}} \frac{1}{R_{12}}\,dz_2 = -\sinh^{-1}\frac{f - 0.5L_2}{\sqrt{-f^2 + g}} + \sinh^{-1}\frac{f + 0.5L_2}{\sqrt{-f^2 + g}} \tag{A.5.2}$$

Index

Madhavan Swaminathan is the John Pippin Chair in Electromagnetics in the School of Electrical and Computer Engineering (ECE) and Director of the Interconnect and Packaging Center, Georgia Tech, USA and the Founder and CTO of E-System Design, a company focusing on the development of CAD tools for achieving signal and power integrity in integrated 3D micro and nano-systems. He is also the co-founder of Jacket Micro Devices, a company that specialized in integrated RF modules and substrates for wireless applications that was acquired by AVX Corporation. He formerly held the position of Joseph M. Pettit Professor in Electronics in ECE and Deputy Director of the NSF Microsystems Packaging Center at Georgia Tech. Prior to joining Georgia Tech, he was with IBM working on packaging for supercomputers. He is the author of more than 400 technical articles, holds 27 patents, is the author of 3 book chapters, primary author and co-editor of 3 books - "Power Integrity Modeling and Design for Semiconductors and Systems", Prentice Hall, Nov 2007, "Introduction to System on Package", McGraw Hill, Mar. 2008 and "Design and Modeling for 3D ICs and Interposers", WSPC, Sep. 2013 in the field of packaging. He is an IEEE Fellow and has served as the Distinguished Lecturer for the IEEE EMC society. Thirty-five Ph.D. and sixteen M.S. students in Electrical Engineering have graduated under his supervision. He received his B.E. degree in Electronics and

Communication from Regional Engineering College, Tiruchirapalli, India (now NITT) in 1985 and M.S. and Ph.D. degrees in Electrical Engineering from Syracuse University in 1989 and 1991, respectively.

Ki Jin Han received the B.S. and M.S. degrees in electrical engineering from Seoul National University in 1998 and 2000, respectively, and the Ph.D. degree in electrical and computer engineering from Georgia Tech in 2009. From 2000 to 2005, he was with the System R&D Laboratory, LG Precision, Korea, where he was involved in the development of wireless transceiver and antenna system for airborne and naval radars. From 2006 to 2009, he was with Mixed Signal Design Group in Georgia Tech as a graduate research assistant. After graduation from Georgia Tech, Dr. Han was with the IBM T. J. Watson Research Center as a postdoctoral researcher. From 2011, he is with the School of ECE, Ulsan National Institute of Science and Technology (UNIST), Korea, as an assistant professor. In UNIST, Dr. Han is collaborating with E-System Design for the development of a 3-D integration design tool. Dr. Han was the recipient of the Samsung Scholarship for graduate study in 2005. He has been a reviewer of journals including IEEE Transactions on Components, Packaging and Manufacturing Technology. Throughout his research career, Dr. Han has been interested in simulation, modeling, and design of various electromagnetic applications, covering microwave circuits, antennas, signal/power integrity for high-speed digital design, and the electromagnetic modeling of electronic packaging and interconnections.